Recent Trends in Industrial Wastewater Treatment

As the global pursuit for environmental sustainability and circular water economy grows ever ambitious, this book explores recent trends, methods, and advancements in industrial wastewater treatment.

With insights from professors and industry experts in the field, this book provides new perspectives on the following themes:

- Cutting-edge treatment technologies like membrane filtration, advanced oxidation processes, anaerobic digestion, process optimisation, and water reuse.
- Adopting a multidisciplinary approach for handling and treatment of industrial wastewater.
- Methods and techniques for the removal and remediation of microplastics from wastewater.
- Recent approaches for treatment of biomedical wastewater from hospitals and diagnostic centres.
- The role of Artificial Intelligence in quality assessment, monitoring, and treatment of industrial wastewater.

This is a valuable resource for environmentalists, scientists, and technologists around the world. It will also be of interest to academics and students within the disciplines of Water Systems Engineering, Environmental Science and Remediation, and Sustainable Development.

Wastewater Treatment and Research

Series Editor: Maulin P. Shah

Wastewater Treatment: Molecular Tools, Techniques, and Applications
Edited by Maulin P. Shah

Advanced Oxidation Processes for Wastewater Treatment: An Innovative Approach
Edited by Maulin P. Shah, Sweta Parimita Bera, and Gunay Yildiz Tore

Emerging Technologies in Wastewater Treatment
Edited by Maulin P. Shah

Bio-Nano Filtration in Industrial Effluent Treatment: Advanced and Innovative Approaches
Edited by Maulin P. Shah

Membrane and Membrane-Based Processes for Wastewater Treatment
Edited by Maulin P. Shah

Phycoremediation Processes in Industrial Wastewater Treatment
Edited by Maulin P. Shah

Recent Trends in Industrial Wastewater Treatment
Edited by Tripti Malik and Maulin P. Shah

For more information, please visit: https://www.routledge.com/Wastewater-Treatment-and-Research/book-series/WASTEWATER

Recent Trends in Industrial Wastewater Treatment

Edited by

Tripti Malik and Maulin P. Shah

CRC Press
Taylor & Francis Group
Boca Raton London New York

CRC Press is an imprint of the
Taylor & Francis Group, an **informa** business
A BALKEMA BOOK

Designed cover image: Shutterstock

First published 2025
by CRC Press/Balkema
4 Park Square, Milton Park, Abingdon, Oxon, OX14 4RN

and by CRC Press/Balkema
2385 NW Executive Center Drive, Suite 320, Boca Raton FL 33431

CRC Press/Balkema is an imprint of the Taylor & Francis Group, an informa business

British Library Cataloguing-in-Publication Data
A catalogue record for this book is available from the British Library

ISBN: 9781032802152 (hbk)
ISBN: 9781032802176 (pbk)
ISBN: 9781003496014 (ebk)

DOI: 10.1201/9781003496014

Typeset in Times
by Newgen Publishing UK

Contents

Preface

The water used-up in different industrial, commercial, and production processes is considered as 'Industrial Wastewater'. Throughout the globe, rapid industrialization has led to increased generation of wastewater. Suspended, colloidal, and dissolved (mineral and organic) solids are invariably present. Additionally, anti-corrosives, fertilizers, flame retardants, gasoline additives, heavy metals, pathogenic microorganisms, pesticides, perfluorinated compounds, and xenobiotics may also be present in wastewater depending on the nature of the industry. The treatment of industrial wastewater has to be legitimate, so as to minimize its environmental impact before its discharge into water bodies or reuse for other processes. Conventional treatment technologies are based on adsorption, coagulation, membrane filtration, ozonation, ion exchange, chemical oxidation, and biological techniques. These methods are often insufficient or inefficient for the removal of toxic and hazardous contaminants. High energy consumption, sludge production, carbon footprint, and limited adaptability to climate change are the key challenges faced by conventional methods.

As per the United Nations' Sustainable Development Goal 6 which calls for universal access to safe water by 2030, newer methods and technologies must be introduced so that industrial waste does not remain a liability, but can be converted into a resource. Industries should follow circular economy models, which aim to recover valuable resources like water, energy, and nutrients from wastewater.

Recent Advances in Industrial Wastewater Treatment has been compiled to assist in this goal. The book describes recent methods such as advanced oxidation processes, electrochemical treatment, nanofiltration technologies, and biological methods which are presently in the nascent stage, but which can be developed with better efficiency for water treatment in the coming years. Hybrid treatment technologies involve combinations of physical, chemical, and biological processes, and new approaches can minimize the costs, energy consumption, and environmental impact associated with wastewater management. The advent of Artificial Intelligence and machine learning methods may further help in optimizing industrial wastewater treatment plants.

Acknowledgements

We extend our grateful thanks to all the contributors for their valuable contributions and timely cooperation. We are thankful to Ms Olivea Sarkar, Assistant Professor, DSMS Institute of Nursing, Durgapur who actively helped in the completion of this book project. Ms Kiran Singh Negi, research scholar, Dolphin PG Institute of Biomedical & Natural Sciences, deserves special thanks for her assistance. Our heartfelt thanks to our family members for their love and support, without which this project could not have been completed.

About the Editors

Tripti Malik is an Associate Professor (Microbiology) at Dolphin PG Institute of Biomedical & Natural Sciences, Dehradun, Uttarakhand, India. She has 22 years of experience teaching and supervising undergraduate, post graduate, and doctorate students. She has received funding from state and central governments for R&D projects. She has authored a number of research articles, review articles, and book chapters in reputable national and international journals and books. She is also actively engaged in editing books and articles for publishers of scientific repute.

Maulin P. Shah is an Environmental Microbiology Consultant at Enviro Technology Ltd., Ankleshwar, Gujarat, India. Dr. Maulin P. Shah has emerged as a prominent researcher with his dedication to research spanning over two decades. He completed his BSc in Microbiology from Gujarat University, Godhra (Gujarat), India, in 1999. He further completed his PhD from Sardar Patel University, Vallabh Vidyanagar (Gujarat), India, in 2005. His expertise and efforts are committed to investigating the complex relation between pollution and microbial diversity of wastewater via both cultivation-dependent and independent analyses. Dr. Shah's prolific scholarly journey is underscored by a staggering publication record, comprising more than 250 research papers across esteemed national and international journals. Beyond his research articles, he holds the distinguished role of editor for over 150 internationally renowned books, and his influence extends beyond publications, as he actively serves on the editorial boards of top-rated journals.

Contributors

Ponnusamy Thillai Arasu
Department of Chemistry, College of Natural and Computational Sciences, Wollega University, Post Box 395, Nekemte, Ethiopia

Hailemariam Assefa
Faculty of Material Science and Engineering, Jimma University, Jimma Institute of Technology, Ethiopia

Reyhan Ata
Çerkezkoy Vocational School of Namık Kemal University, Tekirdag, Turkey

A. Mohamed Azharudeen
Department of Chemistry, Pasumpon Thiru Muthuramalinga Thevar Memorial College, Kamuthi, Ramanathapuram, Tamil Nadu, India

Yogeshwar R. Baste
Department of Chemistry, M.V.P. Samaj's K.S.K.W Arts, Science & Commerce College, Uttam Nagar, CIDCO, Nashik, Maharashtra, India

Jagadeesh Chandra Bose K.
University Institute of Biotechnology, Chandigarh University, Mohali, Punjab, India

Daljeet Singh Dhanjal
Department of Biotechnology, Lovely Professional University, Phagwara, 144411, Punjab, India

Jaskaran Dhiman
Agricultural and Environmental Sciences, Macdonald Campus of McGill University, Quebec, Canada

Manisha Duseja
Department of Chemistry, School of Physical Science, DIT University, Dehradun, Uttarakhand, India

G. Elango
Department of Chemistry, Kalaignar Karunanidhi Government Arts College, Tiruvannamalai, Tamil Nadu 606603, India

G. Kanthimathi
Department of Chemistry, Ramco Institute of Technology, Rajapalayam, Virudhunagar District, Tamil Nadu, India

Harry Kaur
Department of Biosciences and Bioengineering, Indian Institute of Technology, Roorkee, India

Palki Sahib Kaur
Biotechnology Department, Chandigarh College of Technology, CGC, Landran, 140307, Punjab, India

Prasann Kumar
Department of Agronomy, School of Agriculture, Lovely Professional University, Phagwara, Punjab, India

A. V. Senthil Kumar
Department of MCA, Hindusthan College of Arts & Science, Coimbatore, 641 028 India

Anjisha Maharshi
Department of Microbiology, Marwadi University, Rajkot, Gujarat, India

Pramod Malik
Himachal Pradesh University, Shimla, India

Tripti Malik
Department of Microbiology, Dolphin PG Institute of Biomedical and Natural Sciences, Dehradun, India

Komal Mittal
University Institute of Biotechnology, Chandigarh University, Mohali, Punjab, India

Sadhana Y. More
Department of Microbiology, M.V.P. Samaj's K.S.K.W Arts, Science & Commerce College, Uttam Nagar, CIDCO, Nashik, Maharastra, India

Darshan Mulchandani
Department of Microbiology, Marwadi University, Rajkot, Gujarat, India

Arumugam Murugan
Department of Chemistry, North Eastern Regional Institute of Science and Technology, Nirjuli, Itanagar, Arunachal Pradesh, India

S.V. Chinna Swami Naik
Department of Civil Engineering, IIT Roorkee, Uttrakhand, India

Manish Pant
Department of Chemistry, School of Physical Science, DIT University, Dehradun, Uttarakhand, India

S. Arun Prabhu
Department of Chemistry, National College, Tiruchirappalli, Tamilnadu, India

Bhairav Prasad
Biotechnology Department, Chandigarh College of Technology, CGC, Landran, 140307, Punjab, India

Rohit Rai
Faculty of Applied Medical Sciences, Lovely Professional University, Phagwara Punjab, India

Praveen C. Ramamurthy
Interdisciplinary Centre for Water Research (ICWaR), Indian Institute of Science, Bengaluru, Karnataka, India

Natarajan Raman
Department of Chemistry, VHNSN College (Autonomous), Virudhunagar, Tamil Nadu, India

Prerna Rana
Biotechnology Department, Chandigarh College of Technology, CGC, Landran, 140307, Punjab, India

Niladri Shekhar Samanta
Department of Petroleum Engineering, Presidency University, Bangalore, Karnataka, India

Gaurav Sanghvi
Department of Microbiology, Marwadi University, Rajkot, Gujarat, India

Jyoti Sarwan
University Institute of Biotechnology, Chandigarh University, Mohali, Punjab, India

Abhinandan Sharma
Biotechnology Department, Chandigarh College of Technology, CGC, Landran, 140307, Punjab, India

Heenu Sharma
Rayat and Bahra University, Mohali, Punjab, India

Jabrinder Singh
Department of Chemistry, DIT University, Dehradun

Joginder Singh
Department of Botany, Nagaland University, Nagaland, India

Simranjeet Singh
Indian Institute of Science, Bengaluru, Karnataka, India

Vijaya Sood
University Institute of Biotechnology, Chandigarh University, Mohali, Punjab, India

R. V. Suganya
School of Management Studies and Commerce, Vels Institute of Science Technology and Advanced Studies, Chennai 600117, India

Günay Yildiz Töre
Environmental Engineering Department of Çorlu Engineering Faculty, Namik Kemal University, 59860, Tekirdag, Turkey

Nazim Uddin
University Institute of Biotechnology, Chandigarh University, Mohali, Punjab, India

Recent Approaches for Recovery, Re-use and Effective Treatment of Wastewater

1

Niladri Shekhar Samanta

1. INTRODUCTION

The recent surge in industry has been fueled by the population explosion and has resulted in two key issues: the shortage of energy and the emergence of new contaminants (ECs) (Das et al., 2023b; Dhara et al., 2023a; Samanta et al., 2023d; Samanta et al., 2021; Shekhar Samanta et al., 2023b). It is forecast that 7.7 billion people will live in the world in 2019 and 10 billion in 2050, with 70% of them expected to reside in cities. Safe drinking water must be treated before it is released into water bodies or used further since it is essential for humans, animals, and plants. Approximately two billion people currently reside in "high water scarcity" nations in Northern Africa, Central, Western, and Southern Asia. Furthermore, by 2050, the need for clean, safe drinking water will rise from 20% to 30%, according to a United Nations report (2019) (Sharma et al., 2023; Tortajada & Van Rensburg, 2020). Regulatory bodies keep an eye on discharges to ensure that set limits are not crossed.

The Water (Prevention and Control of Pollution) Act of 1974, as amended in 1988, and the Environment (Protection) Act of 1986, which regulates sewage treatment

DOI: 10.1201/9781003496014-1

plant (STP) discharge policy and establishes pollution limitations in several zones, such as water bodies, coastal regions, and land, are two of the initiatives that India has launched on occasion to address the issues of waste-water generation, treatment, and disposal. With the help of modern treatment techniques, improved infrastructure, and legislation like NOM-002-ECOL-1996, NOM-003-SEMARNAT-1997, NOM-001-SEMARNAT-1996, Mexico is striving to enhance wastewater management (Preisner et al., 2020). Sustainable water access in accordance with international standards is the goal of the UAE's Water Security Strategy 2036. Depending on the type of receiver, Dubai's Emirate Government sets quality restrictions for wastewater discharge. The Federal Water Act (WHG), the Ordinance on requirements for contaminated water ejection (AbwV), and mandatory EU laws control wastewater management in Germany. Wastewater quality in British Columbia (BC) is regulated according to receiver type and treatment facility capacity. WWTPs are classified according to their receivers (lakes, seawaters, and streams) and capacity (below 10 m^3/d, 10–50 m^3/d, >50 m^3/d) (Preisner et al., 2020).

It is challenging to abide by the strict government rules and regulations due to the high energy and capital costs associated with treating industrial and municipal wastewater as well as the large land requirements. Untreated wastewater discharge has a significant negative impact on both human and environmental health. If we desire the management of water resources to be sustainable, the water cycle must be closed. On the other hand, industry and agriculture must use water in a sensible and suitable manner due to the global shortage of safe drinking water. Globally, more than a million tons of emerging contaminants are created annually and released into water bodies after receiving little to no treatment (Rambabu et al., 2023).

Numerous conventional technologies, such as photocatalytic treatment, nanofiltration, solvent extraction, activated sludge, coagulation/flocculation, screening and grit trapping, electrocoagulation, ultrafiltration, chemical precipitation, membrane filtration, reverse osmosis, rotating biological contactors, electrochemical treatment, sequencing batch reactors, and anaerobic bed reactors, have been developed as physical, chemical, or biological treatment techniques for the expulsion of emerging contaminants and the treatment of contaminated water (Dhara et al., 2024; Karungamye et al., 2023; Mondal et al., 2019; Samanta et al., 2022a, 2023b, 2023c, 2024). These methods are based on the principles of advanced oxidation process (AOP), degradation, separation, or adsorption (Dhara et al., 2023b; Samanta et al., 2023a; Shafiquzzaman et al., 2021; Shekhar Samanta et al., 2023a). Even though some of these methods are effectively used, have high energy and capital requirements, creation of greenhouse gases (GHGs), low recovery rates, hazardous intermediates, intense aeration, low efficacy, and other factors make them extremely ineffective when used on a large scale (Priyadarshini et al., 2022; Samanta et al., 2022b; Sontakke et al., 2023). For instance, the USA's urban polluted water treatment system uses between 2.8% and 4.1% of the country's total energy consumption, translating into 21 million metric tons of greenhouse gas emissions every year. Similar to this, when an external power source is applied during electrocoagulation, sacrificial electrodes dissolve electrochemically, producing metal coagulants, primarily iron, and aluminum, in situ.

Phytoremediation presents a viable solution to address these limits in wastewater treatment. The minimal requirements for algae growth combined with the absorption of carbon dioxide in the form of biomass make this process highly promising (Zribi et al., 2023). Bioenergy production offers numerous advantages over conventional treatment methods, including the generation of value-added products and resource recovery through circular wastewater treatment. High-rate algal ponds (HRAP) are the most widely utilized algal-contaminated water purification techniques. Since the 1950s, algae have been included in wastewater treatment systems. Algal membrane photobioreactors, or AMBRs, are a new circular economy wastewater treatment technique that has encountered several setbacks. With the application of this technology, pond systems require less area, have shorter hydraulic retention times (HRT), experience less fouling, and offer superior biomass separation, manageable operation, nutrient expulsion, recovery of resources, and good effluent water quality with the possibility of water reuse (Mofijur et al., 2023).

This chapter focuses on the resource recovery from wastewater by conventional as well as advanced techniques for instance membrane filtration, advanced oxidation, ultraviolet radiations, cavitations, etc. This chapter also throws light on energy (heat) recovery, biogas recovery, and bioplastic recovery. Readers of this chapter will be benefitted as it will serve the wastewater treatment strategies, recovery of valuable resources, their utilization, and future outlooks towards the process's sustainability.

2. RESOURCE RECOVERY FROM WASTEWATER BY DIFFERENT TECHNIQUES

2.1. Microalgae-based Resource Recovery from Wastewater Treatment

Because microalgae are capable of photosynthetic conversion, they may effectively utilize solar energy to produce valuable biomasses, which makes wastewater treatment an affordable and environmentally friendly solution (Das et al., 2023a; De La Noüe & Bassères, 1989). Through the bioremediation process, microalgae may absorb nutrients such as phosphorus and nitrogen from contaminated water, lowering treatment costs. They can also be employed as a resource recovery alternative in biorefineries (Clarens et al., 2010).

Microalgae-based resource recovery from wastewater is limited by its low biomass production, contaminants that necessitate additional processing, and intricate nutrient removal methods. Even though numerous studies have produced encouraging results regarding the growth of algae in wastewater and the creation of biodiesel, widespread adoption of the same has not yet taken hold. Figure 1.1 depicts the microalgae-based resource recovery from wastewater (Janssen et al., 2003).

Prokaryotic photosynthetic organisms are known as microalgae. They utilize atmospheric carbon dioxide and solar light by using organic carbon from wastewater. The natural habitat should be duplicated in order to develop microalgae in an artificial setting, however, sunlight access limits the ability to do so (Janssen et al., 2003).

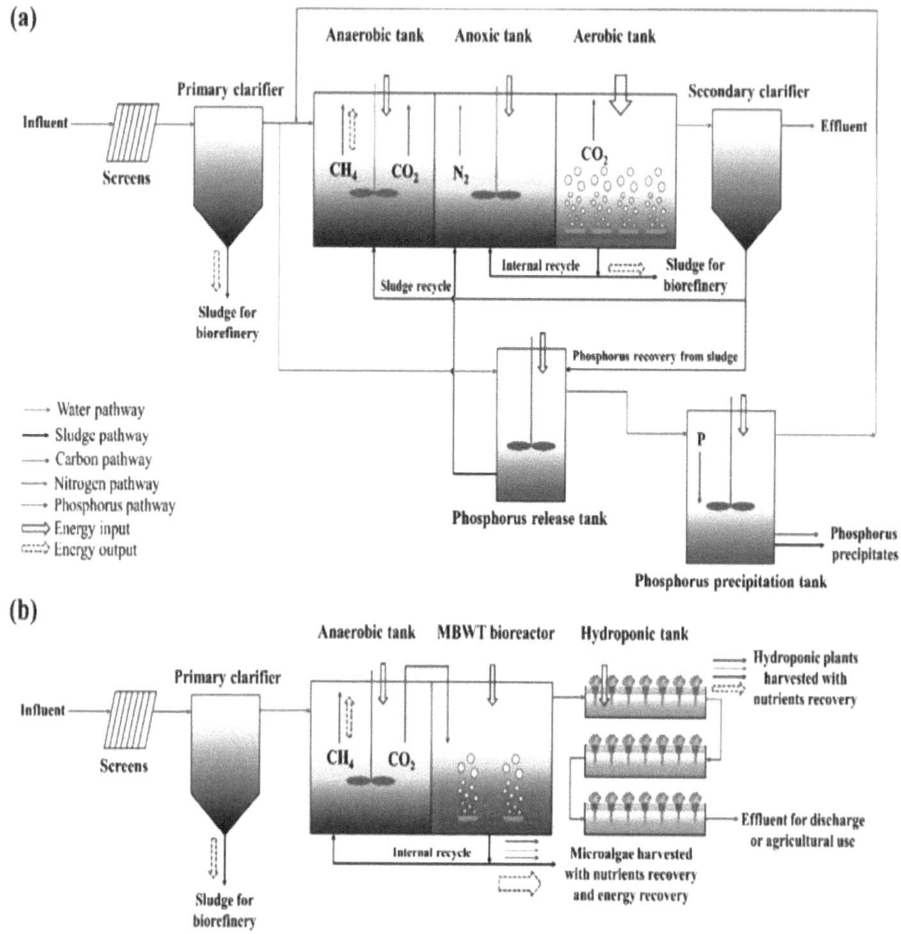

FIGURE 1.1 Microalgae-based resource recovery from wastewater.

Reproduced with permission from Li et al. (2019) ©Elsevier.

An artificial light source, such as fluorescent light, can be employed to make up for the lack of sunshine during the microalgae growing process, however, this is not an economical approach (Molina Grima et al., 2003). Because they are tiny (3 to 30 µm), microalgae are exceedingly challenging to harvest profitably (Kumar et al., 1981). Chemical, biological, electrical, or mechanical methods are to be followed for harvesting of microalgae. All of these strategies have been studied in an effort to lower the cost of harvesting, but none of the strategies has been able to produce an economically viable algal harvesting solution (Kumar et al., 1981). The third generation of biofuels, a feasible alternative energy source are made from the gathered algae. Microalgae are effective in eliminating coliform bacteria from wastewater, lowering BOD and COD levels, eliminating nutrients including phosphorus and nitrogen, and removing heavy metals (Lau et al., 1995).

2.2. Membrane Filtration

Domestic, agricultural, and industrial waste products are growing quickly along with the population, contaminating groundwater and rivers (Mondal et al., 2020). One of the main concerns is the recovery of energy or other resources from wastewater. Wastewater can now have its resources recovered and waste products removed through the use of membrane technology. In essence, membranes are used to selectively block the passage of substances through the membrane, thereby separating two phases (ions and dissolved solids). Pressurized synthetic polymers are utilized in the production of ultrafiltration (UF), nanofiltration (NF), reverse osmosis (RO), and microfiltration (MF) membranes. RO has been shown to be the most effective membrane technology for removing up to 99.5% of tiny waste particles. The configurations are shown in Figure 1.2.

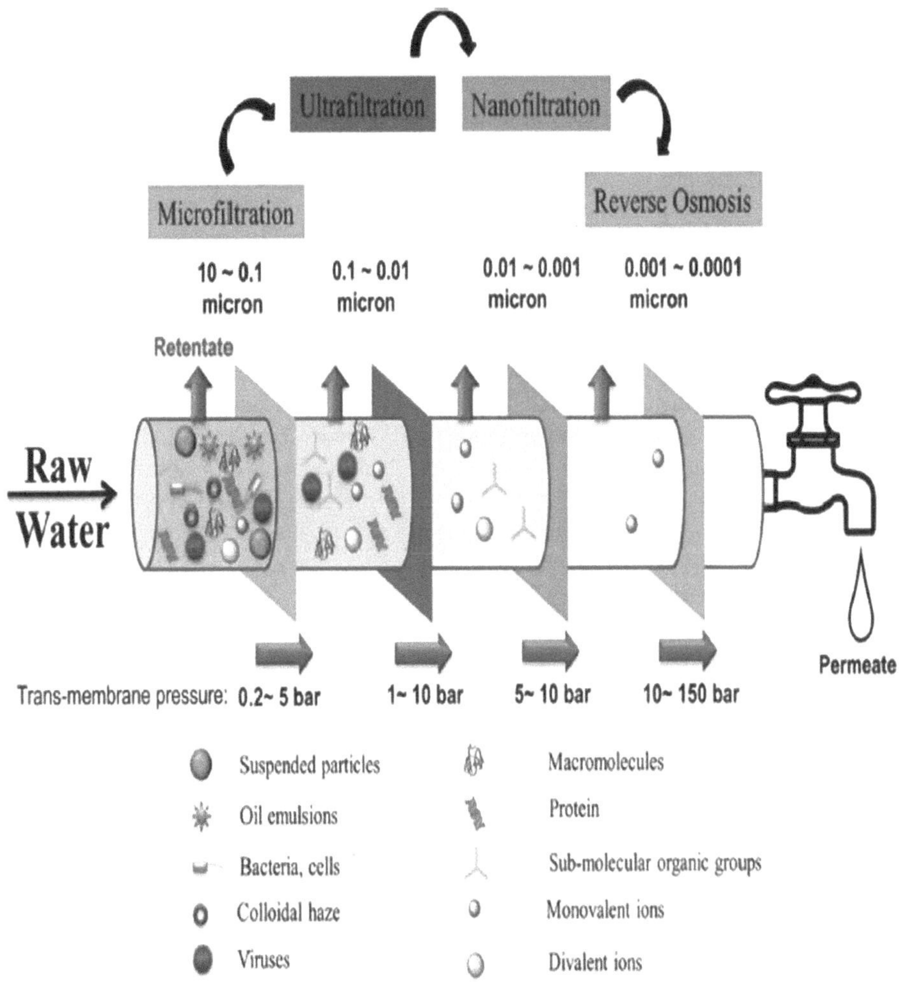

FIGURE 1.2 Reverse osmosis for recovery of heavy metals from wastewater.
Reproduced with permission from Selatile et al. (2018) © RSC Adv.

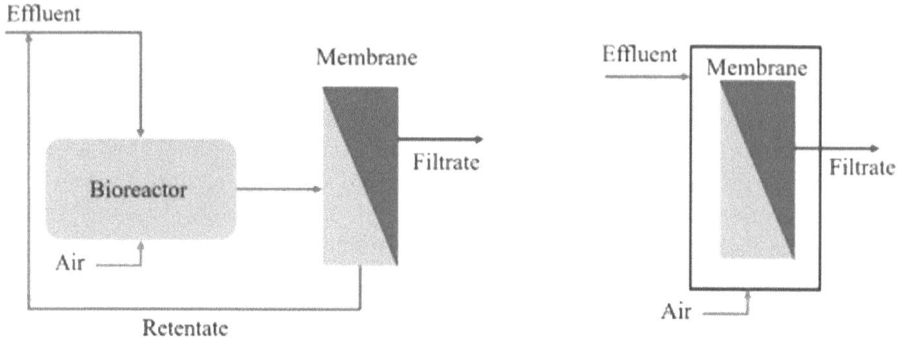

FIGURE 1.3 Configuration of two types of membrane bioreactors (MBR).
Reproduced with permission from Bohra et al. (2022) ©Elsevier.

Sometimes membrane filtration (UF, NF, and MF) acts as a pretreatment step for reverse osmosis, which removes pollutants from wastewater by acting as a multibarrier and recovers resources. The removal/reclaim efficiency will rise when one or more membrane technologies are used in conjunction with other procedures. The membrane bioreactor (MBR), for instance, is an integrated process. When activated sludge, a biological treatment process, is mixed with membrane filtration, MBR becomes a viable treatment method for the extraction of resources from wastewater. There are two types of MBRs that are commonly used: side stream and submerged. The filtration element in the side stream membrane bioreactor is located outside the bioreactor, and the biomass must be transferred for filtration via an intermediary pumping system. In the submerged MBR, the treated permeate is applied after the membranes are directly submerged in the tank containing the biological sludge (Figure 1.3) (Obotey Ezugbe & Rathilal, 2020).
Nevertheless, despite their many advantages, these technologies also have certain disadvantages. Some of the drawbacks of the membrane filtration approach are, for instance, the extremely high equipment operating costs, the need for higher aeration, high energy, and high pressure. Furthermore, it has been discovered that membrane fouling and clogging pose a significant risk at large wastewater flow. Pretreatment of secondary effluents can be used to prevent membrane fouling and blockage. Membrane-based filtering treatment methods need a significant amount of electricity. Therefore, further research in this area is needed in the future given the limited resources in this world. The recovery process from wastewater needs to be carefully examined in order to make the resources available, as resource demand rises along with population growth (Kehrein et al., 2020).

2.3. Advanced Oxidation Process for Water Treatment and Resource Recovery

Hazardous compounds are treated using advanced oxidation processes, or AOPs. These techniques produce a wide range of energy due to their inventiveness. The fundamental idea underlying AOPs is the production of extremely reactive oxidants called hydroxyl radicals ($OH^{\bullet}$), which destroy bacteria, protozoa, and other nonbiodegradable pollutants. Ozonation, UV irradiation, photo-catalysis (UV/TiO_2) electrolysis, and sonolysis are the most often utilized AOPs ("Approaches to energy and resource recovery from municipal wastewater," 2020).

2.3.1. Ultraviolet radiations

Through homoleptic cleavage, the combination of ultraviolet (UV) energy with wavelengths shorter than 280 nm produces the $OH^{\bullet}$ radical. This particular $OH^{\bullet}$ radical is a potent oxidant. When UV radiation is applied, contaminants release electrons that cause them to become unstable. Moreover, UV radiation breaks down pollutants by producing hydroxyl radicals (photochemically). If turbidity, hardness, suspended solids, and chemicals are not taken into consideration, the overall amount of bacteria in the water can return to its initial level in a matter of days (Kehrein et al., 2020).

2.3.2. Cavitation

Cavitation is the process of creating tiny bubbles and then popping them. Much energy is released in this way. The two kinds of cavitation techniques are ultrasonic and hydrodynamic techniques. The energy from ultrasound frequencies (16 Hz to 100 Mhz) oxidizes the substrate. The localized energy of bubbles results in a sharp rise in temperature and pressure. Water and hydroxyl ions separate due to this energy (Kehrein et al., 2020).

2.3.3. Fenton oxidation

When H_2O_2 and Fe^{2+} react with each other as a catalyst, the $OH^{\bullet}$ radical is created. It is this $OH^{\bullet}$ radical that breaks down organic molecules.

$$Fe^{2+} + H_2O_2 \rightarrow Fe^{3+} + OH^- + OH^{\bullet}$$

Exposure of organic molecules to UV-VIS radiation with a wavelength longer than 300 nm accelerates their rate of deterioration (Allison, 1995). Fenton oxidation breaks down compounds like phenols and pesticides. The biodegradability of wastewater containing surfactants is greatly enhanced by this method. This integrated water treatment technique is straightforward, adaptable, and efficient. The ideal parameters for the Fenton procedure are 30 mmol/L of H_2O_2 and 0.5 g/L of $FeSO_4.7H_2O$, with an irradiation period of 60 minutes (Jiang et al., 2019). Fenton oxidation breaks down compounds like phenols and pesticides. The biodegradability of wastewater containing surfactants is greatly enhanced by this method. This integrated water treatment technique is straightforward, adaptable, and efficient. The ideal parameters for the Fenton procedure are 30 mmol/L of H_2O_2 and 0.5 g/L of $FeSO_4.7H_2O$, with an irradiation period of 60 minutes. Although it is a straightforward procedure, there are several drawbacks,

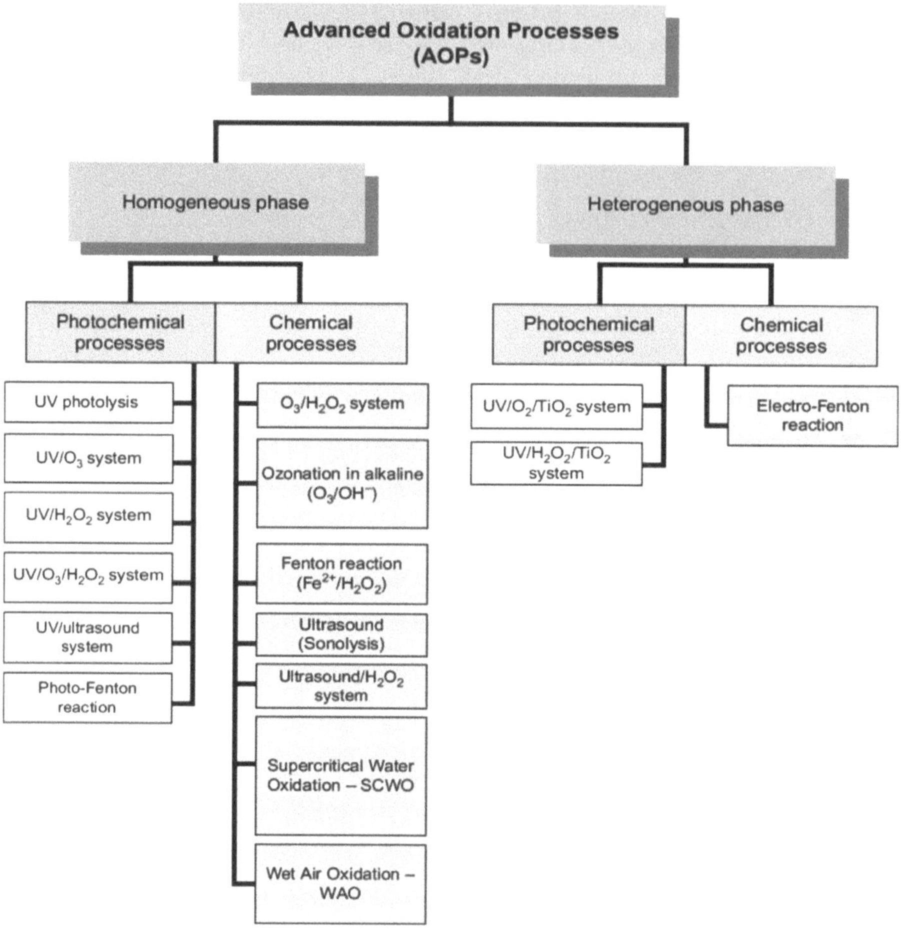

FIGURE 1.4 Advanced oxidation process for the treatment of contaminated water. Reproduced with permission from (Biń & Sobera-Madej, 2012). ©CRC.

such as a rise in cost due to the removal of iron (Fe). The process's low pH requirement raises the expense of operation and maintenance. Sludge from this method has iron hydroxide as a byproduct (Kehrein et al., 2020).

3. ENERGY RECOVERY FROM WASTEWATER

3.1. Recovery of Heat

As wastewater flows from the residential source to the WWTP, some of the thermal energy it contains is retained because hot water used for household chores (such as cleaning and bathing) generates this waste (Cipolla & Maglionico, 2014). It is possible to recover this thermal energy at the source, at the WWTPs, or along the cesspit line. However, due to the increased volumetric flow, which in turn produces a significant amount of recoverable thermal energy, heat recovery in WWTPs receives more attention. To extract heat energy from the filtrate, digester sludge, or effluent, WWTPs use heat exchangers or heat pumps. Heat pumps that use wastewater as their energy source provide thermal energy that is, on average, 3.5 times greater than the electrical energy required to run them (Hu et al., 2016). The recovered thermal energy can be used in the WWTPs for office building heating or process heating (such as sludge drying), which will reduce the amount of electricity needed to meet the heating requirements of the WWTP. District heating gear can also be used to circulate thermal energy for larger-scale heat recovery facilities. Utilizing the calorific value of biogas produced in anaerobic digesters, either in a combined power and heat generation unit or solely for heat production by burning it in a boiler, is the most often used method for heat recovery at a WWTP (Funamizu et al., 2001). The digester is heated concurrently with the recovered thermal energy. Setting up heat balances to account for heat associated with the influent streams and effluent, heat gained or lost from the liquid/air interface, heat produced by biological reactions, and heat dispersed from mechanical actuators, among other factors, is necessary to assess the viability of WWTP for the recovery of thermal energy. Several studies have satisfactorily shown that heat balance permits temperature estimation related to wastewater treatment (Gillot & Vanrolleghem, 2003). These days, the focus must be on the thorough cost analysis to evaluate the commercial viability of heat recovery, which depends on the current energy prices and the balance between heat produced and needed. Furthermore, the recovered heat must not only be limited to on-site use but also be available for use at remote places when needed, creating further difficulties with heat storage and transportation. Heat recovery and storage on-site can be accomplished with the use of aquifer thermal energy storage facilities. In order to control the financial and mechanical viability of a thermal energy production plant, it is necessary to simultaneously study the meteorological and local variables influencing the population's thermal energy requirements (Corbala-Robles et al., 2018).

3.2. Recovery of Biogas

The organic percentage in wastewater is better viewed as an energy reservoir than a pollutant when it comes to resource recovery. The WWTPs incorporate the redistribution of organic carbon for energy generation, taking into account the massive amount of organic materials assessed in terms of chemical oxygen demand (COD). Using methods like high-rate activated sludge systems and enhanced primary treatment can increase the amount of organic matter in wastewater before moving on to the anaerobic digestion stage, where the organic matter is eaten to produce biogas (Wan et al., 2016). Since these fractions are used to produce methane, the amount of biodegradable particulate COD and active biomass fraction released from the activated sludge is what determines how effective anaerobic digestion is. For energy recovery, therefore, no more complex anaerobic digestion is required (Ikumi et al., 2014). Another alternative for increasing biogas output is co-digestion, although feeding food wastes or other substrates into the digester can cause problems related to pH variations once more. Additionally, directing food waste into WWTPs won't exactly increase the efficacy of wastewater recovery; rather, the introduction of an additional energy source-food waste will lead to increases in energy generation. High-rate activated sludge is currently being dug up for carbon redirection because there are notable differences in the solid-liquid separation procedures and conversion efficiency between the various WWTPs. The fact that part of the biodegradable soluble COD is eliminated in the settler rather than being used by bacteria in high-rate activated sludge is one of the most important measures to take care of. The most crucial thing to remember is that a portion of the biodegradable soluble COD fraction is colloidal material, which will be eliminated in the settler instead of being transformed by bacteria in an A-stage. This calls for the accurate characterization of wastewater in addition to the evaluation of other factors and procedures (Jimenez et al., 2015).

3.3. Recovery of Bioplastics

The increasingly pervasive presence of petroleum-based, nonbiodegradable plastic in the environment is one of the most important ecological problems facing humanity today. According to Jambeck et al. (2015), 192 counties disposed of 275 million metric tons of plastic garbage in 2010. If petroleum-based polymers are swapped out for bio-based, biodegradable plastics in this scenario, the amount of plastic pollution in the environment might be reduced and a critical focal print could be made. According to Możejko-Ciesielska and Kiewisz (2016), bioplastics such as polyhydroxyalkanoates (PHAs), which are manufactured using renewable biomass resources, have become a superior alternative to petroleum-based plastics that are routinely utilized. Bioplastics are completely safe for the environment and have a surprising number of applications, including building, agriculture, pharmaceuticals, food packaging, consumable items, and biological purposes (Reddy et al., 2003). PHAs are naturally produced by a variety of microorganisms found in soil, sewage, and marine environments. They build up inside cells and serve as a source of energy. Due to the high expense of using synthetic substrates and pure cultures in its manufacture, PHA has not been given much thought (Salehizadeh & Van Loosdrecht, 2004). Therefore, using composite cultures in activated sludge and a variety of complicated, low-value natural substrates found in waste streams presents another choice. Perhaps the most encouraging stage was producing PHA as a result of the wastewater's biological treatment process. In addition to utilizing the current infrastructure and processes, this method promotes the circular economy and provides a steady supply of carbon that microbes can use to produce PHA. However, studies to increase the yields from these composite cultures are still in progress because it has been observed that composite cultures in sludge can only store 20% to 60% of the dry weight of the sludge, which is significantly less than the almost 90% storage in pure cultures (Puyol et al., 2017; Salehizadeh & Van Loosdrecht, 2004). PHA production at a full-scale municipal WWTP is currently being studied by the PHARIO project in the Netherlands using composite culture. However, their analysis could not find a direct correlation between PHA accumulation potential and storage rate. High PHA buildup can theoretically be attained by optimizing yield through experimentation and making minor adjustments to the kinetic parameters. The assessment of the techno-economic and environmental features of bioplastic production facilities is more important than the biokinetic parameter. This multi-aspect evaluation can assist in developing PHA production value chains for the expansion of current technologies (Fernández-Dacosta et al., 2015).

4. FUTURE OUTLOOK

The resources that can be recovered from wastewater and the technologies that support resource recovery were the main topics of this chapter. The idea is to turn technologies that often eliminate pollution into goods that may be used in the circular economy. This involves reorienting traditional practices—which previously concentrated on removal processes—toward rehabilitation. This might be achieved by making little adjustments or by completely re-imagining the therapeutic process. Nutrients (phosphorus and nitrogen) and energy in the form of carbon are the two main kinds of recoverable resources. The nutrients can be employed in carbon/energy products (generally economically viable if a carbon product is feasible) or as elemental inputs to the circular economy (now unfeasible, but eventually inevitable). Burgess et al. (2015) conducted a thorough financial analysis utilizing commodity items (such as energy, phosphorus, and nitrogen) to compare conventional wastewater treatment systems with emerging alternatives. The study found that emerging next-generation technologies are typically competitive at a larger scale and capital cost neutral when compared to existing technologies. However, because of their high risk and relatively low product value, next-generation process adoption is lagging behind due to these factors. However, the individual worth of carbon and energy in wastewater will also need to be considered in the future. Up until now, the primary focus has been on producing biofuels, namely biomethane, leaving most of the nutrients, which are sometimes lost or stay concentrated in sludge to be recovered.

Eventually, the concentrated carbon is released as CO_2, even if this has led to resource use. There has been a recent shift in the understanding of wastewater's worth, as it is now being used to produce byproducts including liquid biofuels, biopolymers, commodity chemicals, and single-cell protein for animal feed. To truly envision the effects of recovering and recycling these goods into a global circular economy, though, is yet premature. Although recycling nutrients is frequently used for land application, management issues, and technological flaws arise. The public's acceptability of nutrient recycling, the trade-offs of technologies at various sizes, and the framework for comparing systems should be the main topics of future research.

Further research using a decentralized approach is necessary due to the challenges connected with water reuse, which include the high cost of tertiary treatment and the energy required for pumping the water back to the user via a centralized treatment unit. Most of the time, when pilot or demonstration facilities are constructed, the supporting technologies are still not at a stage where they are ready for widespread use. This means that specialized economic assessments, including life cycle assessments, are required.

5. CONCLUSION

The goal of a WWTP is to minimize the harm that human activity causes to the receiving stream and environment while reducing the pollution of wastewater. However, in the process, they also add to environmental pollution by consuming energy to run the treatment system or by employing materials or chemicals that are also linked to emissions. The main issue with WWTP operation is greenhouse gas (GHG) emissions. According to earlier research, a typical WWTP's overall environmental effects are always negative. Onsite energy generation, on the other hand, allows us to counteract the negative environmental effects of wastewater treatment by using organic load (such as water flow, residual heat, and huge areas) and nutrient recycling (as fertilizer).

Resource and energy recovery could even improve the environment as a whole, helping WWTPs get closer to having a net-zero impact. The water itself is the most valuable resource found in wastewater. Its origin cannot be changed, unlike energy, which has various sources. Common water reuse through effluent recycling is insufficient to achieve a net-zero impact of the WWTPs on the entire environment, according to the resource recovery-based assessment. The ecology as a whole can greatly benefit from energy recovery. The main obstacles to resource and energy recovery include high capital costs, unreliability, and climate-specific requirements.

One common issue with nutrient recycling is handling and technological imperfections. When it comes to reducing the environmental impact, phosphorus recovery contributes less than burning ash reuse for building materials. Nevertheless, recovery is required for additional reasons since phosphate is a resource that is actually limited. It is only advised to recover ammonium if the procedure uses less energy than traditional ammonium manufacturing. The wastewater treatment system uses less energy and less of each material when integrated resource recovery is used. Consequently, it is possible to enhance the wastewater treatment system's sustainability.

REFERENCES

Allison, R.P. (1995). Electrodialysis reversal in water reuse applications. *Desalination*, *103*, 11–18. https://doi.org/10.1016/0011-9164(95)00082-8

Biń, A.K., & Sobera-Madej, S. (2012). Comparison of the advanced oxidation processes (UV, UV/H_2O_2 and O_3) for the removal of antibiotic substances during wastewater treatment. *Ozone Sci. Eng.*, *34*, 136–139. https://doi.org/10.1080/01919512.2012.650130

Bohra, V., Ahamad, K. U., Kela, A., Vaghela, G., Sharma, A., & Deka, B. J. (2022). Energy and resources recovery from wastewater treatment systems. In *Clean Energy and Resource Recovery*. Elsevier, pp. 17–36. https://doi.org/10.1016/B978-0-323-90178-9.00007-X

Cipolla, S.S., & Maglionico, M. (2014). Heat recovery from urban wastewater: Analysis of the variability of flow rate and temperature. *Energy Build*, 69, 122–130. https://doi.org/10.1016/j.enbuild.2013.10.017

Clarens, A. F., Resurreccion, E. P., White, M. A., & Colosi, L. M. (2010). Environmental life cycle comparison of algae to other bioenergy feedstocks. *Environ. Sci. Technol*, *44*, 1813–1819. https://doi.org/10.1021/es902838n

Corbala-Robles, L., Ronsse, F., Pieters, J. G., & Volcke, E. I. P. (2018). Heat recovery during treatment of highly concentrated wastewater: Economic evaluation and influencing factors. *Water Sci. Technol*, 78, 2270–2278. https://doi.org/10.2166/wst.2018.507

Das, P.P., Samanta, N.S., Dhara, S., & Purkait, M.K. (2023a). Biofuel production from algal biomass. In *Green approach to alternative fuel for a sustainable future*, Elsevier, pp. 167–179. https://doi.org/10.1016/B978-0-12-824318-3.00009-6

Das, P. P., Sontakke, A. D., Samanta, N. S., & Purkait, M. K. (2023b). Emerging contaminants in wastewater: eco-toxicity and sustainability assessment. In Shah, M.P. (Ed.), *Industrial wastewater reuse.* Springer Nature Singapore, Singapore, pp. 63–87. https://doi.org/10.1007/978-981-99-2489-9_4

De La Noüe, J., & Bassères, A. (1989). Biotreatment of anaerobically digested swine manure with microalgae. *Biol. Wastes*, 29, 17–31. https://doi.org/10.1016/0269-7483(89)90100-6

Dhara, S., Samanta, N.S., Uppaluri, R., & Purkait, M.K. (2023a). High-purity alkaline lignin extraction from Saccharum ravannae and optimization of lignin recovery through response surface methodology. *Int. J. Biol. Macromol.*, 234, 123594. https://doi.org/10.1016/j.ijbiomac.2023.123594

Dhara, S., Shekhar Samanta, N., Das, P. P., Uppaluri, R. V. S., & Purkait, M. K. (2023b). Ravenna grass-extracted alkaline lignin-based polysulfone mixed matrix membrane (MMM) for aqueous Cr(VI) removal. *ACS Appl. Polym. Mater., 5,* 6399–6411. https://doi.org/10.1021/acsapm.3c00999

Dhara, S., Sontakke, A.D., Shekhar Samanta, N., Venkata Sai Uppaluri, R., & Purkait, M.K. (2024). Biowaste-alkaline lignin and GO integrated polysulfone ultrafiltration membrane fabrication for Pb2+ and Eosin Y dye removal. *Sep. Purif. Technol., 341,* 126894. https://doi.org/10.1016/j.seppur.2024.126894

Fernández-Dacosta, C., Posada, J. A., Kleerebezem, R., Cuellar, M. C., & Ramirez, A. (2015). Microbial community-based polyhydroxyalkanoates (PHAs) production from wastewater: Techno-economic analysis and ex-ante environmental assessment. *Bioresour. Technol., 185,* 368–377. https://doi.org/10.1016/j.biortech.2015.03.025

Funamizu, N., Iida, M., Sakakura, Y., & Takakuwa, T.(2001). Reuse of heat energy in wastewater: implementation examples in Japan. *Water Sci. Technol., 43,* 277–285. https://doi.org/10.2166/wst.2001.0640

Gillot, S., & Vanrolleghem, P. A. (2003). Equilibrium temperature in aerated basins—comparison of two prediction models. *Water Res., 37,* 3742–3748. https://doi.org/10.1016/S0043-1354(03)00263-X

Hu, W., Zhou, W., Zhang, K., Zhang, X., Wang, L., Jiang, B., … & Fu, H. (2016). Facile strategy for controllable synthesis of stable mesoporous black TiO_2 hollow spheres with efficient solar-driven photocatalytic hydrogen evolution. *J. Mater. Chem. A., 4,* 7495–7502. https://doi.org/10.1039/C6TA01928E

Ikumi, D. S., Harding, T. H., & Ekama, G. A. (2014). Biodegradability of wastewater and activated sludge organics in anaerobic digestion. *Water Res.*, 56, 267–279. https://doi.org/10.1016/j.watres.2014.02.008

Janssen, M., Tramper, J., Mur, L.R., & Wijffels, R. H. (2003). Enclosed outdoor photobioreactors: Light regime, photosynthetic efficiency, scale-up, and future prospects. *Biotechnol. Bioeng.*, 81, 193–210. https://doi.org/10.1002/bit.10468

Jiang, C., Cui, S., Han, Q., Li, P., Zhang, Q., Song, J., & Li, M. (2019). Study on application of activated carbon in water treatment. IOP Conf. Ser. *Earth Environ. Sci., 237,* 022049. https://doi.org/10.1088/1755-1315/237/2/022049

Jimenez, J., Miller, M., Bott, C., Murthy, S., De Clippeleir, H., & Wett, B. (2015). High-rate activated sludge system for carbon management – Evaluation of crucial process mechanisms and design parameters. *Water Res., 87,* 476–482. https://doi.org/10.1016/j.watres.2015.07.032

Karungamye, P., Rugaika, A., Mtei, K., & Machunda, R. (2023). The removal of ciprofloxacin from synthetic wastewater in constructed wetland. *HydroResearch, 6*, 138–146. https://doi.org/10.1016/j.hydres.2023.04.001

Kehrein, P., Van Loosdrecht, M., Osseweijer, P., Garfí, M., Dewulf, J., & Posada, J.(2020). A critical review of resource recovery from municipal wastewater treatment plants – market supply potentials, technologies and bottlenecks. *Environ. Sci. Water Res. Technol., 6*, 877–910. https://doi.org/10.1039/C9EW00905A

Kumar, H. D., Yadava, P. K., & Gaur, J. P. (1981). Electrical flocculation of the unicellular green alga Chlorella vulgaris Beijerinck. *Aquat. Bot., 11,* 187–195. https://doi.org/10.1016/0304-3770(81)90059-0

Lau, P. S., Tam, N. F. Y., & Wong, Y. S.(1995). Effect of algal density on nutrient removal from primary settled wastewater. *Environ. Pollut., 89*, 59–66. https://doi.org/10.1016/0269-7491(94)00044-E

Li, K., Liu, Q., Fang, F., Luo, R., Lu, Q., Zhou, W., … & Ruan, R. (2019). Microalgae-based wastewater treatment for nutrients recovery: A review. *Bioresour. Technol., 291,* 121934. https://doi.org/10.1016/j.biortech.2019.121934

Liu, Y., Gu, J., & Zhang M. (2020). Approaches to energy and resource recovery from municipal wastewater. In, *A-B Processes: Towards energy self-sufficient municipal wastewater treatment.* IWA Publishing (pp. 29–67). https://doi.org/10.2166/9781789060089_0029

Mofijur, M., Hasan, M. M., Sultana, S., Kabir, Z., Djavanroodi, F., Ahmed, S.F., … & Khan, T. M. Y. (2023). Advancements in algal membrane bioreactors: Overcoming obstacles and harnessing potential for eliminating hazardous pollutants from wastewater. *Chemosphere, 336,* 139291. https://doi.org/10.1016/j.chemosphere.2023.139291

Molina Grima, E., Belarbi, E.-H., Acién Fernández, F.G., Robles Medina, A., & Chisti, Y. (2003). Recovery of microalgal biomass and metabolites: process options and economics. *Biotechnol. Adv., 20*, 491–515. https://doi.org/10.1016/S0734-9750(02)00050-2

Mondal, P., Samanta, N. S., Kumar, A., & Purkait, M. K. (2020). Recovery of H2SO4 from wastewater in the presence of NaCl and KHCO3 through pH responsive polysulfone membrane: Optimization approach. *Polym. Test., 86*, 106463. https://doi.org/10.1016/j.polymertesting.2020.106463

Mondal, P., Samanta, N. S., Meghnani, V., & Purkait, M. K. (2019). Selective glucose permeability in presence of various salts through tunable pore size of pH responsive PVDF-co-HFP membrane. *Sep. Purif. Technol., 221*, 249–260. https://doi.org/10.1016/j.seppur.2019.04.001

Możejko-Ciesielska, J., & Kiewisz, R.(2016). Bacterial polyhydroxyalkanoates: Still fabulous? *Microbiol. Res., 192,* 271–282. https://doi.org/10.1016/j.micres.2016.07.010

Obotey Ezugbe, E., & Rathilal, S. (2020.) Membrane technologies in wastewater treatment: A review. *Membranes, 10,* 89. https://doi.org/10.3390/membranes10050089

Preisner, M., Neverova-Dziopak, E., & Kowalewski, Z.(2020). An analytical review of different approaches to wastewater discharge standards with particular emphasis on nutrients. *Environ. Manage., 66,* 694–708. https://doi.org/10.1007/s00267-020-01344-y

Priyadarshini, M., Ahmad, A., Das, S., & Ghangrekar, M.M. (2022). Application of innovative electrochemical and microbial electrochemical technologies for the efficacious removal of emerging contaminants from wastewater: A review. *J. Environ. Chem. Eng., 10,* 108230. https://doi.org/10.1016/j.jece.2022.108230

Puyol, D., Batstone, D. J., Hülsen, T., Astals, S., Peces, M., & Krömer, J.O. (2017). Resource recovery from wastewater by biological technologies: Opportunities, challenges, and prospects. *Front. Microbiol., 7.* https://doi.org/10.3389/fmicb.2016.02106

Rambabu, K., Avornyo, A., Gomathi, T., Thanigaivelan, A., Show, P.L., & Banat, F.(2023). Phycoremediation for carbon neutrality and circular economy: Potential, trends, and challenges. *Bioresour. Technol., 367,* 128257. https://doi.org/10.1016/j.biortech.2022.128257

Reddy, C.S.K., Ghai, R., Rashmi, & Kalia, V.C. (2003). Polyhydroxyalkanoates: an overview. *Bioresour. Technol., 87,* 137–146. https://doi.org/10.1016/S0960-8524(02)00212-2

Salehizadeh, H., & Van Loosdrecht, M. C. M. (2004). Production of polyhydroxyalkanoates by mixed culture: Recent trends and biotechnological importance. *Biotechnol. Adv., 22,* 261–279. https://doi.org/10.1016/j.biotechadv.2003.09.003

Samanta, N. S., Anweshan, Mondal, P., Bora, U., & Purkait, M. K. (2023a). Synthesis of precipitated calcium carbonate from LD-slag using CO2. *Materials Today Commun, 36,* 106588. https://doi.org/10.1016/j.mtcomm.2023.106588

Samanta, N. S., Anweshan, & Purkait, M. K. (2023b). Techniques in removal of organics and emerging contaminants from wastewater for water reuse application, In, *Development in wastewater treatment research and processes.* Elsevier (pp. 73–96). https://doi.org/10.1016/B978-0-323-95684-0.00009-9

Samanta, N. S., Banerjee, S., Mondal, P., Anweshan, Bora, U., & Purkait, M. K. (2021). Preparation and characterization of zeolite from waste Linz-Donawitz (LD) process slag of steel industry for removal of Fe3+ from drinking water. *Adv. Powder Technol., 32,* 3372–3387. https://doi.org/10.1016/j.apt.2021.07.023

Samanta, N. S., Das, P.P., Mondal, P., Bora, U., & Purkait, M. K. (2022a). Physico-chemical and adsorption study of hydrothermally treated zeolite A and FAU-type zeolite X prepared from LD (Linz–Donawitz) slag of the steel industry. *Int. J. Environ. Anal. Chem.*, 1–23. https://doi.org/10.1080/03067319.2022.2079082

Samanta, N. S., Das, P. P., Mondal, P., Changmai, M., & Purkait, M. K. (2022b). Critical review on the synthesis and advancement of industrial and biomass waste-based zeolites and their applications in gas adsorption and biomedical studies. *J. Indian Chem. Soc., 99,* 100761. https://doi.org/10.1016/j.jics.2022.100761

Samanta, N. S., Mondal, P., & Purkait, M. K. (2023c). Nanofiltration technique for the treatment of industrial wastewater. In Shah, M. P. (Ed.), *Advanced application of nanotechnology to industrial wastewater* (pp. 165–190). Springer Nature Singapore, Singapore. https://doi.org/10.1007/978-981-99-3292-4_9

Samanta, N. S., P. Das, P., Sharma, M., & Purkait, M. K. (2023d). Recycle of water treatment plant sludge and its utilization for wastewater treatment. In, *Resource recovery in drinking water treatment* (pp. 239–264). Elsevier. https://doi.org/10.1016/B978-0-323-99344-9.00010-4

Samanta, N. S., Mondal, P., & Purkait, M. K. (2024). Waste-based zeolites and their advanced composites for wastewater and environmental remediation applications. In Ikhmayies, S. J. (Ed.), *Advanced composites, advances in material research and technology* (pp. 451–490). Springer Nature Switzerland, Cham. https://doi.org/10.1007/978-3-031-42731-2_14

Selatile, M. K., Ray, S. S., Ojijo, V., & Sadiku, R. (2018). Recent developments in polymeric electrospun nanofibrous membranes for seawater desalination. *RSC Adv., 8,* 37915–37938. https://doi.org/10.1039/C8RA07489E

Shafiquzzaman, M. D., Haider, H., & Ashadullah, A. K. M. (2021). Optimization of algal-based membrane bioreactor for greywater treatment. *Process Saf. Environ. Prot., 154,* 81–88. https://doi.org/10.1016/j.psep.2021.08.007

Sharma, M., Samanta, N. S., Chakraborty, A., & Purkait, M. K. (2023). Simultaneous treatment of industrial wastewater and resource recovery using microbial fuel cell. In, *Resource recovery in industrial waste waters* (pp. 621–637). Elsevier. https://doi.org/10.1016/B978-0-323-95327-6.00002-6

Shekhar Samanta, N., Das, P. P., Dhara, S., & Purkait, M. K. (2023a). An overview of precious metal recovery from steel industry slag: Recovery strategy and utilization. *Ind. Eng. Chem. Res., 62,* 9006–9031. https://doi.org/10.1021/acs.iecr.3c00604

Shekhar Samanta, N., Mondal, P., Dhara, S., Bora, U., & Purkait, M. K. (2023b). Fabrication of LD-slag derived zeolite Y coated polysulfone (PSf) membrane for decontamination of groundwater. *Chem. Eng. J., 478,* 147330. https://doi.org/10.1016/j.cej.2023.147330

Sontakke, A. D., Deepti, Samanta, N. S., & Purkait, M. K. (2023). Smart nanomaterials in the medical industry. In, *Advances in Smart nanomaterials and their applications* (pp. 23–50). Elsevier. https://doi.org/10.1016/B978-0-323-99546-7.00025-2

Tortajada, C., & Van Rensburg, P. (2020). Drink more recycled wastewater. *Nature*, *577*, 26–28. https://doi.org/10.1038/d41586-019-03913-6

Wan, J., Gu, J., Zhao, Q., & Liu, Y. (2016). COD capture: A feasible option towards energy self-sufficient domestic wastewater treatment. *Sci. Rep.*, *6*, 25054. https://doi.org/10.1038/srep25054

Zribi, I., Zili, F., Ben Ali, R., Masmoudi, M.A., Karray, F., Sayadi, S., ... & Chamkha, M. (2023). Trends in microalgal-based systems as a promising concept for emerging contaminants and mineral salt recovery from municipal wastewater. *Environ. Res.*, *232,* 116342. https://doi.org/10.1016/j.envres.2023.116342

2 Effective Removal and Remediation of Heavy Metals and Emerging Pollutants from Industrial Wastewater

Ponnusamy Thillai Arasu,
Arumugam Murugan,
A. Mohamed Azharudeen,
G. Kanthimathi, Natarajan Raman,
S. Arun Prabhu and G. Elango

1. INTRODUCTION

Today, environmental contamination poses a severe threat to the entire world. Large-scale urbanization and industrialization have significantly increased environmental pollution. It is nearly impossible to eradicate environmental contamination from the planet, despite the fact that its effects are well recognized by people everywhere. This is a result of people's strong desire to live in a technologically advanced society and reap the rewards of science and technology. Therefore, although pollution in the environment is unavoidable, it can be greatly reduced by using the various pollution management methods, which are offered by science and technology. Adopting pollution control systems is

DOI:10.1201/9781003496014-2

quite challenging for industries because of their high cost. The scientific community is now more interested in developing low-cost pollution management methods as a result of this issue. Many studies are being conducted these days to identify low-cost pollution mitigation methods. Using binary combinations of metal ions solutions, the researcher is attempting to remove heavy metal ions at a low cost using the current study. Because the study's conclusions can be used by manufacturers and industries to treat their effluents more economically, it is highly relevant. The most necessary component for life to exist on Earth is water. The ocean contains 97% of all the water on Earth, which is too salty for agriculture or drinking (Musie & Gonfa, 2023). Three percent is made up of fresh water. Of this, glaciers and ice caps contain 2.997% of the total. As a result, only 0.003% of the water on Earth is readily accessible as water vapor, soil moisture, groundwater, and water from lakes, streams, rivers, and wet areas. Mother Nature has given all living things the pure type of water that exists on Earth. Water pollution results from any external factor altering the ecosystem or from pollutants contaminating it. According to Sharma and Bhattacharya (2017) water pollution is the introduction of any material into water or modifications made to its physical, chemical, and biological properties that could have a negative impact on aquatic biota and humans. The physico-chemical properties of water are governed by several super-governing processes seen in nature (Arasu et al, 2016). The main causes of water pollution are anthropogenic processes like industrial, agricultural, urban, domestic, radioactive, mining sources, use of pesticides and fertilizers by farming communities, and natural processes that bring decomposed vegetable, animal, and weathered products into main water resources. Excessive levels of inorganic minerals, heavy metals, trace metals, organic molecules, and coloured pigments eventually contaminate drinkable water. The pollutants which are hazardous have been classified as (a) Heavy metals: zinc, nickel, lead, copper, chromium, mercury, etc.; (b) trace organics include amines, haloacetic acids, and trihalomethanes; (c) fertilizers and pesticides; and (d) both metallic and azo dyes.

Water-borne illnesses such as dysentery, cholera, typhoid, jaundice, and diarrhoea are mostly caused by microbial pollution as a result of inadequate sanitation (Muthulakshmi et al., 2015). Of all the contaminants found in water, heavy metals are said to be particularly hazardous.

One of the contaminants in wastewater exhausted by the industries that lingers the longest is heavy metals, sometimes referred to as trace metals. Numerous industries pour off top levels of substantial metal discharge into water reservoirs possess large harmful effects on the surroundings and human fitness (Akpor et al., 2014). Various sectors generate heavy metal-containing industrial effluent streams. Significant amounts of effluent comprising heavy metals (such as cadmium, zinc, lead, chromium, nickel, copper, vanadium, platinum, silver, and titanium) are produced by electroplating and metal surface treatment operations from a range of applications. These consist of milling, etching, conversion-coating, anodizing-cleaning, electroplating, and electroless depositions (Saxena et al., 2017). The production of printed circuit boards (PCBs) is a noteworthy further producer of heavy metal wastes. The most fascinating resistant overplates are made of tin, lead, and nickel solder. In addition, the following industries produce metal wastes: the wood processing sector, which uses a chromated copper-arsenate wood treatment to produce wastes containing arsenic; the inorganic pigment manufacturing sector, which creates pigments containing cadmium sulphide and chromium compounds; the petroleum purifying sector, that produces exchange catalysts tainted with Ni, V, and Cr; and the photographic industry, which produces film having extraordinary silver and ferrocyanide concentrations. Each of these generators generates an enormous quantity of residues, sludges, and wastewater that fall under the category of hazardous wastes that need to be thoroughly treated (Odumbe et al., 2023).

Furthermore, people can come into contact with heavy metals through various ways, such as consuming them in food and drinks, inhaling them as dust or fumes of being exposed to them through vaporization (Odumbe et al., 2023). Heavy metals can have detrimental effects on aquatic ecosystems, including causing the death of aquatic species, triggering algal blooms, damaging habitats through settlings and debris, increasing water flow and inducing various short- and long-term toxicities as observed through chemical examination. High levels of heavy metals in soil hinder the plant growth, nutrient absorption and essential metabolic functions ultimately diminishing both the quality and quantity of food produced. Impaired development and expansion, cancer, injury to organs, nervous system destruction, and in severe cases, death, are examples of severe impacts on animals (Halawa & Zabin, 2017; Ince & Ince, 2017).

Because of this, heavy metal ions are most troubling environmental concerns. As a result of heavy metals' detrimental effects on human health, there has been a constant focus on their disposal as they pose a rising hazard to the environment (Halawa & Zabin, 2017; Saxena et al., 2017). Due to their toxicity and non-biodegradability, metal ions pose health dangers to all living organisms. Hence, removing heavy metals from industrial waste is a very significant candidate of discussion. There are numerous remediation techniques available to diminish the harmful effects of heavy metals on the environment, wildlife, and human health (Shakir et al., 2017). They encompass, ion change, membrane filtration, electrodialysis, photocatalysis, advance oxidation process (AOP) technique, and biological treatment (Abdullah et al., 2019; Elsayed, 2018; Karnib et al., 2014; Turkmen et al., 2022; Verma et al., 2013).

2. REMOVAL OF HEAVY METALS METHODS

The elimination of heavy metal toxicity, which involves both an environmental and a technological component, and the recovery of precious metals are the main goals of the heavy metals removal process from industrial effluents and wastewater systems. Currently, a variety of physico-chemical techniques have been employed to remove and recover heavy metals from various materials. These techniques include oxidation and reduction, precipitation, ion-exchange, electrochemical methods, reverse osmosis, electro dialysis, and foam flotation, bio-sorption, and adsorption (Carmona & Abejón, 2021; Kapepula & Luis, 2024; Yadav, et al., 2021).

In addition to competing with several other technologies to meet the demands of the various industries involved, electrochemical approaches have the distinct ability to recover pure metal for recycling. Many other approaches are not viable economically or do not meet the strict guidelines established for industrial effluents. It has been discovered that the precipitation process is inappropriate for effluents with elevated metal concentrations. It also has to do with the sludge's disposal issue. While effective, ion-exchange, reverse osmosis, and electrodialysis are costly and inefficient processes. The traditional approaches to meet the necessary high standards are increasingly becoming inefficient and unfeasible from an economic standpoint. One potential substitute for the current methods for removing and recovering heavy metals from industrial and natural water environments is the adsorption of heavy metals on various adsorbents. It has become a popular and affordable method of metal removal; less problematic and inexpensive. The adsorption technique, which is thought to be the most efficient, quick, and sludge-free approach for treating industrial effluents and removing heavy metal ions from wastewater, is quickly gaining popularity. Furthermore, finding innovative water treatment methods is crucial for reducing the concentration of heavy metals in water. Not only can these metals be harmful to the environment, but they can also pose serious health problems to humans due to their toxicity and tendency to accumulate in the body over time (Topare et al., 2023).

2.1. Ion Exchange Method

The undesirable metal ion can be replaced with a safe, environmentally friendly alternative using a reversible chemical process called the ion exchange procedure (Barman, 2023). As seen in Figure 2.1, a heavy metal ion is extracted from a wastewater solution by bonding it to a stationary solid particle in place of the solid particle cation. Solid ion-exchange particles can be made of naturally occurring materials like inorganic zeolites or artificially made materials like organic resins.

Applying the ion exchange technique to waste fluids is an effective way to safeguard the environment. New or enhanced classical materials have been examined as ion exchangers in recent years (Kumar et al., 2017). Aqueous discharges from industrial to nuclear sources contain a wide range of contaminants, including metal ions like Cd^{2+}, Ni^{2+}, V^{4+}, V^{5+}, Cr^{3+}, Cr^{4+}, Pb^{2+}, and Hg^{2+}; radionuclides like, ^{235}U, ^{137}Cs, ^{89}Sr, ^{59}Fe, ^{65}Zn, and ^{57}Co; and anionic species like CrO_4^{2-} and AsO_4^{3-}. Several removal approaches were investigated for all these metal ions, ranging from precipitation (Barman, 2023), which is a cheap and straightforward procedure rather than other techniques have more expensive and some drawback as precipitation (Narayani & Shetty, 2013), which generates a large amount of sludge.

Ion exchange is recognized as an appropriate and economical method in this area, with tremendous potential for used exchanger regeneration. Typically, exchange materials consist of traditional materials like clay, inorganic zeolites, or organic polymers. For instance, industrial effluents treatment utilizes organic compounds containing sulfonate, amino, thiol, or amide functional groups for removing cadmium and nickel (Chitpong & Husson, 2017). The effectiveness of ion exchange in eliminating these metals hinges on the chemical selectivity of the involved species.

In general, polymers containing carboxylate groups had the same level of efficacy in eliminating residues of cadmium, nickel, copper, and lead from wastewater (Kyzas et al., 2014). Even though these standard materials produce some positive results, there are a few drawbacks that are significant enough to be discussed, including low reuse capacity, kinetics, and thermal and chemical endurance for the reuse of organic resins.

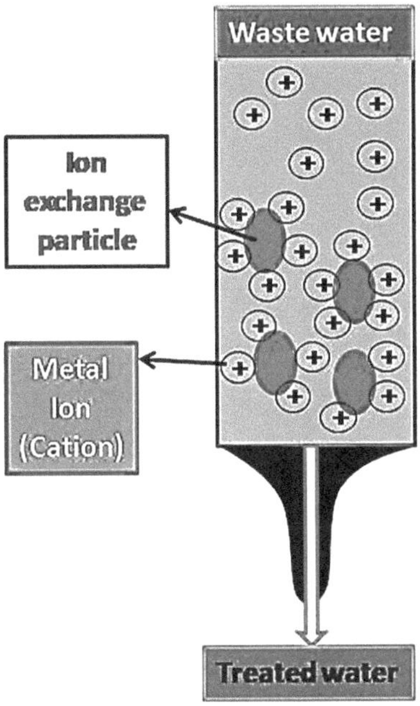

FIGURE 2.1 Ion exchange method for recovery of heavy metals from wastewater. Figure created by author.

This is the rationale behind the seeming need to find and evaluate new substances with ion exchange potential. As a result, obtaining and testing superior materials with large surface area, multiple functional groups, and thermal and chemical resistance has become more popular in recent years (Kumar et al., 2017). In this context, metal-organic ion exchange materials combine conventional materials with high porosity and binding groups (Liu et al., 2014; Rapti et al., 2016). An instance of such materials includes zeolite nanoparticles impregnated with polysulfone, utilized for eliminating lead and mercury from wastewater (Kumar et al., 2017) (Figure 2.1).

2.2. Membrane Filtration

When a driving force moves across a membrane, a separation technique called membrane filtration uses the membrane to separate substrates. Substrates are separated via membrane filtration according to their sizes, membrane pore sizes, solution concentrations, and applied pressures. Membrane methods have drawn more attention in the last few decades for the management of heavy metals (Xiang et al., 2022). Various types of membrane filtration for removing heavy metals exist, each based on the size of the largest particle they can retain. Microfiltration (MF), ultrafiltration (UF), nanofiltration (NF), and reverse osmosis (RO) are among the various varieties. Any of these kinds can be used to filter industrial effluent for heavy metals. It is now essential to search for novel heavy metal removal solutions, such as membrane technology. The ideal solution must be affordable, scalable, adaptable to field settings, and able to reduce the levels of heavy metals to the predetermined level. Recently, membrane filtration is fascinating to offer the prospects and challenges for heavy metal removal.

2.2.1. Microfiltration

Microfiltration is a procedure involving membranes where separation is achieved by sieving the membrane and using the static pressure variation as the driving force. Microfiltration uses a similar separation technique to regular filtration, but it has a better filtration accuracy and can intercept organic macromolecules or particles as small as 0.13 μm to 15 μm. The separation process of microfiltration is variable due to the shape of the microporous membrane; it may be broadly classified into three categories: adsorption interception, mechanical interception, and bridging, with physical interception being the primary mechanism (Sun et al., 2015).

Microfiltration membranes typically have pore sizes between 0.1 μm and 1 μm, which means that while dissolved solids and macromolecular organic molecules can flow through, heavy metal ions cannot be directly intercepted by the membrane, nor can pollutants in the form of dissolved solids smaller than 1 μm be removed. Furthermore, microfiltration does not completely block the virus. Microfiltration, however, can manage these microorganisms in water when combined with disinfection (Torki et al., 2017). To treat contaminated water, microfiltration technology can also be used in conjunction with other procedures. It is crucial to the process of treating industrial wastewater containing heavy metals and radioactive materials.

2.2.2. Ultrafiltration (UF)

For the purification of industrial wastewater, membrane filtration technology, specifically ultrafiltration, is frequently utilized. In this improved treatment of wastewater from the printing and dyeing industry, paper-making industry, greasy wastewater, heavy metal wastewater, and food industry, ultrafiltration technology offers strong technological advantages and a wide range of applications. When ultrafiltration is used in conjunction with membrane filtration technology, wastewater treated by this method can have a higher water flux and meet industry standards for water quality in terms of colorimetric wastewater, chemical oxygen demand (COD), and electrical conductivity (Li et al., 2018).

The removal of heavy metals from industrial wastewater has shown promising results when various materials, including zeolites, metal-organic frameworks (MOFs) and graphene oxide (GO), are incorporated into ultrafiltration membranes (Ursino et al., 2022). Furthermore, the technology of antifouling ultrafiltration membranes, such as those prepared using polyacrylic acid (PAA), has enhanced heavy metal rejection efficiency and fouling resistance. Ultrafiltration membranes present a viable option for the actual heavy metal elimination from water sources (Xiang et al., 2022).

2.2.3. Nanofiltration (NF)

A novel and efficient technique for removing heavy metals from sources of tainted water is nanofiltration (NF). Using semi-permeable membranes with pore sizes between 1 to

10 nm, this advanced filtering technique allows for the specific elimination of heavy metal ions from industrial wastewater while retaining essential minerals and nutrients. Because of this, the nanofiltration system is the greatest way to get rid of heavy metals (Li et al., 2017). The application of nanofiltration to remove heavy metals such as copper, cobalt, zinc, cadmium, mercury, lead, iron, chromium, nickel, manganese, antimony, and arsenic (Bera et al., 2018; Dadari et al., 2022; Kamari & Shahbazi, 2021; Li et al., 2017; Mahmoud et al., 2021) is becoming more and more popular.

2.2.4. Reverse osmosis (RO)

This approach utilizes a semi-permeable membrane that only permits the purified fluid to flow through, separating the contaminant. The procedure is widely employed in the method of desalination and has the potential to eliminate a wide variety of species (Inyang et al., 2016). Several industries can produce wastewater that contains cadmium and copper. The potential of employing reverse osmosis (RO) methods to purify wastewater containing copper and cadmium ions, with the aim of reducing freshwater consumption and mitigating environmental degradation. Samples of synthetic wastewater with varying concentrations of Cu^{2+} and Cd^{2+} ions were created, and RO was used to treat them. Consequently, 98% and 99% of the Cu^{2+} and Cd^{2+}, respectively, were obtained using the RO procedure (Thaci & Gashi, 2019). Therefore, RO plays a dramatic candidate for the treatment of industrial effluent. The schematic diagram of RO is demonstrated below (Figure 2.2).

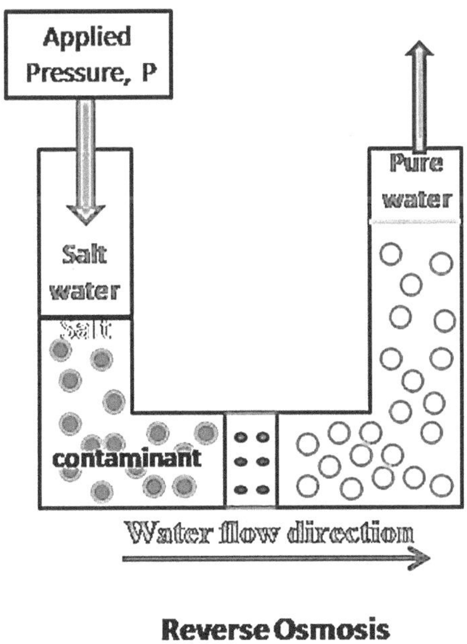

FIGURE 2.2 Reverse osmosis for recovery of heavy metals from wastewater.

Figure created by author.

2.3. Electrodialysis

Electrodialysis (ED) is a kind of membrane separation where the ionized species in the solution move through an ion exchange membrane under the influence of an electric potential. The membranes consist of thin plastic sheets that possess either cationic or anionic properties (Gurreri et al., 2020). The fundamentals of electrodialysis are illustrated in Figure 2.3, where there are ionic species in a solution moving across cell compartments, anions are drawn towards the anode while cations are attracted towards the cathode.

Under comparable operating conditions, two cation-exchange membranes, perfluorosulfonic Nafion 117 and sulfonated polyvinylidene fluoride membrane (SPVDF), were evaluated. Co (II) and Ni (II) had removal efficiencies of 90% and 69%, respectively, with initial concentrations of metals of 0.84 and 11.72 mg/L when perfluorosulfonic Nafion 117 was used (Avci et al., 2020). Using an ED cell, industrial effluent like lead elimination was examined under the effects of temperature, voltage, and flow rate at various concentrations utilizing two different kinds of commercial membranes (Ponomar et al., 2024).

In ED, the industrial effluent of Cd (II) is removed by higher exchange of ions capacity membranes, which produced superior cell functionality (Juve et al., 2022). Throughout the cleanup, an agitated suspension of sludge from wastewater was in the

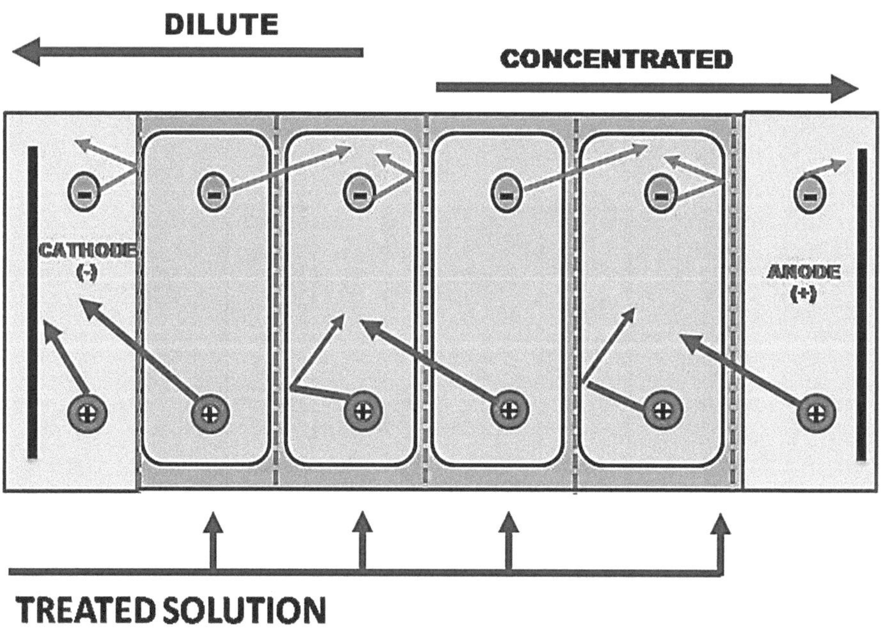

FIGURE 2.3 Application of electrolysis for industrial wastewater treatment.

Figure created by author.

presence of a DC electric field. The ratio of liquid to solid (mL/g fresh sludge) ranged from 1.4 to 2. The experiment was conducted in three sets; the sludge was suspended in distilled water/HNO_3/citric acid. The removal of Cd (II) was determined to be 69%, 67%, and 70%. Basic electrochemistry principles and experimental data on copper ion separation were used to describe the ED process (Alkhadra et al., 2022).

The industrial wastewater containing lead, chromium and zinc ions were also subjected to this experiment. It was discovered that an ED cell's performance is mostly independent of the kind of ions present and is only influenced by the cell's structure and operating conditions. Despite its drawbacks, ED has benefits for treating heavy metal-laden wastewater, including the capacity to reject unwanted contaminants from water and provide a highly concentrated channel for recovery. Important metals like chromium and copper can also be extracted. Being a membrane process, ED needs clean feed, cautious operation, and routine maintenance to avoid stack damage (Avci et al., 2020; Alkhadra et al., 2022).

2.4. Photocatalysis

Several technologies are employed for treating industrial wastewater, with some newer methods being employed as well. One of the newer methods is sophisticated oxidation method. Photocatalysis is a well-liked technique that shows promise in a few domains (Argurio et al., 2018; Gao et al., 2020; Yao et al., 2020a; Yao et al., 2021b). A scheme illustrating the fundamentals of the photocatalytic process is depicted in Figure 2.4. Photocatalysts typically feature semiconductor architectures. When the energy of a photon in incoming light matches or exceeds the bandgap energy of the semiconductor, electrons (e^-) in the valence band (VB) can move to the conduction band. This process can create a positively charged hole (h^+) in the VB when the semiconductor material is exposed to light.

When treating heavy metals, this pair of h^+ and e^- can drift to the semiconductor's surface and experiences an array of reduction and oxidation behaviours, which are represented by the alteration of various valence states (Ullah et al., 2021; Ulyankina et al., 2020; Wang et al., 2021). The wide bandgap's high energy requirements for incident light and the readily recombinant nature of photogenerated carriers (e+ and h+) are the primary limiting factors in the photocatalysis process. The greatest research has been done on TiO_2, which is the most common photocatalyst. Its broad bandgap, however, is exclusively sensitive to UV light (Li et al., 2018; Mukherjee et al., 2014; Safaei et al., 2018). Consequently, a greater number of photocatalysts with appropriate and effective bandgaps have been created and put to use, including bi-based catalysts, ZnO and C_3N_4 (Liang et al., 2020). Numerous studies on photocatalysis for treating organic wastewater processing and splitting of water to produce hydrogen and oxygen, and nitrogen fixation (Li et al., 2020) are now in progress. The generated active species have strong redox capabilities and don't generate extra pollution because photocatalysis only needs light as energy source (Zhu et al., 2019). Due to its distinct benefits, photocatalysis has enormous potential for treating heavy metals (Meng et al., 2019) (Figure 2.4). Consequently, it has also caught the interest of other scholars (Ni et al., 2014).

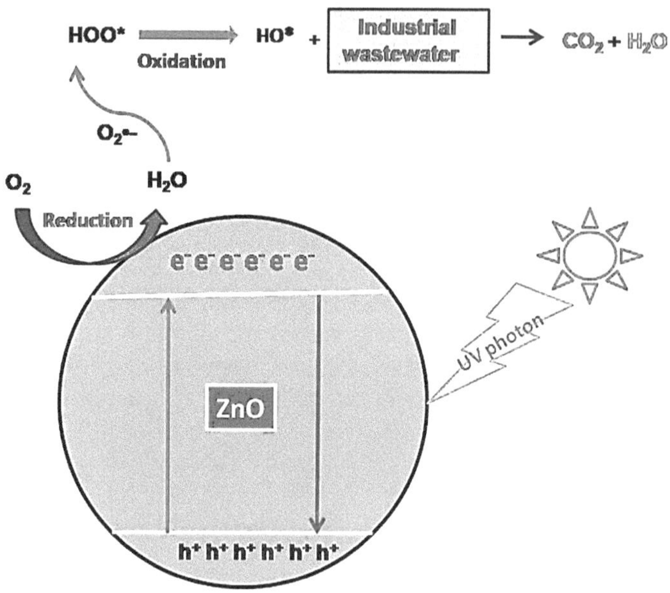

FIGURE 2.4 Photocatalytic mechanism for removal of heavy metals from industrial wastewater. Figure created by author.

2.5. Biological Treatment

Currently, there is an increasing amount of study being done on the application of microorganisms for ecological restoration and remediation. The microbes, which can be used for the removal therapies, perform the best functionalities if they are separated from the same natural environment in which they are intended to be applied (Razzak et al., 2022). Nevertheless, many microorganisms can have their capacities increased through genetic modification approaches. Utilising primarily bacterial cells, biological treatment techniques convert heavy metal wastewaters into stable end products through biochemical breakdown. Sludges, or more microorganisms, are created, and some of the waste is transformed into water, carbon dioxide, and other byproducts. Biological treatment methods are broadly categorized into two groups based on amount of dissolved oxygen content.

i. Aerobic treatment (revival of biological sludge in aeration basins)
ii. Anaerobic treatment

2.5.1. *Aerobic treatment*

Medium rich in minerals is used to acclimatize selected aerobic biomass at pH levels that are ideal for growth; such media serves as a provider of nutrients for biomass development and immobilization only during the initial phase. Yeasts typically have a pH of 3.4, and their medium is made up of the following substances: magnesium sulphate, potassium phosphate, sodium chloride, glucose, and ammonium phosphate. Aerobic reactors are connected in series and are employed to eliminate heavy metals from industrial wastewater.

2.5.2. *Anaerobic treatment*

Anaerobic digestion, a technique which has been utilized extensively in the treatment and stabilization of pollutants with high organic contents, is a metabolic process carried out by anaerobic microbes. The two primary bacterial groups involved in anaerobic digestion are methanogenic microorganisms, which make methane, and acidogenic microbes, which convert organic waste into volatile fatty acids. Methanogenic bacteria are generally thought to be less resilient to outside modifications in their growth environments, such as variations in pH, temperature, and/or the presence of hazardous metals. Additionally, a prior study found that acidogenic bacteria were less sensitive to heavy metal inhibition (Harirchi, 2022).

2.6. Degradation of Heavy Metal Complexes by Advanced Oxidation Processes

By using photochemical or catalytic techniques, recent advancements in industrial wastewater treatment procedures have improved the oxidative degradation process for organic molecules which are dissolved or distributed in aqueous environments. They are commonly known as Advanced Oxidation Processes (AOPs) and are thought to be superior to traditional methods of treating water (Alexandre-Franco & Fernández-González, 2020; Boczkaj & Fernandes, 2017; Iyyappan et al., 2023). Because AOPs may break down refractory organic contaminants into innocuous by-products, they have garnered a lot of fascination in the field of wastewater treatment and are a possible replacement for traditional techniques (Li et al., 2023).

AOPs oxidise organic materials by combining conventional oxidants (H_2O_2 and/or O_3) with extra stimuli like ultraviolet (UV) radiation to produce extremely reactive species (hydroxyl radicals, $^{\bullet}OH$) and super oxide radicals, $O_2^{\bullet-}$. These radicals can oxidize and mineralize nearly any organic molecule, producing CO_2 and inorganic ions in the process. They are extremely reactive, short-lived, non-selective, and simple to create. AOPs may mitigate the negative effects of water treatment on the environment by employing catalysts, creating energy-efficient designs, and transforming contaminants into innocuous byproducts (Yan et al., 2023). The degradation efficiency attained by AOPs is greatly influenced by a number of reaction parameters, including temperature, irradiation wavelength, current density, oxidant concentration, and pH of the treated solution. Based on the kinds of oxidants and radicals generated, as well as the chemicals employed in the process (Aziz & Mustafa, 2024), AOPs are classified into five primary types (Figure 2.5).

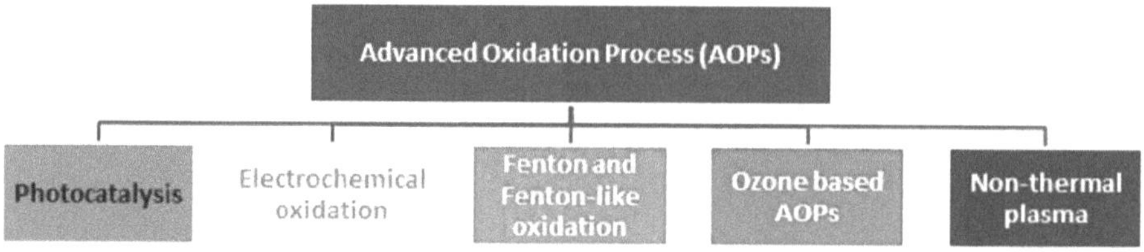

FIGURE 2.5 Major types of Advanced Oxidation Processes (AOPs) for removal of heavy metals from industrial wastewater.

Figure created by author.

3. ADSORPTION PROCESS

The term "adsorption" describes the collection of molecules, atoms, or ions on a surface that is maintained there by chemical forces, electrostatic physical forces (physisorption), or both (Raji et al., 2023). Environmental engineers utilize this procedure to remove different types of contaminants. The most effective method is adsorption because the other methods have drawbacks including producing a lot of sludge, being inefficient, requiring delicate operating conditions, and being expensive to dispose of. Although the adsorption approach is relatively new, it is starting to gain popularity as a possible preferred alternative for the removal of heavy metals due to its ability to offer high-quality treated effluent, flexibility in design, reversibility, and regenerability of the adsorbent (Hussain et al., 2021).

3.1. Adsorption Isotherm

The study of equilibrium and kinetics is crucial for gaining deeper insights into adsorption processes and so many studies have been carried out in order to understand the mechanism of adsorption. The adsorption equilibrium of metal ions is often formulated by two empirical models, the Freundlich and Langmuir (Murphy et al., 2023). The widespread use of the two models stems from their simplicity, requiring only because they are simple and only two parameters for calculations, metal removal capacity (x/m) defined as mass of metal ion removed per gram of adsorbent and Ce, the metal ion concentration at equilibrium.

3.2. Freundlich Adsorption Isotherm

Freundlich adsorption isotherm is purely an empirical one. According to this, the amount of solute adsorbed per unit mass of adsorbent increases with concentration at initial stage and reaches a constant at high concentration; afterwards the adsorption rate is very slow. This is due to the fact that the surface is covered by the adsorbed molecules (unimolecular layer) (Khuchunaev, 2021).

3.3. Langmuir Adsorption Isotherm

Langmuir adsorption isotherm is based on the following assumptions (Heraldy, 2016):

The solid adsorbent surface contains only a fixed number of adsorption sites.

At equilibrium and at constant temperature and pressure, a fraction of the sites Θ is covered by the adsorbed molecules. The fraction of the free (uncovered) site is $(1-\Theta)$.

Each site can hold only one molecule, i.e., unimolecular layer of adsorption is formed.

The adsorption is taking place on the fixed sites, and there is no interaction between the adsorbed molecules on the surface.

Langmuir adsorption isotherm equation is

$$P/(x/m) = 1/(K_1+K_2) + P/K_2 \quad (1.1)$$

The modified form of Langmuir adsorption isotherm equation is given as

$$C_e/q_e = 1/(Qo\ b) + Ce/Qo \quad (1.2)$$

where Ce is the equilibrium concentration expressed in ppm, q_e is the amount adsorbed per unit mass of the adsorbent expresses in mg g^{-1}, Qo and b are Langmuir constants related to adsorption capacity (mg g^{-1}) and energy (L mg-1). A plot of Ce/q_e versus Ce will result in a straight line. Further, the essential characteristics of Langmuir isotherm and the feasibility of the process are expressed in terms of the dimensionless constant described by the separation factor, R which is defined by Weber and Chakravorti and the equation for R is given as

$$R = 1/(1+bCo) \quad (1.3)$$

where Co is the optimum concentration (ppm) of the substrates. The separation factor R indicates the shapes of the isotherm.

VALUES OF R	*TYPES OF ISOTHERM*
$R > 1$	unfavourable
$R = 1$	linear
$0 < R < 1$	favourable
$R = 0$	irreversible

A number of scientists have applied the Langmuir isotherm to test the formation of monolayer of adsorbate as well as the nature and the favourability of the adsorption process using R values (Ayawei et al., 2017; Amrutha et al., 2023).

3.4. BET Adsorption Isotherm

A BET adsorption isotherm is a mathematical or graphical depiction of the relationship between the pressure or concentration of the adsorbate at constant temperature and the amount of adsorbate (such as gas or liquid) adsorbed onto a solid surface. It bears the names of the three scientists who created it in 1938: Stephen Brunauer, Paul H. Emmett, and Edward Teller. The adsorption of gas molecules onto the surface of solid materials, especially porous materials like activated carbon, metal-organic frameworks, and zeolites, is widely described by the BET (Brunauer-Emmett-Teller) hypothesis. According to the BET equation, adsorption takes place in a monolayer, and because of interactions with the initial layer, adsorption in successive layers is said to be reduced. Usually, the isotherm shows discrete regions: (i) As the surface fills with adsorbate molecules, adsorption usually increases linearly at low pressures. (ii) The adsorption rate decreases at intermediate pressures as the adsorbate saturates the surface. (iii) Multilayer adsorption might happen at higher pressures, which would cause the adsorption to grow more sharply. (iv) In order to fit experimental data and determine parameters like the adsorbent material's specific surface area and monolayer adsorption capacity, the BET equation is frequently utilized.

3.5. Types of Adsorbents

Various adsorbents such as Silica gel, Activated Alumina, carbons, zeolites polymer resins, and clay (Luing et al., 2018) are being used in the removal of the heavy metal ions by adsorption technique. A small number of adsorbents, like some zeolites, are naturally occurring, while the majority, like activated carbons, are produced. Every material has unique properties, including pore structure, porosity, and adsorbing surface nature. Adsorbents may have pore diameters that are dispersed throughout the solid. The three broad categories of pore sizes are: macro pores have "diameters" greater than 50 nm, meso pores (sometimes called transitional pores) have "diameters" between 2 and 50 nm, and micro holes have "diameters" less than 2 nm. Numerous adsorbent substances, including silica gels, activated carbons, and activated alumina, are amorphous and comprise intricate webs of interconnected micro, meso, and macropores. Zeolite adsorbents, on the other hand, have precisely defined pores.

3.5.1. Activated carbon adsorbents

Because of its high pore volume and surface area, activated carbon is an effective adsorbent in the treatment of wastewater. Aside from this, unlike other adsorbents, activated carbon can be disposed of anywhere because it won't harm groundwater. Activated carbon can be produced from almost any carbonaceous material (Rajan et al., 2022; Wang et al., 2023). Coconut shell and coal are the main sources of activated carbon generation. A high degree of porosity that develops during the production of activated carbon is linked to a wide surface area and offers several sites for the adsorption of molecules, metal ions, and atoms. This material's interior pore structure also offers a relatively big surface. The molecule needed for adsorption on the surface is provided by the pores' walls. Activated carbon is a well-known industrial adsorbent that has been researched for the removal of heavy metals from detergents, agro-chemical wastes, industrial wastes, and potable water treatment. Granular activated carbon is an eco-friendly way to remove inorganic and organic contaminants before releasing them into the environment since it can be regenerated and reactivated.

3.5.2. Zeolites

Zeolites are commonly used as adsorbents, ion exchangers, and catalysts. They remain cutting edge adsorbents in numerous separation procedures. Zeolites are crystalline microporous aluminosilicates that are frequently employed as ion exchangers, adsorbents, and catalysts. Natural zeolites, which are crystalline hydrated aluminosilicates with a framework structure having pores occupied by water, alkali, and alkaline earth cations, are abundant and inexpensive supplies. Natural zeolites are highly effective adsorbents in separation and purification procedures because of their strong cation-exchange capacity and molecular sieve characteristics (Velarde et al., 2023) (Figure 2.6).

3.5.3. Clay materials

Another possible substitute for activated carbon is clay. Like zeolites, clay minerals are significant inorganic soil constituents. Their large surface area and exchange capacities provide them with sorption capabilities. Clay has the potential to bind metal ions due to the negative charge on its mineral structure (Kumari & Mohan, 2021).

3.5.4. Nanostructured materials

One of the most sophisticated and accurate methods for detecting heavy metals is the use of nanomaterials based on quantum dots. Since quantum dot nanomaterials have high specific surface area, strong reactivity, and unique physicochemical characteristics, they can be used as sensors (Pooja & Chowdhury, 2021). Powerful detection devices that can identify several metals in difficult conditions can be created by linking optical

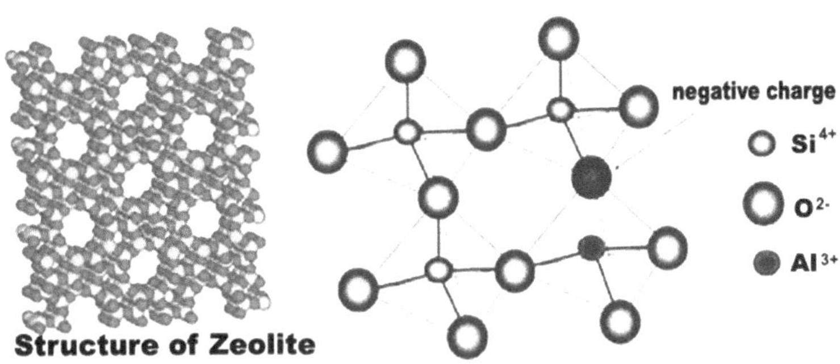

FIGURE 2.6 Structure of zeolites.

Reproduced from *Open Access Journal* under the term of Creative Commons Attribution License (Golomeova & Zendelska, 2016) Copyright © 2016.

or chemical sensor converters to quantum dots nanomaterials (Wang et al., 2021). For heavy metal ion nanosensors, for instance, zero-dimensional graphene dots are used due to their intriguing properties, such as good stability, excellent optical properties, and an easily fabricated and prepared material through doping in optical detector species (Zhang et al., 2018). Additionally, some biomass is utilized in the quantum dot nanomaterial technology. These nanobiosensors are environment friendly and biocompatible, making them useful as eye-catching fluorescent nanosensors. For instance, Fe(III) in effluents is detected using carbon quantum dots that are isolated from waste green algae (Liu et al., 2020). Gold-silica nanocomposites exhibit a heightened surface plasmon resonance bond compared to gold nanoparticles, rendering them more adaptable at detecting trace levels of heavy metals in drinking water. When heavy metal ions clash, gold-silica core-shell nanocomposites are formed, which are then used to detect the presence of these pollutants. Additionally, because of their chemical adsorption on the surface of the gold-containing nanocomposite, these ions cause changes in the position of the plasmon adsorption link (Meng et al., 2018). Metal nanoparticles, like gold and silver, have a high affinity for adsorbing light in the ultraviolet-visible range of the electromagnetic spectrum. This is because of the combined interactions between the metal atoms and electrons. Compared to solid nanoparticles, nanocomposites with a core-shell structure, like gold and silica, have more refined surface plasmon resonance bonds, making them more sensitive to detect trace levels of heavy metals in drinking water. When heavy metal ions clash, gold-silica core-shell nanocomposites are formed, which are used to sense the presence of these pollutants (Beni & Jabbari, 2022; Zhang et al., 2018).

3.6. Low-cost Adsorbents

Materials that are easily available and economically viable for use in adsorption processes are referred to as low-cost adsorbents. Usually, these substances are employed to eliminate impurities from liquids such as water or air. Low-cost adsorbents include:

3.6.1. Byproducts of agriculture

Resources like sawdust, sugarcane bagasse, rice husk, and coconut shell are widely available and reasonably priced. Their huge surface area and porous structure allow them to be treated and used as adsorbents (Khan & Mondal, 2021).

3.6.2. Industrial byproducts

Waste products from a variety of industries, including spent coffee grounds from the processing of coffee, slag from the production of steel, and fly ash from coal-fired power plants, can be used again as adsorbents (De Gisi et al., 2016).

3.6.3. Natural minerals

Activated carbon made from charcoal, zeolites, and clays are examples of naturally occurring minerals that work well as adsorbents and can be less expensive than synthetic substitutes.

3.6.4. Biochar

The pyrolysis of biomass yields biochar, a substance rich in carbon. It can be useful for the adsorption of heavy metals and organic pollutants due to its large surface area.

3.6.5. Nut shells

Because they are porous, waste materials from the nut processing industry, such as peanut, almond, and walnut shells, can be made into adsorbents (Ungureanu et al., 2023).

3.6.6. Agricultural waste

A variety of agricultural wastes can be processed and utilized as inexpensive adsorbents, such as maize cobs, wheat straw, and sugarcane bagasse (Singh et al., 2020).

3.6.7. Municipal solid waste

After the proper processing and treatment, materials like paper, cardboard, and some plastics can be recovered and used as adsorbents.

All these are considered as inexpensive adsorbents and provide an economical and environmentally friendly way to remove pollutants from environmental remediation applications. For particular pollutants and treatment scenarios, their efficacy, selectivity, and regeneration potential should be carefully assessed.

4. CHEMICAL METHODS

Chemical agents are employed in the elimination processes of heavy metals from water or wastewater in order to precipitate, complex, or otherwise immobilize the metals, which facilitates their separation from the solution. The following are some typical chemical techniques for removing heavy metals (Pohl, 2020).

4.1. Precipitation

The process of adding a precipitating agent to create insoluble metal compounds that may be filtered or settled to separate them. Sulfide salts, sodium hydroxide, calcium hydroxide, and sodium carbonate are examples of common precipitating agents (Pohl, 2020).

4.2. Ion Exchange Method

The process of substituting less hazardous ions for heavy metal ions in a solution using ion exchange resins or membranes. Usually, the resins are functionalized with certain groups that bind heavy metal ions only (Ince et al., 2019).

4.3. Adsorption

Through physical or chemical interactions, adsorbent materials such as zeolites, activated carbon, clays, and modified natural materials like chitosan and biosorbents can adsorb heavy metal ions onto their surfaces (Ungureanu et al., 2023).

4.4. Chelation

By forming stable complexes with heavy metal ions, chelating compounds like EDTA (ethylenediaminetetraacetic acid), DTPA (diethylenetriamine pentaacetic acid), and NTA (nitrilotriacetic acid) improve the solubility of the heavy metal ions and make removal easier.

4.5. Oxidation-Reduction (Redox) Reactions

Reducing agents such as ferrous sulfate, sodium bisulfite, or sodium metabisulfite can chemically reduce certain heavy metal ions to insoluble forms or less toxic oxidation states.

4.6. Electrocoagulation

By using an electrical current to destabilize and coagulate heavy metal ions, insoluble precipitates are formed that are easily removed through sedimentation or filtration. This process is known as electrocoagulation (Ince et al., 2020).

4.7. Membrane Processes

By physically rejecting heavy metal ions or selectively permitting their passage via semi-permeable membranes, methods like electrodialysis, nanofiltration, and reverse osmosis can remove them from a mixture efficiently (Vidu et al., 2020).

4.8. pH Adjustment

Adjusting the pH of a solution to maximize the precipitation or complexation of heavy metal ions is known as pH adjustment. For instance, increasing pH might encourage metal hydroxide precipitation, whereas decreasing pH can improve metal ion adsorption onto surfaces or complexation with ligands (Ince et al., 2020).

To effectively remove heavy metals from contaminated water sources, these chemical approaches can be used alone or in conjunction with other treatment strategies. The best approach is designed considering the different criteria such as the presence of specific metal pollutants, their concentration, the chemistry of the water, the treatment goals, and economic concerns.

5. CONCLUSIONS AND FUTURE PROSPECTS

In this chapter, various treatment technologies have been discussed in detail for the removal of heavy metals from industrial wastewater including ion-exchange, membrane filtration, electro dialysis, photocatalysis, biological treatment, and advanced oxidation processes (AOPs). This chapter review sought to explore heavy metals in wastewater effluent. It identified the primary human and natural activity-related sources of heavy metal pollution. However, human activities like industrial growth and technological improvements have accelerated the occurrence of metal ion contamination. The environment and public health are seriously threatened by heavy metal-polluted wastewater and improperly treated effluents. Heavy metal-contaminated wastewater can be treated using a variety of techniques, but it's important to remember that the best treatment depends on a number of fundamental factors, including pH, initial metal concentration, treatment performance overall in comparison to other methods, environmental impact, and economic factors like capital investment and operating costs. This chapter provides a basic assessment of the common approaches in terms of their suitability for the economy, their level of technological maturity, their ability to reduce environmental pollution, their operation, and automatic control.

REFERENCES

Abdullah, N., Yusof, N., Lau, W. J., Jaafar, & Ismail, A. F. (2019). Recent trends of heavy metal removal from water/wastewater by membrane technologies. *J. Ind. Eng. Chem.,76,* 17–38. https://doi.org/10.1016/j.jiec.2019.03.029

Akpor, O. B., Ohiobor, G. O., & Olaolu, T. D. (2014). Heavy metal pollutants in wastewater effluents: Sources, effects and remediation. *Advances in Bioscience and Bioengineering, 2,* 37–43. DOI: 10.11648/j.abb.20140204.11

Alexandre-Franco, M. F., & Fernández-González, C. (2020). Advanced oxidation processes for the removal of antibiotics from water. An overview. *Water, 12*(1), 102. https://doi.org/10.3390/w12010102

Alkhadra, M. A., Su, X., Suss, M. E., Tian, H., Guyes, E. N., Shocron, A. N., … & Bazant, M. Z. (2022). Electrochemical methods for water purification, ion separations, and energy conversion. *Chemical Reviews, 122*(16), 13547–13635. https://doi.org/10.1021/acs.chemrev.1c00396

Amrutha, Jeppu, G., Girish, C. R., Prabhu, B., & Mayer, K. (2023). Multi-component adsorption isotherms: review and modeling studies. *Environmental Processes, 10*(2), 38. https://doi.org/10.1007/s40710-023-00631-0

Arasu, P. T., Murugan, A., & Badusha, M. S. M. (2016). A study of physico-chemical analysis of ground water in and around SIPCOT, Tuticorin, Tamil Nadu, India. *International Journal of Earth Sciences and Engineering*, 2499–2502.

Argurio, P., Fontananova, E., Molinari, R., & Drioli, E. (2018). Photocatalytic membranes in photocatalytic membrane reactors. *Processes, 6*, 162. https://doi.org/10.3390/pr6090162

Avci, A. H., Messana, D. A., Santoro, S., Tufa, R. A., Curcio, E., Di Profio, G., & Fontananova, E. (2020). Energy harvesting from brines by reverse electrodialysis using nafion membranes. *Membranes, 10*(8), 168.

Ayawei, N., Ebelegi, A. N., & Wankasi, D. (2017). Modelling and interpretation of adsorption isotherms. *Journal of Chemistry*. https://doi.org/10.1155/2017/3039817

Aziz, K. H. H., & Mustafa, F. S. (2024). Advanced oxidation processes for the decontamination of heavy metal complexes in aquatic systems: A review. *Case Studies in Chemical and Environmental Engineering, 9,* 100567. https://doi.org/10.1016/j.cscee.2023.100567

Barman, M. K., Bhattarai, A., & Saha, B. (2023). Applications of ion exchange resins in environmental remediation. *Vietnam Journal of Chemistry, 61*(5), 533–550.

Beni, A.A., & Jabbari, H. (2022). Nanomaterials for environmental applications. *Results in Engineering, 15*, 100467. https://doi.org/10.1016/j.rineng.2022.100467

Bera, A., Trivedi, J. S., Kumar, S. B., Chandel, A. K. S., Haldar, S., & Jewrajka, S. K. (2018). Anti-organic fouling and anti-biofouling poly (piperazine amide) thin film nanocomposite membranes for low pressure removal of heavy metal ions. *J. Hazard. Mater, 343*, 86–97. https://doi.org/10.1016/j.jhazmat.2017.09.016

Boczkaj, G., & Fernandes, A. (2017). Wastewater treatment by means of advanced oxidation processes at basic pH conditions: A review. *Chemical Engineering Journal, 320*, 608–633. https://doi.org/10.1016/j.cej.2017.03.084

Carmona, B., & Abejón, R. (2023). Innovative membrane technologies for the treatment of wastewater polluted with heavy metals: perspective of the potential of electrodialysis, membrane distillation, and forward osmosis from a bibliometric analysis. *Membranes, 13*(4), 385. https://doi.org/10.3390/membranes13040385

Chitpong, N., & Husson, S. M. (2017). High-capacity, nanofiber-based ion-exchange membranes for the selective recovery of heavy metals from impaired waters. *Sep. Purif. Technol., 179*, 94–103. https://doi.org/10.1016/j.seppur.2017.02.009

Dadari, S., Rahimi, M., & Zinadini, S. (2022). Novel antibacterial and antifouling PES nanofiltration membrane incorporated with green synthesized nickel-bentonite nanoparticles for heavy metal ions removal. *Chem. Eng. J., 431*, 134116. https://doi.org/10.1016/j.cej.2021.134116

De Gisi, S., Lofrano, G., Grassi, M., & Notarnicola, M. (2016). Characteristics and adsorption capacities of low-cost sorbents for wastewater treatment: A review. *Sustainable Materials and Technologies, 9*, 10–40. https://doi.org/10.1016/j.susmat.2016.06.002

ElSayed, E. E. (2018). Natural diatomite as an effective adsorbent for heavy metals in water and wastewater treatment (a batch study), *Water Science, 32*, 32–43. https://doi.org/10.1016/j.wsj.2018.02.001

Gao, X., Yao, Y., & Meng, X. (2020). Recent development on BN-based photocatalysis: A review. *Mater. Sci. Semicond. Process., 120*, 105256. https://doi.org/10.1016/j.mssp.2020.105256

Golomeova, M., & Zendelska, A. (2016). Application of some natural porous raw materials for removal of lead and zinc from aqueous solutions. *Microporous and Mesoporous Materials,* 21–49. InTech. http://dx.doi.org/10.5772/62347

Gurreri, L., Tamburini, A., Cipollina, A., & Micale, G. (2020). Electrodialysis applications in wastewater treatment for environmental protection and resources recovery: A systematic review on progress and perspectives. *Membranes, 10*(7), 146.

Halawa, R. A. E., & Zabin, S. A. (2017). Removal efficiency of Pb, Cd, Cu and Zn from polluted water using dithiocarbamate ligands. *Journal of Taibah University for Science, 11*, 57–65. https://doi.org/10.1016/j.jtusci.2015.07.002

Harirchi, S., Wainaina, S., Sar, T., Nojoumi, S. A., Parchami, M., Parchami, M., & Taherzadeh, M. J. (2022). Microbiological insights into anaerobic digestion for biogas, hydrogen or volatile fatty acids (VFAs): A review. *Bioengineered, 13*(3), 6521–6557.

Heraldy, E., Hidayat, Y., & Firdaus, M. (2016). *The langmuir isotherm adsorption equation: the monolayer approach.* In IOP Conference Series: Materials science and engineering (Vol. 107, No. 1, p. 012067). IOP Publishing.

Hussain, A., Madan, S., & Madan, R. (2021). Removal of heavy metals from wastewater by adsorption. In M. Khaled Nazal & H. Zhao (Eds.), *Heavy metals—their environmental impacts and mitigation*. IntechOpen. https://doi.org/10.5772/intechopen.95841

Ínce, M. (Ed.). (2020). Biochemical toxicology: Heavy metals and nanomaterials. IntechOpen.

İnce, M., & İnce, O. K. (2017). An overview of adsorption technique for heavy metal removal from water/wastewater: A critical review. *International Journal of Pure and Applied Sciences, 3*, 10–19. https://doi.org/10.29132/ijpas.358199

Ince, M., & Ince, O. K. (2019). Heavy metal removal techniques using response surface methodology: water/wastewater treatment. *Biochemical toxicology-heavy metals and nanomaterials*. IntechOpen.

Inyang, M. I., Gao, B., Yao, Y., et al. (2016). A review of biochar as a low-cost adsorbent for aqueous heavy metal removal. *Crit. Rev. Environ. Sci. Technol., 46*, 406–433. https://doi.org/10.1080/10643389.2015.1096880

Iyyappan, J., Gaddala, B., Gnanasekaran, R., Gopinath, M., Yuvaraj, D., & Kumar, V. (2023). Critical review on wastewater treatment using photo catalytic advanced oxidation process: Role of photocatalytic materials, reactor design and kinetics. *Case Studies in Chemical and Environmental Engineering*, 100599. https://doi.org/10.1016/j.cscee.2023.100599

Juve, J. M. A., Christensen, F. M. S., Wang, Y., & Wei, Z. (2022). Electrodialysis for metal removal and recovery: A review. *Chemical Engineering Journal, 435*, 134857.

Kamari, S., & Shahbazi, A. (2021). High–performance nanofiltration membrane blended by Fe_3O_4@SiO_2–CS bionanocomposite for efficient simultaneous rejection of salts/heavy metals ions/dyes with high permeability, retention increase, and fouling decline. *J. Chem. Eng., 417*, 127930. https://doi.org/10.1016/j.cej.2020.127930

Kapepula, V. L., & Luis, P. (2024). Removal of heavy metals from wastewater using reverse osmosis. *Frontiers in Chemical Engineering, 6,* 1334816. https://doi.org/10.3389/fceng.2024.1334816

Karnib, M., Kabbani, A., Holail, H., & Olama, Z. (2014). Heavy metals removal using activated carbon, silica and silica activated carbon composite. *Energy Procedia., 50*, 113–120. https://doi.org/10.1016/j.egypro.2014.06.014

Khan, A. A., & Mondal, M. (2021) Low-cast adsorbents, removal techniques, and heavy metal removal efficiency. In M. P. Shah, S. R. Couto, & V. Kumar, (Eds.), *New trends in removal of heavy metals from industrial wastewater* (pp. 83–103). Elsevier. https://doi.org/10.1016/B978-0-12-822965-1.00004-0.

Khuchunaev, B., Baysiev, K., Gekkieva, S., & Budaev, A. (2021). *Researches of ice-forming efficiency of products of sublimation of pyrotechnic compositions consisting of silver iodide AgI particles and zinc oxide*. In IOP Conference Series: Materials Science and Engineering.

Kumar, P., Pournara, A., Kim, K. -H., Bansal, V., Rapti, S., & Manos, M. J. (2017). Metal-organic frameworks: Challenges and opportunities for ion-exchange/sorption applications. *Prog. Mater. Sci., 86*, 25–74. https://doi.org/10.1016/j.pmatsci.2017.01.002

Kumari, N., & Mohan, C. (2021). Basics of clay minerals and their characteristic properties. In G. Morari Do Nascimento (Ed.), *Clay and clay minerals*. IntechOpen. https://doi.org/10.5772/intechopen.97672

Kyzas, G. Z., Siafaka, P. I., Lambropoulou, D. A., Lazaridis, N. K., & Bikiaris, D. N. (2014). Poly(itaconic acid)-grafted chitosan adsorbents with different cross-linking for Pb(II) and Cd(II) uptake. *Langmuir, 30*, 120–131. https://doi.org/10.1021/la402778x

Li, M., Lv, Z., Zheng, J., et al. (2017). Positively charged nanofiltration membrane with dendritic surface for toxic element removal. *ACS Sustain. Chem. Eng., 5*, 784–792. https://doi.org/10.1021/acssuschemeng.6b02119

Li, X., Jiang, L., & Li, H. (2018). *Application of Ultrafiltration Technology in Water Treatment*. IOP Conf. Series: Earth and Environmental Science. doi:10.1088/1755-1315/186/3/012009

Li, Y., Dong, H., Xiao, J., et al. (2023). Advanced oxidation processes for water purification using percarbonate: insights into oxidation mechanisms, challenges, and enhancing strategies. *J. Hazard Mater., 442,* 130014. https://doi.org/10.1016/j.jhazmat.2022.130014

Li, Z., Meng, X., & Zhang, Z. (2018). Recent development on MoS_2-based photocatalysis: A review. *J. Photochem. Photobiol. C Photochem. Rev., 35*, 39–55. https://doi.org/10.1016/j.jphotochemrev.2017.12.002

Li, Z., Zhang, Z., Wang, L., & Meng, X. (2020). Bismuth chromate (Bi_2CrO_6): A promising semiconductor in photocatalysis. *J. Catal., 382,* 40–48. https://doi.org/10.1016/j.jcat.2019.12.0

Liang, H., Li, T., Zhang, J., et al. (2020). 3-D hierarchical Ag/ZnO@CF for synergistically removing phenol and Cr(VI): Heterogeneous vs. homogeneous photocatalysis. *J. Colloid Interface Sci., 558,* 85–94. https://doi.org/10.1016/j.jcis.2019.09.105

Liu, F., Ma, B.-R., Zhou, D., Xiang, Y., & Xue, L. (2014). Breaking through tradeoff of Polysulfone ultrafiltration membranes by zeolite 4A. *Microporous and Mesoporous Materials*, 186, 113–120. https://doi.org/10.1016/j.micromeso.2013.11.044

Liu, F., Zhu, S., Li, D., Chen, G., & Ho, S.H. (2020). Detecting ferric iron by microalgal residue derived fluorescent nanosensor with an advanced kinetic model. *iScience, 23*, 101174. https://doi.org/10.1016/j.isci.2020.101174.

Luing, W.S., Ngadi, N., Inuwa, I. M., & Onn Hassan, O. H. (2018). Recent advances in applications of activated carbon from biowaste for wastewater treatment: A short review. *Journal of Cleaner Production 175*, 361–375.

Mahmoud, A. E. D., Al-Qahtani, K. M., Alflaij, S. O., Al-Qahtani, S. F., & Alsamhan, F. A. (2021). Green copper oxide nanoparticles, for lead, nickel, and cadmium removal from contaminated water. *Sci. Rep., 15*, 12547. https://doi.org/10.1038/s41598-021-91093-7

Meng, C., Zhikun, W., Qiang, L., Chunling, L., Shuangqing, S., & Songqing, H. (2018). Preparation of amino-functionalized Fe_3O_4@mSiO_2 core-shell magnetic nanoparticles and their application for aqueous Fe^{3+} removal. *J. Hazard Mater., 341*, 198–206. https://doi.org/10.1016/j.jhazmat.2017.07.062

Meng, X., Yun, N., & Zhang, Z. (2019). Recent advances in computational photocatalysis: A review. *Can. J. Chem. Eng., 97,* 1982–1998. https://doi.org/10.1002/cjce.23477

Mukherjee, D., Barghi, S., & Ray, A., K. (2014). Preparation and characterization of the TiO_2 immobilized polymeric photocatalyst for degradation of aspirin under UV and solar light. *Processes, 2*, 12–23. https://doi.org/10.3390/pr2010012

Murphy, O. P., Vashishtha, M., Palanisamy, P., & Kumar, K. V. (2023). A review on the adsorption isotherms and design calculations for the optimization of adsorbent mass and contact time. *ACS Omega, 8*(20), 17407–17430.https://doi.org/10.1021/acsomega.2c08155

Musie, W., & Gonfa, G. (2023). Fresh water resource, scarcity, water salinity challenges and possible remedies: A review. *Heliyon*. doi: 10.1016/j.heliyon.2023.e18685

Muthulakshmi, L., Arasu, P. T., Ilavarasi, R., & Selvanayagam, S. (2015). Hydrochemical trend of ground water Rajapalayam town, Tamil Nadu, India. *J Mater Environ Sci, 6*(4), 1004–1008.

Narayani, M., & Shetty, K. V. (2013). Chromium-resistant bacteria and their environmental condition for hexavalent chromium removal: A review. *Crit. Rev. Environ. Sci. Technol., 43,* 955–1009. https://doi.org/10.1080/10643389.2011.627022

Ni, Y., Wang, W., Huang, W., Lu, C., & Xu, Z. (2014). Graphene strongly wrapped TiO_2 for high-reactive photocatalyst: A new sight for significant application of grapheme. *J. Colloid Interface Sci., 428*, 162–169. doi: 10.1016/j.jcis.2014.04.022

Odumbe, E., Murunga, S., & Ndiiri, J. (2023). Heavy metals in wastewater effluent: Causes, effects, and removal technologies [Internet]. In *Trace Metals in the Environment.* IntechOpen. http://dx.doi.org/10.5772/intechopen.1001452

Pohl, A. (2020). Removal of heavy metal ions from water and wastewaters by sulfur-containing precipitation agents. *Water Air Soil Pollut., 231*, 503. https://doi.org/10.1007/s11270-020-04863

Ponomar, M., Ruleva, V., Sarapulova, V., Pismenskaya, N., Nikonenko, V., Maryasevskaya, A., ... & Améduri, B. (2024). Structural characterization and physicochemical properties of functionally porous proton-exchange membrane based on PVDF-SPA graft copolymers. *International Journal of Molecular Sciences, 25*(1), 598.

Pooja, & Chowdhury, P. (2021). Functionalized CdTe fluorescence nanosensor for the sensitive detection of water borne environmentally hazardous metal ions. *Opt. Mater. 111*, 110584, https://doi.org/10.1016/j.optmat.2020.110584

Rajan, M. J., & Anish, C. I. (2022). Role of activated carbon in water treatment. In *Water Quality-New Perspectives*. IntechOpen. doi: 10.5772/intechopen.108349

Raji, Z., Karim, A., Karam, A., & Khalloufi, S. (2023). Adsorption of heavy metals: Mechanisms, kinetics, and applications of various adsorbents in wastewater remediation—A review. *Waste, 1*(3), 775–805. https://doi.org/10.3390/waste1030046

Rapti, S., Pournara, A., Sarma, D., et al. (2016). Selective capture of hexavalent chromium from an anion-exchange column of metal organic resin-alginic acid composite. *Chem. Sci., 7,* 2427–2436. https://doi.org/10.1039/C5SC03732H

Razzak, S. A., Faruque, M. O., Alsheikh, Z., et al. (2022). A comprehensive review on conventional and biological-driven heavy metals removal from industrial wastewater. *Environmental Advances, 7,* 100168. https://doi.org/10.1016/j.envadv.2022.100168

Safaei, J., Ullah, H., Mohamed, N. A., Mohamad Noh, M. F., Soh, M. F., Tahir, A. A., ... & Mat Teridi, M. A. (2018). Enhanced photoelectrochemical performance of Z-scheme g-C3N4/BiVO4 photocatalyst. *Applied Catalysis B: Environmental, 234,* 296–310. https://doi.org/10.1016/j.apcatb.2018.04.056

Saxena, A., Bhardwaj, M., Allen, T., Kumar, S., & Sahney, R. (2017). Adsorption of heavy metals from wastewater using agricultural–industrial wastes as biosorbents. *Water Science, 31,* 189–197. https://doi.org/10.1016/j.wsj.2017.09.002

Shakir, E., Zahraw, Z., & Al-obaidy, A. H. M. J. (2017). Environmental and health risks associated with reuse of wastewater for irrigation. *Egyptian Journal of Petroleum, 26,* 95–102. https://doi.org/10.1016/j.ejpe.2016.01.003

Sharma, S., & Bhattacharya, A. (2017). Drinking water contamination and treatment techniques. *Applied Water Science, 7*(3), 1043–1067. https://doi.org/10.1007/s13201-016-0455-7

Sheng, P. X., Ting, Y. P., Chen, J. P., & Hong, L. (2004). Sorption of lead, copper, cadmium, zinc, and nickel by marine algal biomass: characterization of biosorptive capacity and investigation of mechanisms. *J. Colloid Interface Sci., 275,* 131–141. https://doi.org/10.1016/j.jcis.2004.01.036

Singh, S., Wasewar, K. L., & Kansal, S. K. (2020). Low-cost adsorbents for removal of inorganic impurities from wastewater. In *Inorganic pollutants in water* (pp. 173–203). Elsevier. https://doi.org/10.1016/B978-0-12-818965-8.00010-X

Sun, Q. J., Sun, C. S., & Jui, H. J. (2015). The application of microfiltration-reverse osmosis process in printing and dyeing wastewater reuse. *Shandong Chemical Industry, 44,* 141–142.

Thaçi, B. S., & Gashi, S. T. (2019). Reverse osmosis removal of heavy metals from wastewater effluents using biowaste materials pretreatment. *Pol. J. Environ. Stud., 28,* 337–341. doi: https://doi.org/10.15244/pjoes/81268

Topare, N. S., & Wadgaonkar, V. S. (2023). A review on application of low-cost adsorbents for heavy metals removal from wastewater. *Material Today Proceedings, 77,* 8–18. https://doi.org/10.1016/j.matpr.2022.08.450

Torki, M., Nazari, N., & Mohammadi, T. (2017). Evaluation of biological fouling of RO/MF membrane and methods to prevent it. *European Journal of Advances in Engineering and Technology, 4,* 707–710.

Türkmen, D., Bakhshpour, M., Akgonullu, S., Aşır, S., & Denizli, A. (2022). Heavy metal ions removal from wastewater using cryogels: A review. *Front. Sustain., Sec. Sustainable Chemical Process Design, 3,* 765592. doi: 10.3389/frsus.2022.765592

Ullah, H., Viglašová, E., & Galamboš, M. (2021). Visible light-driven photocatalytic rhodamine b degradation using CdS nanorods. *Processes, 9,* 263. https://doi.org/10.3390/pr9020263

Ulyankina, A., Mitchenko, S., & Smirnova, N. (2020). Selective photocatalytic oxidation of 5-HMF in water over electrochemically synthesized TiO_2 nanoparticles. *Processes, 8,* 647. https://doi.org/10.3390/pr8060647

Ungureanu, E. L., Mocanu, A. L., Stroe, C. A., Panciu, C. M., Berca, L., Sionel, R. M., & Mustatea, G. (2023). Agricultural byproducts used as low-cost adsorbents for removal of potentially toxic elements from wastewater: A comprehensive review. *Sustainability, 15*(7), 5999. https://doi.org/10.3390/su15075999

Ursino, C., Santoro, S., Ounifi, I., Hafiane, A., & Figoli, A. (2022). A novel cellulose acetate fouling-resistant ultrafiltration membranes for heavy metal remediation. EGU General Assembly 2022, EGU22–12154. https://doi.org/10.5194/egusphere-egu22-12154

Varma, G. V. P., Singh, R. K., & Sahu, V. (2013). A comparative study on the removal of heavy metals by adsorption using fly ash and sludge: A review. *International Journal of Application or Innovation in Engineering and Management, 2,* 45–56. https://api.semanticscholar.org/CorpusID:212526709

Velarde, L., Nabavi, M. S., Escalera, E., Antti, M. L., & Akhtar, F. (2023). Adsorption of heavy metals on natural zeolites: A review. *Chemosphere,* 138508. https://doi.org/10.1016/j.chemosphere.2023.138508

Vidu, R., Matei, E., Predescu, A. M., Alhalaili, B., Pantilimon, C., Tarcea, C., & Predescu, C. (2020). Removal of heavy metals from wastewaters: A challenge from current treatment methods to nanotechnology applications. *Toxics, 8*(4), 101. doi10.3390/toxics8040101

Wang, B., Lan, J., Bo, C., Gong, B., & Ou, J. (2023). Adsorption of heavy metal onto biomass-derived activated carbon. *RSC Advances, 13*(7), 4275–4302.

Wang, C., Sun, X., Shan, H., Zhang, H., & Xi, B. (2021). Degradation of landfill leachate using UV-TiO_2 photocatalysis combination with aged waste reactors. *Processes, 9,* 946. https://doi.org/10.3390/pr9060946

Wang, X., Kong, L., Zhou, S., et al. (2022). Development of QDs-based nanosensors for heavy metal detection: A review on transducer principles and in-situ detection. *Talanta, 239,* 122903. https://doi. org/10.1016/j.talanta.2021.122903, 122903.

Xiang, H., Min, X., Tang, C. -J., Sillanpää, M., & Zhao, F. (2022). Recent advances in membrane filtration for heavy metal removal from wastewater: A mini review, *Journal of Water Process Engineering, 49,* 103023. https://doi.org/10.1016/j.jwpe.2022.103023

Yadav, M., Singh, G., & Jadeja, R. N. (2021). Physical and chemical methods for heavy metal removal. *Pollutants and Water Management* (pp. 377–397). https://doi.org/10.1002/9781119693635.ch15

Yan, L., Yang, X., Zeng, H., et al. (2023). Nanocomposite hydrogel engineered hierarchical membranes for efficient oil/water separation and heavy metal removal. *J. Membr. Sci., 668,* 121243. https://doi.org/10.1016/j.memsci.2022.121243

Yao, Y., Gao, X., Li, Z., & Meng, X. (2020a). Photocatalytic reforming for hydrogen evolution: A review. *Catalysts, 10,* 335. https://doi.org/10.3390/catal10030335

Yao, Y., Gao, X., & Meng, X. (2021b). Recent advances on electrocatalytic and photocatalytic seawater splitting for hydrogen evolution. *Int. J. Hydrogen Energy, 46,* 9087–9100. https://doi.org/10.1016/j.ijhydene.2020.12.212

Zhang, L., Peng, D., Liang, R. P., & Qiu, J. D. (2018). Graphene-based optical nanosensors for detection of heavy metal ions. *TrAC Trends Anal. Chem., 102,* 280–289. https://doi.org/10.1016/j.trac.2018.02.010.

Zhu, C., Liu, C., Fu, Y., et al. (2019). Construction of CDs/CdS photocatalysts for stable and efficient hydrogen production in water and seawater. *Appl. Catal. B Environ., 242,* 178–185. https://doi.org/10.1016/j.apcatb.2018.09.096

Development of Robust Strategies for Wastewater Treatment for Agro-Foods and Beverage Industries

3

Prasann Kumar and Joginder Singh

The agro-foods and beverage industries encounter substantial obstacles in treating wastewater, primarily because of their liquid waste's varied and intricate characteristics. Creating solid and practical plans for treating wastewater in these industries is crucial to reduce environmental pollution and meet regulatory requirements. It comprehensively summarises the critical factors and methods in developing efficient wastewater treatment plans for agro-food and beverage sectors. The chapter emphasises the distinctive attributes of wastewater produced in these sectors, including elevated organic content, fluctuating pH levels, and nutrients and contaminants. These factors require customised treatment solutions that effectively eliminate pollutants while minimising operational costs and resource consumption. The following chapter explores a range of technologies frequently used in wastewater treatment from agro-foods and beverage industries. These technologies encompass physical, chemical, and biological processes. The approach highlights integrating multiple treatment stages to thoroughly remove pollutants and guarantee compliance with regulatory standards for effluent. Moreover, it examines how advanced treatment techniques, such as membrane filtration, oxidation processes, and anaerobic digestion, improve treatment efficiency and sustainability. Additionally,

DOI:10.1201/9781003496014-3

it emphasises the possibility of extracting resources from wastewater streams, such as generating energy through biogas production and recovering nutrients for use in agriculture. It highlights the significance of embracing comprehensive wastewater management strategies, including process optimisation, water reuse, pollution prevention measures, and technological solutions. The text underscores the necessity of cooperation among industry stakeholders, regulatory agencies, and research institutions to stimulate innovation and tackle emerging obstacles in wastewater treatment. This chapter highlights the importance of creating solid plans for treating wastewater in agro-foods and beverage industries. This is crucial for protecting the environment, promoting sustainability, and ensuring the long-term success of these sectors.

1. INTRODUCTION

The agro-foods and beverage industries play a crucial role in the global economies, supplying indispensable products for consumption and pleasure worldwide (Abdelfattah et al., 2023; Franco-Morgado et al., 2023). Nevertheless, the manufacturing procedures in these sectors unavoidably produce substantial quantities of wastewater, distinguished by an intricate combination of organic substances, nutrients, and diverse pollutants. Efficient treatment of this wastewater is essential to reduce environmental contamination, for compliance with regulatory requirements, and sustainable acquisitions.This introduction provides a foundation for effective strategies for treatment of wastewater specifically designed to address the distinct challenges and needs of agro-food and beverage industries (Alcántara et al., 2020; Kumar et al., 2020; Kumar & Pathak, 2019; Sarangi et al., 2023). The wastewater produced by agro-food and beverage industries presents unique difficulties as compared to other industrial sectors, primarily because of the varied nature of their production processes and the composition of their wastewater. Within agro-food processing, including meat, dairy, and crops, the wastewater frequently contains elevated levels of organic substances, fats, proteins, and solid particles suspended in the water. For instance, slaughterhouse wastewater can have a high concentration of blood, fats, and animal by-products. In contrast, effluents from dairy processing facilities may consist of proteins, lactose, and milk fats. Likewise, the process of producing beverages, such as brewing, wine-making, and soft drink manufacturing, generates wastewater streams containing different acidity (pH), sugars, and organic acids (Jha et al., 2024; Varadavenkatesan et al., 2021). Furthermore, agricultural practices, including applying fertilisers, pesticides, and animal excrement, add to the intricacy of wastewater in these sectors. Effluent from farm fields can transport nutrients, such as nitrogen and phosphorus, into wastewater streams, thereby exacerbating the complexity of treatment procedures. Additionally, the fluctuation in wastewater flow rates and compositions caused by seasonal variations in production and farming activities can challenge the stability and effectiveness of treatment systems. Efficient wastewater treatment from agro-foods and beverage industries is crucial for multiple reasons.

To begin with, untreated or insufficiently treated wastewater can present substantial environmental and public health hazards. Sewage contains various contaminants, including organic matter, nutrients, and pathogens, which can cause a decline in water quality, harm aquatic ecosystems, and pose a risk to drinking water sources (Adetunji et al., 2022; Jha et al., 2024; Kee et al., 2022; Park et al., 2024). In addition, releasing untreated waste into water bodies can result in eutrophication, the growth of harmful algal blooms, and the reduction of oxygen levels, which can cause ecological imbalances and the destruction of habitats. Furthermore, regulatory authorities enforce stringent restrictions on releasing contaminants from industrial establishments to safeguard environmental integrity and promote human welfare. Failure to comply with these regulations may lead to financial penalties, legal repercussions, and harm to a company's image. Hence, agro-foods and beverage industries must formulate resilient approaches for wastewater treatment to fulfil their ethical obligations and legal mandates, ensuring adherence to regulations and mitigating adverse environmental consequences (Arya et al., 2022; Jornet-Martínez et al., 2019). Furthermore, the implementation of sustainable wastewater management strategies can yield economic advantages for businesses through the reduction of water usage, the depreciation of waste production, and the extraction of valuable resources from wastewater streams. Technologies like anaerobic digestion can enable the production of biogas from organic waste, offering renewable energy sources and decreasing greenhouse gas emissions. Similarly, nutrient recovery technologies can retrieve valuable resources like nitrogen and phosphorus from wastewater to be reused in agricultural fertilisers. This process helps to close nutrient cycles and advance the principles of a circular economy.

This chapter offers valuable insights and guidance to industry stakeholders, policymakers, researchers, and practitioners engaged in developing and implementing wastewater treatment strategies for agro-foods and beverage industries. Its main objective is to address these goals. The ultimate goal is to advance sustainable practices that protect environmental well-being, guarantee adherence to regulations, and bolster the long-term sustainability of these vital industries (Bangar et al., 2024; Cruz et al., 2023; Kee et al., 2021; López-Gámez et al., 2023).

2. CHARACTERISTICS OF AGRO-FOODS AND BEVERAGE WASTEWATER

The agro-foods and beverage industries produce wastewater with distinct characteristics due to the diverse processes used in food production, processing, and agricultural activities. Understanding the composition and variability of waste liquids is crucial for developing effective strategies to treat wastewater tailored to these industries' specific needs and challenges (Bangar et al., 2024; Jha et al., 2024; Kee et al., 2022; López-Gámez et al., 2023). The composition of wastewater generated by agro-foods and beverage industries can vary significantly due to multiple factors, including the characteristics of the products being processed, the techniques employed in production, and the fluctuations in agricultural practices across different seasons. However, these effluents generally display standard components and characteristics. Agro-foods and beverage wastewater frequently contain high concentrations of organic compounds derived from raw materials, residual food processing, and microbial activities (Gehlot et al., 2022; Kee et al., 2022). The presence of organic compounds in wastewater, including carbohydrates, proteins, fats, oils, and cellulose, contributes to the levels of biochemical oxygen demand (BOD) and chemical oxygen demand (COD). An abundance of organic matter can decrease oxygen levels in water bodies, causing oxygen depletion and detrimental impacts on aquatic ecosystems. The wastewater from agro-food and beverage industries may contain significant amounts of nutrients, such as nitrogen (N), phosphorus (P), and potassium (K), which come from the use of fertilisers, pesticides, and organic materials. Nitrogen and phosphorus are essential nutrients for plant growth. However, if these substances are excessively discharged into aquatic ecosystems, they can trigger eutrophication, characterised by algae's rapid proliferation and ecological equilibrium disruption. Regulating nutrient concentrations in wastewater is essential to prevent water quality degradation and protect aquatic ecosystems. Suspended solid particles, such as organic matter, sediments, and particulate matter, contribute to the turbidity and opacity of sewage (Kee et al., 2021; Kumar et al., 2021a; Mishra et al., 2023; Udaiyappan et al., 2017). Suspended solids can hinder treatment processes, reduce effectiveness, and cause aesthetic issues in the receiving waters. Efficient solid-liquid separation is essential for removing suspended solids from wastewater and improving treatment effectiveness.

Effluent from agro-foods and beverage industries may contain various contaminants, including organic pollutants, pesticides, heavy metals, and pathogens. The presence of these contaminants can be attributed to agricultural inputs, processing additives, and microbial activities. Discharging contaminants into water bodies without treatment can adversely impact the environment and human health (Dey et al., 2023; Kumar et al., 2018; Kumar & Pathak, 2019; Reddy et al., 2022; Srivastav et al., 2023). Therefore, it is crucial to devise effective treatment methods to eradicate or mitigate the repercussions of pollutants in wastewater. Fluctuations in wastewater characteristics are due to agricultural activities such as planting, harvesting, and livestock management. For example, the intensified use of pesticides during planting seasons and the elevated production of animal waste at specific times of the year can lead to fluctuations in the concentrations of nutrients and pollutants in wastewater. Different food processing and beverage production methods generate wastewater with distinct compositions. Meat processing facilities produce high protein and fat wastewater, while dairy processing plants generate effluents rich in lactose and milk solids (Hernández-Pinto et al., 2024; Song et al., 2020). For an accurate assessment of wastewater composition, it is imperative to thoroughly comprehend the distinct processes and inputs linked to each sector within the industry. The composition of wastewater is influenced by the characteristics of the raw materials, ingredients, and additives used in the manufacturing of food and beverages. The properties of brewery wastewater are affected by the particular grains, hops, and adjuncts used during brewing. On the other hand, winery effluents vary based on the specific types of grapes and fermentation techniques employed. The wastewater from agro-foods and beverage production contains a substantial quantity of organic substances, rendering it highly susceptible to decomposition by microorganisms (Diaz et al., 2018; Hasan et al., 2023; Okaiyeto et al., 2024; Udaiyappan et al., 2017). The decomposition of organic substances present in wastewater result in variations in treatment efficacy and treated water quality of treated water. Vigilant monitoring and regulation of the process is crucial to ensure consistent treatment protocols and efficient removal of organic substances. Elevated levels of nutrients, notably nitrogen and phosphorus, can induce eutrophication and degrade water quality in the receiving bodies. Adopting effective strategies for managing nutrients, such as employing biological nutrient removal (BNR) processes and chemical precipitation, is crucial for reducing the number of nutrients in wastewater and avoiding harmful environmental effects. Wastewater from agro-foods and beverage industries can contain various pollutants, including organic pollutants, pesticides, and pathogens (López-Gámez et al., 2023; Park et al., 2024). These pollutants require specific treatment methods to be removed or degraded. To adhere to stringent effluent quality standards and meet regulatory obligations, sophisticated treatment methodologies, such as activated carbon adsorption, ozonation, and membrane filtration, may be imperative.

The fluctuation in the composition and flow rates of wastewater during various seasons presents challenges for the operation and efficacy of treatment plants. It is imperative to build and oversee treatment facilities that can effectively accommodate variations in wastewater characteristics and adapt to changing conditions throughout the year (Ahmad et al., 2020; Franco-Morgado et al., 2023; Jornet-Martínez et al., 2019; Okaiyeto et al., 2024; Park et al., 2024; Sarangi et al., 2023; Varadavenkatesan et al., 2021). To address these challenges, it is imperative to develop and implement efficient strategies for wastewater treatment that integrate diverse treatment techniques, optimise resource utilisation, and prioritise environmental conservation. By understanding the characteristics and variations of agro-foods and beverage wastewater, stakeholders can develop tailored treatment strategies that effectively remove pollutants, minimise environmental impacts, and ensure compliance with regulations. Moreover, ongoing research and innovative efforts are essential for advancing treatment technologies, improving treatment effectiveness, and addressing emerging contaminants and challenges in wastewater management for these industries (Table 3.1).

TABLE 3.1 Advantages and disadvantages of different strategies for agro-foods and beverage wastewater treatment.

TREATMENT STRATEGY	*ADVANTAGES*	*DISADVANTAGES*
I. Physical Treatment		
Sedimentation and Filtration	– Simple and cost-effective	– Ineffective in removing dissolved contaminants
	– Effective for removing large particles and suspended solids	– Requires periodic maintenance of filters
	– No chemicals required	– Limited efficiency for organic and nutrient removal
Centrifugation	– Efficient for separating solids from liquids	– High operational costs and energy consumption

(*Continued*)

TABLE 3.1 (Continued)

TREATMENT STRATEGY	*ADVANTAGES*	*DISADVANTAGES*
Flotation	– Can handle large volumes of wastewater – No chemical additives required – Effective in removing oils, greases, and fine suspended particles – Rapid treatment process – Can be combined with chemical coagulants to improve efficiency	– Not effective for dissolved pollutants – Equipment is expensive – High energy requirements – Requires careful control of operating conditions – Potential for chemical residues in treated water
II. Chemical Treatment		
Coagulation and Flocculation	– Effective in removing suspended solids, colloids, and some dissolved organic matter – Can significantly reduce turbidity and colour – Relatively fast process	– Requires chemical additives (coagulants and flocculants) – Generates chemical sludge that requires further disposal – Operational costs associated with chemical usage
pH Adjustment	– Simple and widely used method to optimise conditions for other treatment processes – Effective in reducing heavy metals and enhancing other chemical treatments	– Requires careful monitoring and control to avoid over-acidification or over-alkalisation – Potential environmental impact from improper neutralisation
Oxidation (e.g., Ozone, UV)	– Effective in breaking down organic pollutants, pathogens, and certain inorganic substances – Can reduce the biological oxygen demand (BOD) and chemical oxygen demand (COD) of wastewater – Does not generate sludge	– High operational costs, especially with advanced oxidation processes (AOPs) – Potential formation of by-products that may require additional treatment – Requires skilled operators
III. Biological Treatment		
Aerobic Treatment (e.g., Activated Sludge,	– Highly effective for removing organic matter (BOD, COD) – Produces stable effluent with low levels of organics	– Requires large land areas for treatment facilities – High energy requirements for aeration

TABLE 3.1 (Continued)

TREATMENT STRATEGY	*ADVANTAGES*	*DISADVANTAGES*
Trickling Filters, Biofilters)	– Can be combined with other treatments for enhanced performance	– Sensitive to fluctuations in wastewater composition and environmental conditions
Anaerobic Treatment	– Energy-efficient process with low operational costs	– Slower process compared to aerobic treatment
(e.g., Anaerobic Digestion)	– Generates biogas (methane) that can be used for energy production	– Requires careful control of operating conditions to prevent system failure
	– Effective for high-strength wastewater	– Produces a more complex and less stable effluent than aerobic processes
	– Lower sludge production	– Long start-up period required for microbial communities to become fully functional
Constructed Wetlands	– Natural and aesthetically pleasing solution	– Requires large land areas
	– Low operational and maintenance costs	– Limited capacity for treating large volumes of wastewater
	– Effective in removing nutrients, BOD, and some metals	– Seasonal variability in treatment efficiency
	– Can support biodiversity	– Slower treatment process
IV. Advanced Treatment		
Membrane Filtration	– High efficiency in removing suspended solids, microorganisms, and some dissolved pollutants	– High initial capital and operational costs
(e.g., Reverse Osmosis,	– Produces high-quality effluent suitable for reuse	– Membrane fouling can reduce efficiency and increase maintenance costs
Ultrafiltration)	– Compact system requiring less space	– High energy consumption
	– Can be integrated with other treatment processes	– Disposal of concentrated waste streams (brine) can be challenging
Ion Exchange	– Effective in removing specific ions and heavy metals	– High cost of ion exchange resins and regeneration chemicals

(Continued)

TABLE 3.1 (Continued)

TREATMENT STRATEGY	*ADVANTAGES*	*DISADVANTAGES*
	– Can achieve very low levels of contaminants in treated water	– Generates concentrated waste streams that require further treatment
	– Regenerable resins can be reused multiple times	– Not effective for treating large volumes of wastewater
Electrochemical Treatment	– Effective in removing a wide range of contaminants including heavy metals, organics, and pathogens	– High operational costs due to electricity usage
	– Produces minimal sludge	– Requires specialised equipment and maintenance
	– Can be controlled and optimised with precision	– May require post-treatment for complete contaminant removal
V. Hybrid Systems		
**Integrated Biological-Chemical	– Combines the strengths of biological and chemical treatments for enhanced efficiency	– Complexity in system design and operation
Treatment**	– Can handle a wide range of contaminants and fluctuating wastewater characteristics	– Higher capital and operational costs compared to single treatment methods
	– Can be tailored to specific wastewater characteristics	– Requires skilled operators and regular monitoring
Anaerobic-Aerobic Treatment	– Maximises organic matter removal and energy recovery through biogas production	– Complex operation requiring careful balancing of anaerobic and aerobic conditions
	– Reduces overall treatment time and land requirements	– Sensitive to variations in wastewater composition
	– Can handle high-strength wastewater effectively	– Potential for odour issues and methane emissions if not properly managed
Advanced Oxidation Processes	– Capable of degrading recalcitrant organic pollutants and achieving complete mineralisation	– High energy consumption and operational costs
(AOPs)	– Can be combined with other treatments for improved performance	– Potential formation of harmful by-products

TABLE 3.1 (Continued)

TREATMENT STRATEGY	*ADVANTAGES*	*DISADVANTAGES*
	– Effective for treating a wide range of contaminants	– Requires specialised equipment and expertise
VI. Emerging Technologies		
Algal Treatment	– Can capture CO_2 and produce biomass for biofuels or other valuable products	– Requires significant land and light for algal growth
	– Effective in nutrient removal (N, P)	– Limited by climatic and seasonal conditions
	– Can be integrated with other treatment processes	– Potential for algal blooms and associated management challenges
Microbial Fuel Cells (MFCs)	– Generates electricity while treating wastewater	– High initial costs and technological complexity
	– Reduces sludge production	– Low power output and efficiency
	– Can treat a wide range of organic pollutants	– Still in experimental stages for large-scale applications
Photocatalysis	– Uses light energy to degrade organic pollutants	– Limited by light availability and efficiency of photocatalysts
	– No chemical additives required	– High cost of photocatalyst materials
	– Can be combined with solar energy for sustainable treatment	– Potential formation of incomplete degradation products

3. TRADITIONAL WASTEWATER TREATMENT TECHNOLOGIES

Wastewater treatment is crucial for reducing environmental pollution and safeguarding public health by eliminating contaminants and pollutants from wastewater before its release into water bodies. Conventional wastewater treatment methods usually blend physical, chemical, and biological processes for efficient treatment (Adetunji et al., 2022; Rodrigues et al., 2024). This section offers a comprehensive introduction to conventional treatment technologies, their utilisation in the treatment of agro-foods and beverage wastewater, and their performance and constraints. Physical processes in wastewater treatment encompass the separation of solids and liquids via sedimentation, filtration, and flotation. Sedimentation, or settling, utilises the force of gravity to separate suspended solids from wastewater by enabling them to settle at the bottom of a sedimentation tank (Kumar et al., 2018, 2019, 2022; Kumar & Dwivedi, 2020). Filtration employs porous materials, such as sand or membranes, to eliminate suspended solids and particles from wastewater by directing them through a bed of filters. Flotation is when air bubbles are attached to suspended solids, causing them to rise to the surface and be removed. Chemical processes entail the addition of chemicals to wastewater to aid in eliminating impurities using precipitation, coagulation, flocculation, or oxidation-reduction reactions. Coagulation and flocculation entail introducing coagulants and flocculants, respectively, into wastewater to disrupt and gather together suspended particles, making it easier to eliminate them through sedimentation or filtration. Precipitation involves the creation of insoluble precipitates by adding chemicals that react with dissolved contaminants, causing them to separate from the solution. Oxidation-reduction reactions, such as chlorination or ozonation, entail the introduction of oxidising agents into wastewater to break down organic pollutants or eliminate pathogens through disinfection (Maurya et al., 2022; Okaiyeto et al., 2024).

Biological processes employ microorganisms, including bacteria, fungi, and algae, to break down organic matter and eliminate pollutants from wastewater using biochemical reactions. Common natural treatment methods involve aerobic processes, such as activated sludge and trickling filters, which use oxygen to decompose organic matter. Additionally, anaerobic processes like digestion and lagoons function without oxygen to generate biogas and stabilise organic sludge. Conventional wastewater treatment methods are extensively employed in the agro-food and beverage sectors to purify wastewater generated from diverse production operations, such as food processing, beverage manufacturing, and agricultural practices. Agro-foods and beverage wastewater treatment often employ physical processes, such as sedimentation and filtration, to eliminate suspended solids, fats, oils, and grease (FOG) from effluents. Sedimentation tanks and clarifiers facilitate the separation of suspended solids from wastewater through settling (Adetunji et al., 2022; Ahmad et al., 2020; Herrmann et al., 2024; Kumar et al., 2018, 2020; Kumar & Pathak, 2019; Sharma et al., 2023; Udaiyappan et al., 2017). Simultaneously, filtration systems, such as sand filters or membrane filtration, are employed to eliminate minute particles and solids further. Chemical processes, such as coagulation, flocculation, and chemical oxidation, are used on agro-foods and beverage wastewater to improve the elimination of organic matter, nutrients, and contaminants. Coagulants and flocculants are introduced into wastewater to gather suspended particles and make eliminating them easier through sedimentation or filtration processes. Chemical oxidation techniques, such as chlorination or ozonation, can disinfect wastewater and break down organic pollutants. Biological treatment methods from agro-foods and beverage industries are commonly used in wastewater treatment to break down soil organic matter, nutrients, and contaminants using microbial metabolism (Bangar et al., 2024, Gecim et al., 2021). Aerobic treatment techniques, such as activated sludge and aerobic lagoons, employ oxygen to facilitate the proliferation of aerobic bacteria, which decompose organic substances and eliminate nutrients from wastewater. Anaerobic treatment techniques, such as anaerobic digestion and lagoons, function without oxygen to generate biogas and stabilise organic sludge via microbial fermentation. Although traditional wastewater treatment technologies have successfully eliminated contaminants and pollutants from wastewater, they possess inherent performance limitations and drawbacks, particularly when utilised for agro-foods and beverage wastewater (Bangar et al., 2024; Diaz et al., 2018; Franco-Morgado et al., 2023; Hoang et al., 2023; Jornet-Martínez et al., 2019; Kumar et al., 2021b; López-Gámez et al., 2023; Maurya et al., 2022; Song et al., 2020). Physical, chemical, and biological processes for the treatment of sewage lead to the efficient elimination of suspended solids, organic matter, and nutrients and also substantially reduce the pollutants. These processes are firmly established and extensively utilised in wastewater treatment plants globally, showcasing dependable efficacy in treating various wastewater streams (Hasan et al., 2023; Herrmann et al., 2024). Nevertheless, conventional treatment approaches have constraints, particularly for agro-foods and beverage wastewater containing substantial amounts of organic matter, diverse compositions, and intricate matrices.

Physical processes may not effectively remove colloidal and fine particles alone, necessitating additional treatment steps or advanced filtration methods. Chemical processes can produce chemical sludges or residual by-products that require disposal or additional treatment, resulting in higher operational expenses and a larger environmental impact. Changes can influence biological processes in wastewater properties, such as pH, temperature, and nutrient levels, which can impact the effectiveness and stability of treatment (Kumar et al., 2022; Kumar & Mistri, 2020; Kumari et al., 2022; Naik & Kumar, 2020; Pankaj et al., 2012b; Pathak et al., 2017). In addition, conventional treatment techniques may not consistently meet strict standards for wastewater quality released nor effectively deal with emerging pollutants found in agro-food and beverage wastewater, such as pharmaceuticals, personal care products, and microplastics. Hence, it is imperative to employ novel methodologies and cutting-edge treatment technologies to supplement conventional approaches and enhance these sectors' efficacy, sustainability, and adherence to regulations (Table 3.2).

TABLE 3.2 Advantages and disadvantages of different processes for agro-foods and beverage wastewater involving traditional technology.

PROCESS	*ADVANTAGES*	*DISADVANTAGES*
I. Primary Treatment		
Screening	– Simple and cost-effective process for removing large solids and debris	– Ineffective for removing dissolved or fine suspended solids
	– Protects downstream equipment from damage	– Requires frequent maintenance and cleaning of screens
	– Easy to implement and operate	– Generates solid waste that requires disposal
Sedimentation	– Effective in removing settleable solids and reducing suspended solids	– Ineffective for colloidal and dissolved pollutants

(Continued)

TABLE 3.2 (Continued)

PROCESS	*ADVANTAGES*	*DISADVANTAGES*
	– Low operational costs and energy requirements	– Requires large settling tanks and land area
	– Simple operation and maintenance	– Sludge generated needs further treatment and disposal
II. Secondary Treatment		
Activated Sludge Process	– Highly effective in removing organic matter (BOD, COD)	– High energy consumption for aeration
	– Produces high-quality effluent suitable for discharge or reuse	– Requires large land area for aeration tanks
	– Can be easily modified or expanded	– Sensitive to fluctuations in wastewater composition
	– Well-established and widely used	– Produces large volumes of sludge that require treatment and disposal
Trickling Filters	– Low operational costs and energy requirements	– Less efficient than activated sludge in removing organic matter
	– Simple design and relatively easy to maintain	– Requires large land area
	– Resistant to shock loads and fluctuations in wastewater composition	– Potential for clogging and maintenance issues
	– Produces less sludge compared to activated sludge	– Limited flexibility in handling high-strength wastewater
Rotating Biological Contactors	– Efficient in removing organic matter with low energy requirements	– High initial capital costs
	– Compact design, requiring less land area compared to trickling filters	– Sensitive to temperature fluctuations
	– Low sludge production	– Requires regular maintenance of rotating disks
	– Can be easily integrated into existing systems	– Potential for mechanical failure of rotating components
Anaerobic Lagoons	– Effective in treating high-strength wastewater	– Requires large land area
	– Low energy consumption	– Long retention time needed for effective treatment
	– Generates biogas that can be captured and used for energy	– Potential for odour issues and methane emissions
	– Low operational costs	– Difficult to control and maintain anaerobic conditions consistently

TABLE 3.2 (Continued)

PROCESS	*ADVANTAGES*	*DISADVANTAGES*
III. Tertiary Treatment		
Sand Filtration	– Effective in removing residual suspended solids and turbidity	– Requires regular backwashing and maintenance
	– Simple and reliable process	– Limited capacity for handling high solids loads
	– Low operational costs	– Not effective in removing dissolved organics and nutrients
	– Can improve the quality of effluent from secondary treatment processes	– Requires large filter beds and land area
Chemical Precipitation	– Effective in removing heavy metals, phosphates, and other specific contaminants	– Requires chemical additives, increasing operational costs
	– Can be tailored to target specific pollutants	– Generates chemical sludge that requires further treatment and disposal
	– Relatively fast process	– Sensitive to variations in wastewater composition
	– Can be integrated with other treatment processes	– Potential for residual chemicals in treated effluent
**Disinfection (e.g., Chlorination, UV Radiation) **	– Effective in destroying pathogens and reducing microbial load	– Potential formation of harmful disinfection by-products (e.g., trihalomethanes)
	– Widely used and well-established techniques	– Requires careful control and monitoring to ensure effectiveness
	– Can be implemented at various scales	– High operational costs, particularly for chemical disinfection
	– Provides a high level of safety in treated water	– UV systems require regular maintenance and bulb replacement
IV. Sludge Management		
Sludge Thickening	– Reduces the volume of sludge for further treatment and disposal	– High operational costs for mechanical thickening equipment
	– Can improve the efficiency of subsequent sludge treatment processes	– Requires skilled operators and regular maintenance

(Continued)

TABLE 3.2 (Continued)

PROCESS	*ADVANTAGES*	*DISADVANTAGES*
	– Simple gravity thickening has low operational costs	– Generates concentrated sludge that still requires stabilisation and disposal
Anaerobic Digestion	– Effective in stabilising sludge and reducing its volume	– High capital costs and long start-up time
	– Produces biogas that can be used for energy generation	– Requires careful control of operational conditions
	– Reduces pathogens in sludge	– Slower process compared to aerobic digestion
	– Can be integrated with other waste management processes	– Odor issues and methane emissions need to be managed
Composting	– Converts organic sludge into stable, nutrient-rich compost	– Requires large land area for composting operations
	– Reduces the volume of waste and produces a valuable end product	– Long processing time
	– Can be combined with other organic waste streams	– Requires careful monitoring of temperature, moisture, and aeration
	– Low operational costs	– Potential for odour and leachate issues if not managed properly
Land Application	– Cost-effective and beneficial use of treated sludge as fertiliser	– Requires large land area and careful management
	– Enhances soil fertility and structure	– Potential for contamination of soil and groundwater if not properly managed
	– Reduces the need for chemical fertilisers	– Limited to specific types of wastewater and sludge
	– Simple and natural process	– Regulatory restrictions and public perception issues
V. Natural Treatment Systems		
Constructed Wetlands	– Low operational and maintenance costs	– Requires large land area
	– Provides natural treatment of wastewater	– Seasonal variability in treatment efficiency
	– Enhances biodiversity and supports wildlife	– Slower treatment process

TABLE 3.2 (Continued)

PROCESS	*ADVANTAGES*	*DISADVANTAGES*
	– Effective in removing nutrients, organic matter, and some metals	– Limited capacity for treating high volumes of wastewater
Overland Flow	– Simple and cost-effective process	– Requires large land area and appropriate topography
	– Enhances soil and vegetation health	– Seasonal and climatic variability can affect performance
	– Effective in reducing BOD, suspended solids, and nutrients	– Limited to specific types of wastewater
	– Can be integrated into agricultural practices	– Potential for groundwater contamination if not properly managed
Stabilisation Ponds	– Simple design and operation	– Requires large land area and long retention time
(e.g., Lagoons)	– Low operational and maintenance costs	– Potential for odour and vector control issues
	– Effective in removing BOD, nutrients, and pathogens	– Limited capacity for treating high-strength or toxic wastewater
	– Can be used for multiple wastewater streams	– Seasonal variability in performance

4. ADVANCED TREATMENT TECHNIQUES

Within the agro-foods and beverage industries, these advanced techniques provide superior elimination of contaminants, heightened retrieval of resources, and enhanced environmental sustainability (Chakraborty et al., 2021; Dey et al., 2023; Dubey et al., 2002; Hidangmayum et al., 2023). This section examines four crucial advanced treatment techniques: membrane filtration (MF/UF/NF/RO), advanced oxidation processes (AOPs), anaerobic digestion and biogas production, and nutrient recovery and resource recovery.

Membrane filtration technologies employ semi-permeable membranes to segregate impurities physically and particles from wastewater, relying on size, shape, and molecular weight (Cruz et al., 2023; Kalra et al., 2021). Microfiltration (MF) involves using membranes with pore sizes ranging from 0.1 to 10 μm. These membranes are highly effective in eliminating suspended solids, bacteria, and specific pathogens present in wastewater. Ultrafiltration (UF) is a process that utilises membranes with smaller pore sizes (ranging from 0.001 to 0.1 μm) to effectively eliminate suspended solids, colloids, viruses, and macromolecules from wastewater. Nanofiltration (NF) membranes possess narrower spore diameters (0.001-0.01 μm) and efficiently eliminate divalent ions, organic compounds, and specific trace contaminants. Reverse osmosis (RO) membranes possess petite pore sizes (<0.001 μm) and are capable of eliminating dissolved ions, salts, organic compounds, and nearly all impurities from wastewater (Dwivedi & Kumar, 2011; Kumar et al., 2018; Kumar et al., 2020; Kumar & Dwivedi, 2018; Kumar et al., 2021c; Sharma et al., 2023). Membrane filtration provides numerous benefits for the treatment of agro-foods and beverage wastewater. These advantages include exceptional removal efficiencies, a small physical footprint, and minimal chemical reliance. Nevertheless, membrane fouling, scaling, and the need for high energy consumption pose considerable obstacles in membrane filtration processes. These challenges necessitate the implementation of appropriate pretreatment, fouling control, and energy optimisation strategies. Advanced oxidation processes (AOPs) utilise the production of highly reactive hydroxyl radicals (•OH) to chemically oxidise and break down organic pollutants, pathogens, and emerging contaminants in wastewater. Ozonation, produces hydroxyl radicals in wastewater by reacting with organic compounds. Consequently, this results in the breakdown of harmful substances and wastewater purification (Cruz et al., 2023; Gecim et al., 2021; Gehlot et al., 2022; Kee et al., 2021; López-Gámez et al., 2023; Maurya et al., 2022; Udaiyappan et al., 2017). UV/H_2O_2 is a process that uses ultraviolet (UV) irradiation and hydrogen peroxide (H_2O_2) to generate hydroxyl radicals. These radicals undergo a chemical reaction with organic contaminants, degrading into non-hazardous by-products. Fenton's reagent, consisting of ferrous iron (Fe^{2+}) and hydrogen peroxide (H_2O_2), produces hydroxyl radicals by reacting with organic compounds. This leads to the quick oxidation and breakdown of pollutants. AOPs effectively eliminate persistent organic pollutants, micro-contaminants, and pathogens in agricultural and food industry wastewater. They help treat intricate wastewater streams containing much organic matter and many newly discovered pollutants. Nevertheless, successfully implementing AOPs necessitates meticulous optimisation of operational parameters such as pH, temperature, and dosing rates to enhance treatment efficacy and minimise energy usage (Udaiyappan et al., 2017; Phiri et al., 2024). Anaerobic digestion is a biological process that employs anaerobic microorganisms to break down organic matter in wastewater, producing biogas (methane and carbon dioxide) as a secondary product. The process consists of four stages: hydrolysis, acidogenesis, acetogenesis, and methanogenesis. The anaerobic digestion process produces biogas, the organic sludge gets stabilised and reduced in its volume; thus, minimises disposal expenses and mitigates environmental consequences. Anaerobic digestion generates digestate which is abundant in nutrients; as an agricultural substitute for fertilisers. This process helps to close nutrient cycles and advance circular economy principles (Chatterjee & Mazumder, 2024; Hernández-Pinto et al., 2024; Jha et al., 2024; Kumar et al., 2018; Mishra et al., 2023; Okaiyeto et al., 2024).

Nevertheless, anaerobic digestion necessitates meticulous control of operational parameters, including temperature, pH, and substrate composition, to sustain optimal microbial activity and biogas generation. Furthermore, the presence of inhibitors, such as toxic substances and metallic elements, can impede the efficiency of anaerobic digestion and necessitate pre-treatment or measures to control inhibition (Kumar et al., 2017, 2018, 2022; Pankaj et al., 2012b). Nutrient recovery and resource recovery technologies

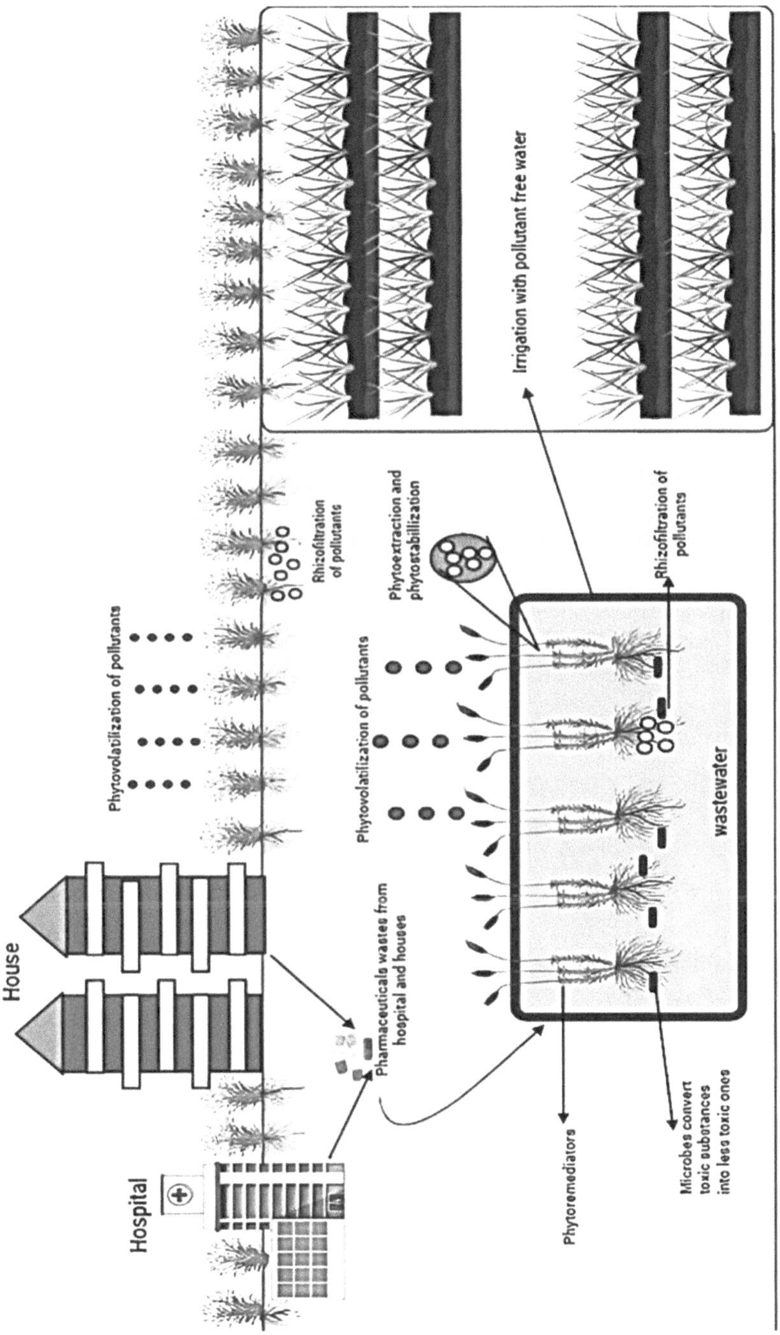

FIGURE 3.1 Remediation of wastewater by natural processes.

are designed to extract valuable resources, such as nitrogen's, phosphorus, and organic matter, from wastewater to reuse or recycle them. Struvite precipitation introduces magnesium and phosphate ions into wastewater to create struvite ($MgNH_4PO_4 \cdot 6H_2O$), used as a slow-release fertiliser containing nitrogen and phosphorus. Enhanced Biological Phosphorus Removal (EBPR) is a process that involves the use of specific microorganisms to gather phosphorus in biomass. This phosphorus-rich biomass can then be collected and reused as a fertiliser that is high in nutrients (Contreras-Jácquez et al., 2022; Díaz et al., 2022; Kee et al., 2021). Algal cultivation involves using algae to absorb nutrients, such as nitrogen and phosphorus, from wastewater to produce biomass, generate bioenergy, and recover nutrients. Resource recovery technologies provide the chance to extract valuable resources from wastewater produced by the agro-food and beverage industries. This helps to decrease the need for new materials and supports sustainability. Nevertheless, to fully exploit the benefits of nutrient and resource recovery in wastewater treatment, it is crucial to tackle obstacles such as nutrient imbalance, scalability, and economic viability. Advanced treatment methods provide cutting-edge solutions to improve wastewater treatment efficiency, sustainability, and resilience in agro-food and beverage sectors. By incorporating membrane filtration, advanced oxidation processes, anaerobic digestion, and nutrient/resource recovery technologies, stakeholders can enhance treatment efficiency, reduce environmental harm, and optimise resource utilisation, promoting a circular and sustainable economy. Continual research, development, and collaboration are crucial for further advancing these technologies and tackling the changing challenges of wastewater treatment in the agro-food and beverage industries (Chakraborty et al., 2021; Devi et al., 2023; Kumar et al., 2023; Upadhyay et al., 2023) (Figure 3.1).

5. INTEGRATED APPROACHES TO WASTEWATER MANAGEMENT

Efficient wastewater management is crucial for ensuring environmental sustainability, complying with regulations, and optimising operations in the agro-food and beverage sectors. Integrated approaches to wastewater management entail coordinating diverse strategies and technologies to enhance treatment processes, minimise resource utilisation, and mitigate environmental repercussions (Kumari et al., 2022; Sharma et al., 2023). The following section examines three essential elements of integrated wastewater management: strategies for optimising and controlling processes, the reuse and recycling of water, and measures for preventing pollution and reducing sources of contamination. Process optimisation and control strategies aim to enhance wastewater treatment processes' efficiency, dependability, and effectiveness using systematic monitoring, analysis, and adjustment of operational parameters (Kumar et al., 2018, 2021a, 2021b; Kumar et al., 2022; Kumar & Naik, 2020; Kumar & Pathak, 2019; Reddy et al., 2022). Consistent observation of crucial performance indicators (KPIs), such as the quality of discharged waste, the effectiveness of removing pollutants, and the amount of energy used, offers a valuable understanding of the performance of the treatment process and highlights areas that need enhancement.

Advanced data analytics techniques, such as statistical analysis, machine learning, and modelling, can facilitate the detection of patterns, trends, and correlations in wastewater quality and treatment performance. This, in turn, allows for predictive maintenance and proactive process optimisation. Operators can utilise performance data and analytics to promptly modify process parameters, including chemical dosing, flow, and aeration levels (Ahmad et al., 2020; Alcántara et al., 2020; Diaz et al., 2018). Moreover, the treatment efficiency, reduction of energy consumption, and adherence to regulatory standards are also optimised. Automated control systems and process optimisation algorithms enhance the dependability and steadiness of wastewater treatment processes by constantly monitoring and adjusting operational parameters to variations in conditions and fluctuations in wastewater properties. By incorporating energy-efficient technologies such as variable frequency drives (VFDs), high-efficiency motors, and energy recovery systems, energy usage and operating expenses can be reduced while treatment performance and environmental sustainability are improved. Through process optimisation and control strategies, agro-food and beverage industries can greatly enhance the efficiency of wastewater treatment, utilisation of resources, and environmental performance, all while reducing operational expenses and regulatory liabilities (Dey et al., 2023; Dubey et al., 2002; Kumar et al., 2022, 2023; Kumari et al., 2022; Saini et al., 2023). Water reuse and recycling strategies encompass retrieving and utilising treated wastewater for non-drinking purposes, such as irrigation, cooling, industrial processes, and sanitation. Advanced treatment methods, such as membrane filtration, UV disinfection, and reverse osmosis, can create reclaimed water of excellent quality that is appropriate for a range of non-drinking uses. These methods also guarantee adherence to regulatory requirements and minimise potential health hazards. Implementing specialised distribution systems and infrastructure for reclaimed water, such as using separate piping networks and storage tanks, guarantees the consistent delivery and availability of recycled water to consumers while preventing mixing or contamination with drinking water sources (Hoang et al., 2023; Mishra et al., 2023; Sarangi et al., 2023). Risk management strategies such as hazard analysis and critical control points (HACCP), microbial monitoring, and water quality testing can identify and reduce potential risks associated with reclaimed water. Informing stakeholders, encompassing employees, customers, and the general public, about the advantages and security of utilising reclaimed water encourages the acceptance and endorsement of water reuse and recycling initiatives, thereby advancing sustainable water management practices and the conservation of resources. These benefits include decreased usage of freshwater, reduced volumes of discharged wastewater, and improved ability to withstand water scarcity and drought. Pollution prevention and source reduction measures seek to minimise the production of pollutants and contaminants in wastewater by implementing optimal practices, technologies, and management strategies (Kumar et al., 2011b, 2021a, 2022; Naik & Kumar, 2020; Pankaj et al., 2012a; Srivastav et al., 2023). By adhering to Good Manufacturing Practices (GMP), which include appropriate facility design, equipment upkeep, and hygiene protocols, the likelihood of product contamination is diminished, and the production of process-related wastewater pollutants, such as organic substances, fats, and oils, is minimised. Implementing waste minimisation strategies, such as optimising processes, substituting materials, and utilising by-products, decreases the creation of waste streams and reduces the environmental consequences of agro-food and beverage production processes. Implementing Zero Liquid Discharge (ZLD)

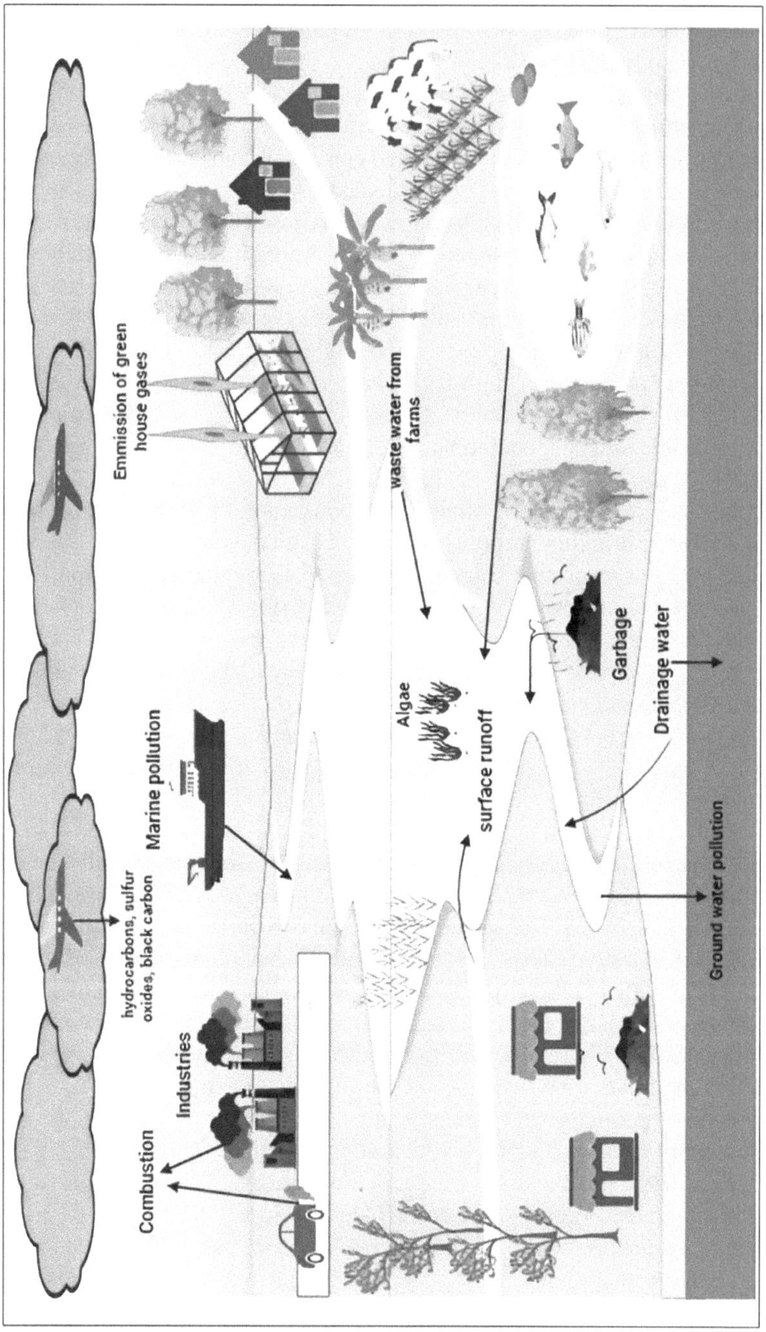

FIGURE 3.2 Remediation of wastewater by natural processes.

systems, such as evaporation, crystallisation, and membrane technologies, allows for the maximum reuse and recycling of water while minimising the volume of wastewater discharged and the emission of pollutants (Chatterjee & Mazumder, 2024; Kumar et al., 2021c; Kundu et al., 2022). This results in the achievement of near-zero liquid discharge from industrial operations. Advocating for product stewardship initiatives, such as extended producer responsibility (EPR), product lifecycle assessment (LCA), and eco-design principles, motivates manufacturers to create products with minimal environmental impact throughout their lifespan. Agro-food and beverage industries can minimise environmental pollution, decrease regulatory compliance costs, and improve sustainability performance throughout the value chain by implementing pollution prevention and source reduction measures. Integrated approaches to wastewater management provide comprehensive solutions to the intricate challenges that agro-food and beverage industries encounter in effectively and sustainably managing wastewater. Through implementing process optimisation and control strategies, water reuse and recycling initiatives, pollution prevention, and source reduction measures, stakeholders can greatly enhance treatment efficiency, resource utilisation, and environmental performance, all while reducing operational costs and regulatory risks. Continual research, innovation, and collaboration are crucial for furthering integrated wastewater management practices and tackling emerging challenges in the agro-food and beverage industries. By adopting comprehensive strategies for managing wastewater, agro-foods and beverage industries can improve their ability to withstand challenges, increase their competitiveness, and promote sustainability in an ever-evolving global context (Dey et al., 2023; Kumar et al., 2020; Kumar & Dwivedi, 2018) (Figure 3.2).

6. CASE STUDIES AND BEST PRACTICES

Analysing adequate wastewater treatment initiatives in the agro-food and beverage sectors offers valuable knowledge on practical approaches, cutting edge technologies, and optimal methods for handling intricate wastewater flows (Kumar et al., 2013, 2018, 2022). By emphasising these specific instances of real-world applications and the insights gained from them, individuals with a vested interest can acquire valuable information and motivation to implement environmentally friendly strategies for managing wastewater effectively. A brewery in a rural area encountered difficulties effectively handling its wastewater, which contained elevated concentrations of organic substances, suspended particles, and nutrients.

The brewery implemented an anaerobic digester to process concentrated organic wastewater, generating biogas as a sustainable energy source and decreasing the organic burden on subsequent treatment procedures. Following anaerobic digestion, the wastewater underwent aerobic treatment utilising activated sludge processes to eliminate organic matter and nutrients, attaining stringent effluent quality standards. The last stage of treatment involved membrane filtration, specifically ultrafiltration (UF) and reverse osmosis (RO), to generate reclaimed water of an exceptional quality suitable for non-potable uses like cleaning and irrigation. Integrating various treatment technologies, such as anaerobic digestion, aerobic treatment, and membrane filtration, enhances treatment efficiency and facilitates resource recovery. Anaerobic digestion treats highly-strength organic wastewater and generates biogas, a renewable energy source. This process improves sustainability and lowers operational expenses (Dwivedi & Kumar, 2011; Kumar & Pathak, 2019; Sharma et al., 2023). Membrane filtration allows reclaimed water to be reused in different non-potable applications, thereby preserving freshwater resources and minimising environmental harm. A dairy processing facility installed a wastewater treatment system to handle its effluent, which contained elevated fats, proteins, and nutrients from milk processing operations. Grease traps and screens were implemented to eliminate fats, oils, grease (FOG), and large particles from the wastewater. This prevents blockages and contamination of subsequent treatment equipment (Kotia et al., 2021; Kumar et al., 2021b; Kumar & Pathak, 2019). The wastewater was subjected to biological treatment, employing aerobic and anaerobic processes, such as activated sludge and anaerobic lagoons, to break down organic matter and nutrients. This treatment successfully ensured compliance with the effluent standards set by regulatory authorities. The facility incorporated nutrient recovery technologies, including struvite precipitation and biological phosphorus removal, to extract nitrogen and phosphorus from the wastewater. These nutrients are then repurposed as fertilisers in agricultural settings. Implementing appropriate pre-treatment measures, such as grease traps and screens, safeguards the functionality of subsequent treatment equipment and mitigates potential operational complications. Using biological treatment methods, such as aerobic and anaerobic digestion, successfully eliminates organic substances and nutrients from dairy wastewater, ensuring adherence to regulations and promoting environmental sustainability. Nutrient recovery technologies provide the means to extract valuable resources, specifically nitrogen and phosphorus, from wastewater to be reused as agricultural fertilisers (Das et al., 2022; Goud et al., 2022; Kumar et al., 2017; Kumar & Mistri, 2020). This practice aligns with circular economy principles and diminishes reliance on synthetic fertilisers. The fruit and vegetable processing plant adopted a wastewater treatment system to handle its effluent, which contained elevated concentrations of organic matter, sugars, and suspended solids resulting from processing activities. The wastewater underwent screening and sedimentation to eliminate large solids and suspended particles, thus preventing obstruction of downstream treatment equipment and enhancing treatment effectiveness (Aley et al., 2022; Kumar et al., 2018; Pankaj et al., 2012b; Pathak et al., 2017; Sharma & Kumar, 1999). The wastewater passed through biological filtration units, such as trickling filters or rotating biological contactors (RBCs), where aerobic microorganisms adhered to media surfaces and decomposed organic matter and pollutants. The treated wastewater was released into constructed wetlands as the last treatment step. In these wetlands, natural processes like sedimentation, filtration, and biological uptake enhanced the water quality further before it was discharged into the receiving water bodies. Integrating physical, biological, and natural treatment methods maximises the effectiveness of wastewater treatment and improves environmental sustainability. Biological filtration units, such as trickling filters and RBCs, efficiently eliminate organic matter and pollutants with minimal energy and chemical requirements. They constructed wetlands as economical and ecologically sound units that refine water quality and enhance the natural habitat for aquatic organisms (Kumar et al., 2018). These case studies exemplify the varied strategies and optimal methods that agro-foods and beverage industries employs to tackle the difficulties of intricate wastewater streams in wastewater treatment. Stakeholders can greatly enhance treatment efficiency, resource recovery, and environmental sustainability by implementing integrated treatment systems that combine multiple treatment technologies and adopt innovative approaches. Furthermore, sharing acquired knowledge and experiences from accomplished wastewater treatment projects promotes cooperation, the exchange of information, and the ongoing enhancement of wastewater management practices within the agro-foods and beverage sectors. To achieve sustainable wastewater management and ensure a cleaner, healthier environment for future generations, industries must utilise case studies and best practices as they innovate and evolve.

7. CHALLENGES AND FUTURE DIRECTIONS

Wastewater treatment is crucial in preserving environmental integrity, safeguarding water reserves, and upholding public welfare. Novel difficulties have arisen due to the progress of industries and environmental problems. The emerging issues, trends, opportunities for innovation and research, and recommendations for industry stakeholders and policymakers in wastewater treatment need to be addressed (Kumar et al., 2022; Kumar & Dwivedi, 2020; Kumar & Naik, 2020). Emerging contaminants in wastewater, such as pharmaceuticals, personal care products, microplastics, and per- and polyfluoroalkyl substances (PFAS), poses challenges to traditional treatment methods. The broad implementation of water reuse practices, especially in water-intensive industries like agro-foods and beverage sectors, is impeded by regulatory obstacles, public opinion, and technological constraints (Kandpal et al., 2018; Kumar et al., 2011b; Naik & Kumar, 2020). To address the effects of climate change on wastewater treatment systems, such as heightened stormwater runoff and shifting precipitation patterns, it is necessary to employ inventive engineering solutions and resilient planning strategies. There is an increasing focus on extracting resources from wastewater, such as energy, nutrients, and valuable materials. Nevertheless, technological feasibility, economic viability, and regulatory constraints impede the adoption of resource recovery technologies, thereby impeding the shift towards a circular economy and sustainable wastewater management practices. Exploration and advancement in cutting-edge treatment methods, such as membrane filtration, advanced oxidation processes (AOPs), and electrochemical treatment, present prospects for improving pollutant removal effectiveness, decreasing energy usage, and enhancing the ability to handle emerging contaminants (Kumar et al., 2011a; Kumar & Dwivedi, 2018; Reddy et al., 2022). Biological treatment processes, such as anaerobic digestion, bioaugmentation, and microbial fuel cells (MFCs) can enhance the breakdown of organic matter, removal of nutrients, and production of biogas from wastewater. This can help achieve energy self-sufficiency and recover valuable resources. Implementing intelligent monitoring and control systems, utilising sensors, Internet of Things (IoT) technology, and data analytics, allows for continuous monitoring of wastewater quality, process performance, and energy usage. This enables proactive decision-making and optimisation of treatment processes in real-time (Kumar et al., 2021a, 2022; Pankaj et al., 2012a). Water quality, biodiversity, and ecosystem regeneration can be improved by incorporating nature-based solutions into wastewater treatment systems, including constructed wetlands, green infrastructure, and ecological engineering. This approach also offers cost-effective and sustainable alternatives to traditional treatment methods. Promoting community support for sustainable wastewater management practices, water conservation, and resilience to water-related challenges is achieved by enhancing public engagement and awareness through education campaigns, outreach initiatives, and participatory decision-making processes. Industry stakeholders, academia, research institutions, and government agencies have to promote cooperation and information exchange among industry stakeholders, academia, research institutions, and government agencies for effective wastewater treatment. Favourable policies, regulations, and incentives have to be enacted so as to encourage investment in infrastructure for treating wastewater, adopting new technologies, and implementing initiatives for recovering resources. Training and capacity-building programs for wastewater treatment have to be provided to professionals, operators, and policymakers to enhance technical skills, knowledge of emerging technologies, and regulatory compliance requirements, ensuring effective operation and maintenance of treatment facilities. Allocation of financial resources and necessary support for research and development projects specifically target emerging issues in wastewater treatment. These initiatives should advance technological innovation, enhance treatment efficiency, and promote resilience and sustainability. Involve local communities, stakeholders, and residents in decision-making processes, project planning, and policy development concerning wastewater treatment, guaranteeing transparency and accountability in decision-making and governance. By embracing innovation, collaboration, and proactive approaches, industry stakeholders and policymakers can overcome the challenges posed by evolving wastewater treatment needs and shape a more sustainable and resilient future for wastewater management. By combining focused endeavours and collaborative measures, the potential of wastewater can be exploited as a valuable asset, mitigate its negative effects on the environment, and guarantee the enduring well-being and vigour of our water systems and ecosystems (Kumar et al., 2019; Kumar et al., 2018; Srivastav et al., 2023). Future investigations and advancements in wastewater treatment technologies are expected to result in more sophisticated and practical approaches for the agro-food and beverage sectors. These may encompass advancements for enhanced treatment efficiencies and reduced operational expenses such as membrane bioreactors, sophisticated oxidation techniques, and electrocoagulation. Potential future strategies could aim to optimise the recovery of resources from wastewater streams and convert waste materials into valuable resources. Technological advancements such as anaerobic digestion and nutrient recovery systems have the potential to facilitate the extraction of biogas for energy generation and the production of fertilisers from nutrient rich wastewater; would contribute to the establishment of circular economy within these sectors. Incorporating IoT devices and automation into wastewater treatment systems can transform wastewater management's efficacy in these industries. Automated control systems with real-time monitoring of crucial parameters such as pH, temperature, and pollutant concentrations can improve process efficiency, optimise resource allocation, and mitigate environmental consequences. Given the growing significance of sustainability for both consumers and regulators, agro-foods and beverage industries would embrace eco-friendly methods for treating wastewater. This may entail the implementation of green infrastructure solutions, such as the establishment of constructed wetlands and the utilisation of phytoremediation techniques. Establishing partnerships among industry stakeholders, research institutions, and government agencies will play a pivotal role in fostering innovation and facilitating the exchange of optimal methodologies in the field of wastewater treatment. Collaborative research projects, industry consortia, and knowledge-sharing platforms can expedite the exchange of ideas and the implementation of efficient wastewater treatment strategies in the agro-foods and beverage sectors. Adopting robust wastewater treatment strategies within these industries would be driven by increasingly stringent environmental regulations and standards. Organisations that allocate resources towards the adoption of cutting-edge treatment technologies and exhibit a steadfast dedication to ecological stewardship have the potential to attain a competitive edge within the market, all the while ensuring adherence to regulatory obligations. The allocation of resources towards educational and training initiatives specifically targeting wastewater treatment technologies and optimal methodologies can contribute to development of a proficient labour force which can implement and update efficient treatment systems within the agro-foods and beverage sectors. In the agro-foods and beverage industries, the future of wastewater treatment is marked by a focus on innovation, sustainability, and collaboration. To achieve their wastewater treatment goals and contribute to a more sustainable future, these industries can achieve their goals by embracing advanced technologies, adopting green practices, and fostering collaboration across the value chain.

8. CONCLUSION

In conclusion, substantial guidelines have to be implemented for the treatment of wastewater in agro-food and beverage industries for environmental sustainability. The distinctive attributes of sewage produced in these sectors, such as its elevated organic content, fluctuating composition, and intricate matrices, pose considerable obstacles which necessitate customised solutions and inventive methodologies. By incorporating cutting-edge treatment technologies like membrane filtration, advanced oxidation processes, anaerobic digestion, process optimisation, water reuse, and pollution prevention measures, agro-food and beverage industries can substantially enhance treatment efficiency, resource recovery, and environmental performance. Moreover, it is imperative to cooperate with various entities in the industry, policymakers, researchers, and communities to stimulate creativity, exchange exemplary methods, and cultivate a sustainable approach to wastewater management. To achieve a more resilient, sustainable, and prosperous future for wastewater treatment in agro-foods and beverage industries, we can invest in research and development, establish supportive policies and regulations, and encourage public engagement and education. In conclusion, creating solid and practical plans for treating wastewater safeguards our water resources, ecosystems, and public health. It also promotes the growth and competitiveness of agro-food and beverage industries, ensuring their long-term sustainability and contribution to a sustainable global economy. Through coordinated endeavours and collaborative measures, we can tackle the intricate dilemmas of wastewater management and establish a more healthful and enduring world for subsequent generations.

REFERENCES

Abdelfattah, A., Ali, S. S., Ramadan, H., El-Aswar, E. I., Eltawab, R., Ho, S.-H., … & Sun, J. (2023). Microalgae-based wastewater treatment: Mechanisms, challenges, recent advances, and prospects. *Environmental Science and Ecotechnology*, *13*, 100205. https://doi.org/https://doi.org/10.1016/j.ese.2022.100205

Adetunji, C. O., Olaniyan, O. T., Anani, O. A., Bodunrinde, R. E., Osemwegie, O. O., & Ubi, B. E. (2022). Chapter 17 - Integrated processes for the production of pharmaceutical products from agro-wastes. In S. Varjani, A. Pandey, T. Bhaskar, S. V. Mohan, & D. C. W. B. T.-B. Tsang (Eds.), *Biomass, biofuels, biochemicals* (pp. 439–461). Elsevier. https://doi.org/10.1016/B978-0-323-89855-3.00021-2

Ahmad, T., Belwal, T., Li, L., Ramola, S., Aadil, R. M., Abdullah, Xu, Y., & Zisheng, L. (2020). Utilization of wastewater from edible oil industry, turning waste into valuable products: A review. *Trends in Food Science & Technology*, *99*, 21–33. https://doi.org/https://doi.org/10.1016/j.tifs.2020.02.017

Alcántara, C., Acién-Fernández, F. G., Fernández-Sevilla, J. M., Riaño, B., Hernández, D., García-González, M. C., … & Muñoz, R. (2020). Chapter 40 - Exploring the potential of microalgae for the bioremediation of agro-industrial wastewaters. *Handbook of Algal Science, Technology and Medicine* (pp. 641–658). Academic Press. https://doi.org/10.1016/B978-0-12-818305-2.00040-1

Aley, P., Singh, J., & Kumar, P. (2022). Chapter 23 - Adapting the changing environment: microbial way of life. In A. Kumar, J. Singh, & L. F. R. B. T.-M. U. C. C. Ferreira (Eds.), *Microbiome under changing climate* (pp. 507–525). Woodhead Publishing. https://doi.org/https://doi.org/10.1016/B978-0-323-90571-8.00023-7

Arya, P. S., Yagnik, S. M., Rajput, K. N., Panchal, R. R., & Raval, V. H. (2022). Valorization of agro-food wastes: Ease of concomitant-enzymes production with application in food and biofuel industries. *Bioresource Technology*, *361*, 127738. https://doi.org/https://doi.org/10.1016/j.biortech.2022.127738

Chakraborty, S., Kumar, P., Sanyal, R., Mane, A. B., Arvind Prasanth, D., Patil, M., & Dey, A. (2021). Unravelling the regulatory role of miRNAs in secondary metabolite production in medicinal crops. *Plant Gene*, *27*, 100303. https://doi.org/https://doi.org/10.1016/j.plgene.2021.100303

Chatterjee, B., & Mazumder, D. (2024). A critical review of the advances in valorizing agro-industrial wastes through mixed culture fermentation. *Journal of Environmental Chemical Engineering*, *12*(1), 111838. https://doi.org/https://doi.org/10.1016/j.jece.2023.111838

Contreras-Jácquez, V., Virgo-Cruz, J. M., García-Fajardo, J., Obregón-Solís, E., Mateos-Díaz, J. C., & Asaff-Torres, A. (2022). Pilot-scale nanofiltration vibratory shear enhanced processing (NF-VSEP) for the improvement of the separation and concentration of compounds of biotechnological interest from tortilla industry wastewater (nejayote). *Separation and Purification Technology*, *300*, 121921. https://doi.org/https://doi.org/10.1016/j.seppur.2022.121921

Cruz, R. M. S., Albertos, I., Romero, J., Agriopoulou, S., & Varzakas, T. B. T.-A. in F. and N. R. (2023). *Innovations in food packaging for a sustainable and circular economy*. Academic Press. https://doi.org/https://doi.org/10.1016/bs.afnr.2023.10.003

Das, T., Saha, S. C., Sunita, K., Majumder, M., Ghorai, M., Mane, A. B., … & Dey, A. (2022). Promising botanical-derived monoamine oxidase (MAO) inhibitors: pharmacological aspects and structure-activity studies. *South African Journal of Botany*, *146*, 127–145. https://doi.org/https://doi.org/10.1016/j.sajb.2021.09.019

Devi, P., Dey, S. R., & Kumar, P. (2023). Integration of water resources management in rural areas. In *Water resources management for rural development: Challenges and mitigation* (pp. 139–152). Elsevier. https://doi.org/10.1016/B978-0-443-18778-0.00013-1

Devi, P., Dey, S. R., Saini, L., Kumar, P., Panigrahi, S., & Dwivedi, P. (2023). Toward sustainable agriculture: Strategies involving phytoprotectants against reactive oxygen species. In *Reactive oxygen species: Prospects in plant metabolism* (pp. 229–247). Springer Nature. https://doi.org/10.1007/978-981-19-9794-5_13

Devi, P., Kumar, P., Sharma, K., & Singh, J. (2023). Is there a relationship between myconanotechnology and sustainable development? In *Myconanotechnology and application of nanoparticles in biology: Fundamental concepts, mechanism and industrial applications* (pp. 175–207). Elsevier. https://doi.org/10.1016/B978-0-443-15262-7.00009-7

Dey, S. R., Devi, P., & Kumar, P. (2023). Organic contaminants in aquatic environments: Sources and impact assessment. In *Metal organic frameworks for wastewater contaminant removal* (pp. 299–317). Wiley. https://www.scopus.com/inward/record.uri?eid=2-s2.0-85174103623&partnerID=40&md5=dbfb41dc5474b67aaaca5dec64c73718

Dey, S. R., Sharma, M., & Kumar, P. (2023). Effects and responses of chromium on plants. In *Environmental science and engineering: Vol. Part F1975* (pp. 385–427). Springer Science and Business Media Deutschland GmbH. https://doi.org/10.1007/978-3-031-44029-8_14

Diaz, A. B., Blandino, A., & Caro, I. (2018). Value added products from fermentation of sugars derived from agro-food residues. *Trends in Food Science & Technology*, *71*, 52–64. https://doi.org/https://doi.org/10.1016/j.tifs.2017.10.016

Díaz, V., Leyva-Díaz, J. C., Almécija, M. C., Poyatos, J. M., del Mar Muñío, M., & Martín-Pascual, J. (2022). Microalgae bioreactor for nutrient removal and resource recovery from wastewater in the paradigm of circular economy. *Bioresource Technology*, *363*, 127968. https://doi.org/https://doi.org/10.1016/j.biortech.2022.127968

Dubey, A., Kumar, P., & Yati, V. (2002). Growth performance and susceptibility of Solanum melongena subjected to different concentration of opium effluent. *Pollution Research*, *21*(1), 89–90. https://www.scopus.com/inward/record.uri?eid=2-s2.0-0036252801&partnerID=40&md5=415ec25e0df7d8852aeffe44c0027b95

Dwivedi, P., & Kumar, P. (2011). Anti-diabetic medicinal plants and their conservation: Waging green war on diabetes. *Medicinal Plants*, *3*(3), 181–189. https://doi.org/10.5958/j.0975-4261.3.3.031

Franco-Morgado, M., Amador-Espejo, G. G., Pérez-Cortés, M., & Gutiérrez-Uribe, J. A. (2023). Microalgae and cyanobacteria polysaccharides: Important link for nutrient recycling and revalorization of agro-industrial wastewater. *Applied Food Research*, *3*(1), 100296. https://doi.org/https://doi.org/10.1016/j.afres.2023.100296

Gecim, G., Aydin, G., Tavsanoglu, T., Erkoc, E., & Kalemtas, A. (2021). Review on extraction of polyhydroxyalkanoates and astaxanthin from food and beverage processing wastewater. *Journal of Water Process Engineering*, *40*, 101775. https://doi.org/https://doi.org/10.1016/j.jwpe.2020.101775

Gehlot, P., Vivekanand, V., & Pareek, N. (2022). Chapter 17 - Cyanobacterial and microalgal bioremediation: an efficient and eco-friendly approach toward industrial wastewater treatment and value-addition. In *Microbial biodegradation and bioremediation (2nd Ed.)* (pp. 343–362). Elsevier. https://doi.org/10.1016/B978-0-323-85455-9.00002-3

Goud, E. L., Singh, J., & Kumar, P. (2022). Climate change and their impact on global food production. In *Microbiome under changing climate* (pp. 415–436). Elsevier.

Hasan, H. A., Muhamad, M. H., Ji, B., Nazairi, N. A., Jiat, K. W., Sim, S. I. S. W. A., & Poh, A. F. M. S. (2023). Revolutionizing wastewater treatment with microalgae: Unveiling resource recovery, mechanisms, challenges, and future possibilities. *Ecological Engineering*, *197*, 107117. https://doi.org/10.1016/j.ecoleng.2023.107117

Hernández-Pinto, F. J., Miranda-Medina, J. D., Natera-Maldonado, A., Vara-Aldama, Ó., Ortueta-Cabranes, M. P., Vázquez del Mercado-Pardiño, J. A., … & Castro-Muñoz, R. (2024). Arabinoxylans: A review on protocols for their recovery, functionalities and roles in food formulations. *International Journal of Biological Macromolecules*, *259*, 129309. https://doi.org/10.1016/j.ijbiomac.2024.129309

Herrmann, L. W., Letti, L. A. J., Penha, R. de O., Soccol, V. T., Rodrigues, C., & Soccol, C. R. (2024). Bacillus genus industrial applications and innovation: First steps towards a circular bioeconomy. *Biotechnology Advances*, *70*, 108300. https://doi.org/https://doi.org/10.1016/j.biotechadv.2023.108300

Hidangmayum, A., Dwivedi, P., Kumar, P., & Upadhyay, S. K. (2023). Seed priming and foliar application of chitosan ameliorate drought stress responses in mungbean genotypes through modulation of morpho-physiological attributes and increased antioxidative defense mechanism. *Journal of Plant Growth Regulation*, *42*(10), 6137–6154. https://doi.org/10.1007/s00344-022-10792-1

Hoang, A. T., Pandey, A., Huang, Z., Nižetić, S., Le, A. T., & Nguyen, X. P. (2023). Chapter 3 - Biofuels an option for agro-waste management. In *Environmental sustainability of biofuels* (pp. 27–47). Elsevier. https://doi.org/10.1016/B978-0-323-91159-7.00011-4

Jha, R., Zhang, K., He, Y., Mendler-Drienyovszki, N., Magyar-Tábori, K., Quinet, M., … & Zhou, M. (2024). Global nutritional challenges and opportunities: Buckwheat, a potential bridge between nutrient deficiency and food security. *Trends in Food Science & Technology*, *145*, 104365. https://doi.org/https://doi.org/10.1016/j.tifs.2024.104365

Jornet-Martínez, N., Bocanegra-Rodríguez, S., González-Fuenzalida, R. A., Molins-Legua, C., & Campíns-Falcó, P. (2019). Chapter 9 - In situ analysis devices for estimating the environmental footprint in beverages industry. In A. M. Grumezescu & A. M. B. T.-P. & S. of B. Holban (Eds.), *Processing and sustainability of beverages* (pp. 275–317). Woodhead Publishing. https://doi.org/10.1016/B978-0-12-815259-1.00009-4

Kalra, R., Gaur, S., & Goel, M. (2021). Microalgae bioremediation: A perspective towards wastewater treatment along with industrial carotenoids production. *Journal of Water Process Engineering*, *40*, 101794. https://doi.org/https://doi.org/10.1016/j.jwpe.2020.101794

Kee, S. H., Chiongson, J. B. V, Saludes, J. P., Vigneswari, S., Ramakrishna, S., & Bhubalan, K. (2021). Bioconversion of agro-industry sourced biowaste into biomaterials via microbial factories – A viable domain of circular economy. *Environmental Pollution*, *271*, 116311. https://doi.org/https://doi.org/10.1016/j.envpol.2020.116311

Kee, S. H., Ganeson, K., Rashid, N. F. M., Yatim, A. F. M., Vigneswari, S., Amirul, A.-A. A., Ramakrishna, S., & Bhubalan, K. (2022). A review on biorefining of palm oil and sugar cane agro-industrial residues by bacteria into commercially viable bioplastics and biosurfactants. *Fuel*, *321*, 124039. https://doi.org/https://doi.org/10.1016/j.fuel.2022.124039

Kotia, A., Rutu, P., Singh, V., Kumar, A., Dhoke, S., Kumar, P., & Singh, D. K. (2021). *Rheological analysis of rice husk-starch suspended in water for sustainable agriculture application.* 2nd International Conference on Functional Materials, Manufacturing and Performances, ICFMMP 2021, 50, 1962–1966. https://doi.org/10.1016/j.matpr.2021.09.325

Kumar, D., Rameshwar, S. D., & Kumar, P. (2019). Effect of intergated application of inorganic and organic fertilizers on the roots of chickpea. *Plant Archives*, *19*(1), 857–860. https://www.scopus.com/inward/record.uri?eid=2-s2.0-85068434345&partnerID=40&md5=4d9e2d8c35a83b31a09556e919e9745f

Kumar, P., Devi, P., & Dey, S. R. (2021a). Fungal volatile compounds: A source of novel in plant protection agents. In *Volatiles and metabolites of microbes* (pp. 83–104). Elsevier.

Kumar, P., & Dwivedi, P. (2018a). Putrescine and glomus mycorrhiza moderate cadmium actuated stress reactions in zea mays l. By means of extraordinary reference to sugar and protein. *Vegetos*, *31*(3), 74–77. https://doi.org/10.5958/2229-4473.2018.00076.9

Kumar, P., & Dwivedi, P. (2020). Lignin estimation in sorghum leaves grown under hazardous waste site. *Plant Archives*, *20*, 2558–2561. https://www.scopus.com/inward/record.uri?eid=2-s2.0-85090326255&partnerID=40&md5=2482a876d9ae56ac04904bd264180711

Kumar, P., Goud, E. L., Devi, P., Dey, S. R., & Dwivedi, P. (2022). Heavy metals: Transport in plants and their physiological and toxicological effects. In *Plant metal and metalloid transporters* (pp. 23–54). Springer Nature. https://doi.org/10.1007/978-981-19-6103-8_2

Kumar, P., Goud, E. L., Devi, P., & Koul, B. (2022). Metal pollutants in the environment. In *Environmental microbiology: Emerging technologies* (pp. 291–323). De Gruyter. https://doi.org/10.1515/9783110727227-012

Kumar, P., Harshavardhan, M., Kumar, P. S., Yumnam, J., Jyoti, N., Naik, M., … & Kumar, S. (2018). Effect on chlorophyll a/b ratio in cadmium contaminated maize leaves treated with putrescine and mycorrhiza. *Annals of Biology*, *34*(3), 281–283. https://www.scopus.com/inward/record.uri?eid=2-s2.0-85075538775&partnerID=40&md5=aa4f2234012e98022d0cb99a02963635

Kumar, P., Koul, B., & Sharma, M. (2022). Phytoremediation of heavy metals. In *Heavy Metals in plants: Physiological to molecular approach* (pp. 369–388). CRC Press. https://doi.org/10.1201/9781003110576-17

Kumar, P., Kumar, S., Harshavardhan, M., Naik, M., Yumnam, J., Kumar, P. S., … & Purnima. (2018). Evaluation of plant height and leaf length of sorghum grown under different sources of nutrition. *Annals of Biology*, *34*(3), 284–286. https://www.scopus.com/inward/record.uri?eid=2-s2.0-85075511816&partnerID=40&md5=570ec8983a32abffaa81d117e0acae1e

Kumar, P., Kumar, T., Singh, S., Tuteja, N., Prasad, R., & Singh, J. (2020). Potassium: A key modulator for cell homeostasis. *Journal of Biotechnology*, *324*, 198–210.

Kumar, P., Mandal, B., & Dwivedi, P. (2011a). Heavy metal scavenging capacity of Mentha spicata and Allium cepa. *Medicinal Plants*, *3*(4), 315–318. https://doi.org/10.5958/j.0975-4261.3.4.053

Kumar, P., Mandal, B., & Dwivedi, P. (2011b). Heavy metals scavenging of soils and sludges by ornamental plants. *Journal of Applied Horticulture*, *13*(2), 144–146. https://www.scopus.com/inward/record.uri?eid=2-s2.0-84860310792&partnerID=40&md5=41f3f6036e261fad9ff5a655f9aef1f5

Kumar, P., Mandal, B., & Dwivedi, P. (2013). Combating heavy metal toxicity from hazardous waste sites by harnessing scavenging activity of some vegetable plants. *Vegetos*, *26*(2), 416–425. https://doi.org/10.5958/j.2229-4473.26.2.106

Kumar, P., & Mistri, T. K. (2020). Transcription factors in SOX family: Potent regulators for cancer initiation and development in the human body. *Seminars in Cancer Biology*, *67*, 105–113. https://doi.org/https://doi.org/10.1016/j.semcancer.2019.06.016

Kumar, P., & Naik, M. (2020). Biotic symbiosis and plant growth regulators as a strategy against cadmium and lead stress in chickpea. *Plant Archives*, *20*, 2495–2500. https://www.scopus.com/inward/record.uri?eid=2-s2.0-85090455388&partnerID=40&md5=e0a3e27f167a9f50afc1c5954d71b2f0

Kumar, P., & Pathak, S. (2019). Responsiveness index of sorghum (Sorghum bicolor (L.) Moench) grown under cadmium contaminated soil treated with putrescine and mycorrhiza. *Bangladesh Journal of Botany*, *48*(1), 139–143. https://www.scopus.com/inward/record.uri?eid=2-s2.0-85063424662&partnerID=40&md5=7b83ecbe705e28da69bc206c34136726

Kumar, P., Pathak, S., Amarnath, K. S., Veerendra Brahma Teja, P., Dileep, B., Kumar, K., ... & Siddique, A. (2018). Effect of growth regulator on morpho-physiological attributes of chilli: A case study. *Plant Archives*, *18*(2), 1771–1776. https://www.scopus.com/inward/record.uri?eid=2-s2.0-85060849603&partnerID=40&md5=8146fd6f6db1681549099575e6a18cdb

Kumar, P., Pathak, S., Kumar, M., & Dwivedi, P. (2018). Role of secondary metabolites for the mitigation of cadmium toxicity in sorghum grown under mycorrhizal inoculated hazardous waste site. *Biotechnological Approaches for Medicinal and Aromatic Plants: Conservation, Genetic Improvement and Utilization*, 199–212.

Kumar, P., Saini, L., & Sharma, M. (2023). Assessment of the phytochemical constituents and metabolites in the medicinal plants and herbal medicine used in the treatment and management of respiratory diseases. In *Reference series in phytochemistry* (Vol. 2023). Springer Science and Business Media B.V. https://doi.org/10.1007/978-3-031-21973-3_19-1

Kumar, P., Sharma, K., Saini, L., & Dey, S. R. (2021b). Chapter 8 - Role and behavior of microbial volatile organic compounds in mitigating stress. In A. Kumar, J. Singh, & J. B. T.-V. and M. of M. Samuel (Eds.), (pp. 143–161). Academic Press. https://doi.org/10.1016/B978-0-12-824523-1.00010-9

Kumar, P., Singh, B. N., & Dwivedi, P. (2017). Plant growth regulators, plant adaptability and plant productivity: A review on abscisic acid (ABA) signaling in plants under emerging environmental stresses. In *Sustaining future food security in changing environments* (pp. 81–97). Nova Science Publishers, Inc. https://www.scopus.com/inward/record.uri?eid=2-s2.0-85022015328&partnerID=40&md5=4d99a190e7c92792cec87af94a7d53ee

Kumari, P., Singh, J., & Kumar, P. (2022). Chapter 21 - Impact of bioenergy for the diminution of an ascending global variability and change in the climate. In A. Kumar, J. Singh, & L. F. R. B. T.-M. U. C. C. Ferreira (Eds.), *Microbiome under changing climate* (pp. 469–487). Woodhead Publishing. https://doi.org/https://doi.org/10.1016/B978-0-323-90571-8.00021-3

Kumar, V., Dwivedi, P., Kumar, P., Singh, B. N., Pandey, D. K., Kumar, V., & Bose, B. (2021c). Mitigation of heat stress responses in crops using nitrate primed seeds. *South African Journal of Botany*, *140*, 25–36. https://doi.org/https://doi.org/10.1016/j.sajb.2021.03.024

Kundu, D., Dutta, D., Samanta, P., Dey, S., Sherpa, K. C., Kumar, S., & Dubey, B. K. (2022). Valorization of wastewater: A paradigm shift towards circular bioeconomy and sustainability. *Science of The Total Environment*, *848*, 157709. https://doi.org/https://doi.org/10.1016/j.scitotenv.2022.157709

López-Gámez, G., Soliva-Fortuny, R., & Elez-Martínez, P. (2023). 4.11 - Green technologies for sustainable food production: Pulsed electric fields. In P. B. T.-S. F. S.-A. C. A. Ferranti (ed.), *Microbiome under changing climate* (pp. 198–217). Elsevier. https://doi.org/https://doi.org/10.1016/B978-0-12-823960-5.00026-3

Maurya, R., Zhu, X., Valverde-Pérez, B., Ravi Kiran, B., General, T., Sharma, S., ... & Angelidaki, I. (2022). Advances in microalgal research for valorization of industrial wastewater. *Bioresource Technology*, *343*, 126128. https://doi.org/https://doi.org/10.1016/j.biortech.2021.126128

Mishra, B., Mohanta, Y. K., Reddy, C. N., Reddy, S. D. M., Mandal, S. K., Yadavalli, R., & Sarma, H. (2023). Valorization of agro-industrial biowaste to biomaterials: An innovative circular bioeconomy approach. *Circular Economy*, *2*(3), 100050. https://doi.org/https://doi.org/10.1016/j.cec.2023.100050

Mohd Udaiyappan, A. F., Abu Hasan, H., Takriff, M. S., & Sheikh Abdullah, S. R. (2017). A review of the potentials, challenges and current status of microalgae biomass applications in industrial wastewater treatment. *Journal of Water Process Engineering*, *20*, 8–21. https://doi.org/https://doi.org/10.1016/j.jwpe.2017.09.006

Naik, M., & Kumar, P. (2020). Role of growth regulators and microbes for metal detoxification in plants and soil. *Plant Archives*, *20*, 2820–2824. https://www.scopus.com/inward/record.uri?eid=2-s2.0-85090334398&partnerID=40&md5=d501ee461b077a20ac3dfe8ab537a3ee

Okaiyeto, S. A., Sutar, P. P., Chen, C., Ni, J.-B., Wang, J., Mujumdar, A. S., ... & Xiao, H.-W. (2024). Antibiotic resistant bacteria in food systems: Current status, resistance mechanisms, and mitigation strategies. *Agriculture Communications*, 100027. https://doi.org/https://doi.org/10.1016/j.agrcom.2024.100027

Pankaj, P. K., Kumar, P., Nigam, R. C., & Mishra, P. K. (2012a). Monitoring and surveillance of synthetic pyrethroids and organophosphate in different brands of soft drinks. *Journal of Chemical and Pharmaceutical Research*, *4*(8), 3939–3943. https://www.scopus.com/inward/record.uri?eid=2-s2.0-84867504985&partnerID=40&md5=0078e81af00e4b6849c087c3ae5c7591

Pankaj, P. K., Kumar, P., Nigam, R. C., & Mishra, P. K. (2012b). To studies organochlorine insecticides in different brands of cold drinks. *Journal of Chemical and Pharmaceutical Research*, *4*(8), 3934–3938. https://www.scopus.com/inward/record.uri?eid=2-s2.0-84867537578&partnerID=40&md5=f2db0a7b022cd9e4dd96e1dcc49250a5

Park, H., He, H., Yan, X., Liu, X., Scrutton, N. S., & Chen, G.-Q. (2024). PHA is not just a bioplastic! *Biotechnology Advances*, *71*, 108320. https://doi.org/https://doi.org/10.1016/j.biotechadv.2024.108320

Pathak, S., Kumar, P., Mishra, P. K., & Kumar, M. (2017). Mycorrhiza assisted approach for bioremediation with special reference to biosorption. *Pollution Research*, *36*(2), 329–332. https://www.scopus.com/inward/record.uri?eid=2-s2.0-85028617468&partnerID=40&md5=da296311c0f3d29ea94d21631fe01d69

Paul, A., Sharma, S. R., Sresty, T. V. S., Devi, S., Bala, S., Kumar, P. S., ... & Kumar, P. A. (2005). Transgenic cabbage *(Brassica oleracea var. capitata)* resistant to Diamondback moth *(Plutella xylostella). Indian Journal of Biotechnology, 4*, 72 77.

Phiri, R., Mavinkere Rangappa, S., & Siengchin, S. (2024). Agro-waste for renewable and sustainable green production: A review. *Journal of Cleaner Production*, *434*, 139989. https://doi.org/10.1016/j.jclepro.2023.139989

Punia Bangar, S., Chaudhary, V., Kajla, P., Balakrishnan, G., & Phimolsiripol, Y. (2024). Strategies for upcycling food waste in the food production and supply chain. *Trends in Food Science & Technology*, *143*, 104314. https://doi.org/10.1016/j.tifs.2023.104314

Reddy, S. S., Kumar, P., & Dwivedi, P. (2022). Heavy metal transporters, phytoremediation potential, and biofortification. In *Plant metal and metalloid transporters* (pp. 387–405). Springer.

Rodrigues, D. M., da Silva, M. F., Almeida, F. L. C., de Mélo, A. H. F., Forte, M. B. S., Martín, C., … & Goldbeck, R. (2024). A green approach to biomass residue valorization: Bacterial nanocellulose production from agro-industrial waste. *Biocatalysis and Agricultural Biotechnology*, *56*, 103036. https://doi.org/10.1016/j.bcab.2024.103036

Saini, L., Sharma, K., & Kumar, P. (2023). Water efficiency in agriculture as a vital approach toward water management in rural areas. In *Water Resources Management for Rural Development: Challenges and Mitigation* (pp. 127–138). Elsevier. https://doi.org/10.1016/B978-0-443-18778-0.00020-9

Sarangi, P. K., Singh, A. K., Sonkar, S., Shadangi, K. P., Srivastava, R. K., Gupta, V. K., ... & Govarthanan, M. (2023). Biorefinery solutions for food processing wastes: A sustainable bioeconomic perspective. *Industrial Crops and Products*, *205*, 117488.https://doi.org/10.1016/j.indcrop.2023.117488

Sharma, K., Devi, P., & Kumar, P. (2023). Metal-organic frameworks for wastewater contaminants removal. In *Metal organic frameworks for wastewater contaminant removal* (pp. 95–117). Wiley. https://www.scopus.com/inward/record.uri?eid=2-s2.0-85174095287&partnerID=40&md5=00a15587f7d58d45b161b04281a93a57

Sharma, K., Devi, P., Kumar, P., Dey, A., & Dwivedi, P. (2023). Hazardous phytotoxic nature of reactive oxygen species in agriculture. In *Reactive oxygen species: Prospects in plant metabolism* (pp. 135–146). Springer Nature. https://doi.org/10.1007/978-981-19-9794-5_8

Sharma, M., Dey, S. R., & Kumar, P. (2023). Antioxidant defence: a key mechanism of chromium tolerance. In *Environmental science and engineering: Vol. Part F1975* (pp. 91–116).

Springer Science and Business Media Deutschland GmbH. https://doi.org/10.1007/978-3-031-44029-8_5

Sharma, P. C., & Kumar, P. (1999). Alleviation of salinity stress during germination in Brassica juncea by pre-sowing chilling treatments to seeds. *Biologia Plantarum, 42*(3), 451–455. https://doi.org/10.1023/A:1002481709121

Song, W., Xie, B., Huang, S., Zhao, F., & Shi, X. (2020). Chapter 6 - Aerobic membrane bioreactors for industrial wastewater treatment. In *Current developments in biotechnology and bioengineering* (pp. 129–145). Elsevier. https://doi.org/10.1016/B978-0-12-819809-4.00006-1

Srivastav, A. L., Patel, N., Rani, L., Kumar, P., Dutt, I., Maddodi, B. S., & Chaudhary, V. K. (2023). Sustainable options for fertilizer management in agriculture to prevent water contamination: A review. *Environment, development and sustainability*. https://doi.org/10.1007/s10668-023-03117-z

Upadhyay, S. K., Devi, P., Kumar, V., Pathak, H. K., Kumar, P., Rajput, V. D., & Dwivedi, P. (2023). Efficient removal of total arsenic (As3+/5+) from contaminated water by novel strategies mediated iron and plant extract activated waste flowers of marigold. *Chemosphere, 313*, 137551. https://doi.org/https://doi.org/10.1016/j.chemosphere.2022.137551

Varadavenkatesan, T., Pai, S., Vinayagam, R., Pugazhendhi, A., & Selvaraj, R. (2021). Recovery of value-added products from wastewater using Aqueous Two-Phase Systems – A review. *Science of The Total Environment, 778*, 146293. https://doi.org/https://doi.org/10.1016/j.scitotenv.2021.146293

4

Bio-catalytic Mitigation for Removal of Microplastics from Water Contaminated with Industrial Effluents

Darshan Mulchandani, Gaurav Sanghvi and Anjisha Maharshi

1. INTRODUCTION

Bakelite is world's the first synthetic plastic discovered and synthesized in 1907 in New York. In 1920, with the advent of theory of polymers including nylon, polyethylene, polyvinyl chloride, and polystyrene were eventually introduced to the world (Morris, 2005). The invention of injection-molding technologies allowed to mold plastic in any shape and size entered into day-to-day life. Further, to meet global demand on a daily basis, plastics have been manufactured at a high rate. The excellent benefits of plastics,

DOI:10.1201/9781003496014-4

including convenience, erosive resistance, processing ease, durability, and low cost, are driving up demand for plastics (Meikle, 1995). In 2018, 360 million tons of plastic were produced, and 500 million tons of plastic are anticipated to be produced globally by 2025, increasing at a compound annual growth rate of 8.4% since 1950. With its usage in daily life, approximately 60% of plastic waste ends up in the environment (Geyer et al., 2017). Their exposure to the environment causes mechanical and physiochemical degeneration, as well as the formation of plastic debris. This debris is regarded as microplastics (small pieces less than 5 mm in size). These microplastics then get inside land animals and sea animals, potentially harming their health (Esteban et al., 2016). Oxidative stress, immune system dysfunction can occur due to microplastic exposure. However, the effect of microplastics on humans is not well understood. Fish were observed with neurotoxicity when exposed to microplastics (Usman et al., 2021). Thermal treatment, microbial degradation, and physical techniques like adsorption and filtration are all used in microplastic remediation. Scientists have become interested in using bacteria, fungi, and algae as a tool for treating microplastics. Microplastic degradation involves enzymatic reactions which are produced by microorganisms, such as cutinase, Lipase, PETEase, etc., that are involved in microplastic degradation (Othman et al., 2021).

Additional research aided by advancements in the fields of genomics, proteomics, gene regulation, enzyme immobilization, protein design, and nanotechnology is anticipated to significantly increase our understanding and will further lead to improvement in the enzymatic efficiency and affordability of degradation (Carmen, 2021). As compared to other techniques, enzyme-assisted biodegradation of microplastics is the most efficient.

2. MICROPLASTICS AND THEIR ENVIRONMENTAL CONCERNS

2.1. Type of Microplastics

Plastic fragments are categorized in the form of megaplastics, macroplastics, mesoplastics, and microplastics based on their presence in the environment. Microplastics are defined as plastic fragments smaller than 5 mm, a term Thompson first proposed in 2004 (Thompson et al., 2004) (Figure 4.1). There are two types of microplastics: primary and secondary. Primary microplastics are made for commercial use such as clothing, textiles, cosmetics, and so on; whereas secondary microplastics are formed from the breakdown of large microplastics such as a plastic bottle (Figure 4.1).

2.2. Microplastics on Environment

Microplastics can increase the impact of global warming. Most of the microplastics enter into soil and water through sewage, the major role in the spreading of microplastics.

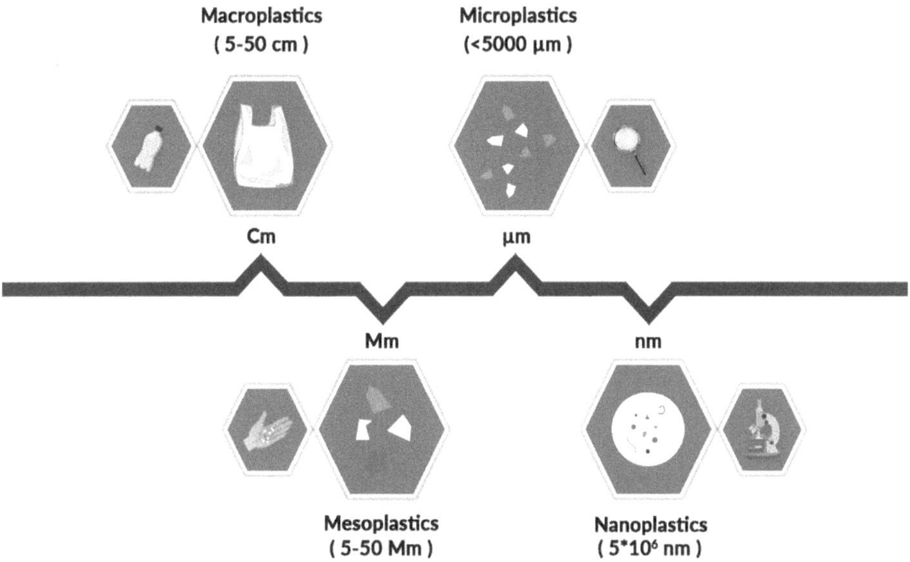

FIGURE 4.1 Classification of microplastics on the basis of size.
Figure created by author.

Around 80% to 90% of plastic debris from clothing fibers are found in sewage in the form of sludge. Each year 1,000 tons of microplastics enter in agricultural fields and soils due to the use of sewage sludge as fertilizer. Even in tap water, micro-plastics are present in trace amounts (Tong et al., 2020). Additionally, the surfaces of microscopic plastic fragments may contain pathogens and act as a vector for environmental diseases. Microplastics may have an effect on the fertility of the soil (Figure 4.2) as well as on the health of the soil's fauna, including earthworms, nematodes, and protozoa. A recent article published by *Science Daily* states, "Earthworms, for example, make their burrows differently when microplastics are present in the soil, affecting the earthworm's fitness and the soil condition." Another statement adds, "When earthworms and chickens eat microplastic that has formed and degraded in the soil, it can lead to oxidative stress and neurotoxicity" (Wang et al., 2022). Microplastics can also take up in inorganic (Tagg et al., 2020) and organic pollutants like dichlorodiphenyltrichloroethane (DDT), polychlorinated biphenyls (PCBs), dioxins, and polycyclic aromatic hydrocarbons (PAHs) (Jiménez-Skrzypek et al., 2021).

Microplastics are difficult to remove from the environment because of their small size and stability (Malankowska et al., 2021). Microplastics can act as a vector for contaminants to be transported both within and between different ecosystem compartments due to their surface interactions (Bank & Hansson, 2019). Microplastic interaction with various elements can have an impact on crop production (Figure 4.2).

FIGURE 4.2 Effect of microplastics on soil and growth of plants.

Currently, the main sources of marine microplastics are land flow. Wind and rain carry land-based plastic wastes like plastic bags, foam, and fibers into rivers and the ocean, where they contribute to marine plastic pollution. Land flow is the primary contributor to microplastics in the marine environment (Li et al., 2021). Mishandled plastic waste finds its way onto land, into rivers, waterways, and coastal waters, where it contributes to the global pollution of oceans and beaches caused by an increasing amount of marine litter. Certain reports suggest that 6–15 million tons of plastics, representing 2–4% of global production, enter the environment every year (Papari et al., 2021). Unfortunately, very little plastic is recycled, and energy recovered from it. Due to low recycling efficiency, the majority of plastics are discarded into the environment through a variety of channels. Concentrations of these particles in the ocean's surface waters appear to have significantly increased over the past 40 years. Over the past few years, plenty of plastics are dumped into the environment through various channels due to low effectiveness in recycling.

Estimating and measuring the amount of microplastics released into the environment is difficult. Due to the numerous primary and secondary sources of microplastics and the absence of standardized sampling and measurement techniques, estimates of the amounts of microplastics released and formed are unknown. According to research, the oceans are currently home to at least 14 million tons of microplastics (Barrett et al., 2020), and an additional 1.5 million tons are believed to enter each year (Boucher & Friot, 2017). Despite the short-term advantages of plastics, scientists are now focusing more on their long-term impacts on crop productivity and soil quality (Forster et al., 2020).

Evidence suggests that snow also contains microplastic fragments. A group of researchers from the University of Canterbury in New Zealand have collected samples from 19 different sites in Antarctica. The presence of black carbon can increase the melting of ice fields and glaciers as it can absorb sunlight, accelerating melting processes (Parolini et al., 2021). Larger plastic also concentrates in ocean gyres (Moore et al., 2002). There is some preliminary evidence that microplastics are transferred to the circulatory system based on analyses with mussels (Zarfl et al., 2011). However, tens of thousands of tons of plastic, far less than anticipated, were thought to be present on the open ocean's surface globally (Andrés Cózar et al., 2014).

Surface runoff and sewage/wastewater treatment plant effluent discharge are the main sources of microplastics (primary and secondary) in water (Iyare et al. 2020). Numerous industrial sectors employ (primary) microplastics in many applications, such as cosmetics and pharmaceuticals (European Commission Report, 2017; Lechner & Ramler, 2015). Following their residential or commercial use, these major microplastics wash off and end up in the wastewater or sewage. Conversely, dumping grounds, landfills, and seas are frequently the areas where large-sized plastic particles break down into smaller pieces, or secondary microplastics, which contributes to the amount of microplastics in water bodies (Kyrikou & Briassoulis, 2007; Galloway & Lewis, 2016). Large amounts of microplastics are present in the effluent emit by sewage/wastewater treatment plants since these facilities are not accurately equipped to eliminate microplastics completely (Amrutha & Warrier, 2020). Mixing of this effluent with the freshwater sources, microplastics enter the freshwater and drinking water supply chain (Okoffo et al., 2019).

Fresh water may accumulate microplastics, however, compared to seawater, fresh water microplastics have received less attention from researchers. Fresh water may be a source, transfer medium, or sink for microplastics. Microplastics in sewage can be heavily contaminated but microplastics in freshwater have less contamination and less organic compounds (Jingyi et al., 2018).

Lakes, groundwater, and rivers where there is a high amount of plastic was observed with higher aggregation of microplastics. However, appropriate sampling and quantification are required. Considering the effects microplastics have on human health, effective wastewater treatment of these particles is imperative. Furthermore, drinking water is a major way for humans to consume microplastics, which can cause a number of health problems, including cancer and mutagenic and teratogenic disorders. Investigating methods for removing microplastics from wastewater is crucial since microplastic contamination of water is a growing public health concern.

3. MICROPLASTICS AND HEALTH

Microplastics are hydrophobic and have a large surface area; are ubiquitous and present in all size and shape; and are present in the soil, atmosphere, freshwater, marine water, food, drinking water, etc. Microplastics may contain different kinds of chemicals: environmental chemicals absorbed from the environment, and polymeric raw materials derived from the plastics (viz. monomers or oligomers) and additives. In the manufacturing process, plastic is intentionally given additives to give it properties like color and transparency. These additives also help plastic products to perform better by increasing their resistance to bacteria, mold, humidity, ozone, light radiation, temperature, as well as mechanical, thermal, and electrical resistance (Hahladakis et al., 2018). Additives such as dyes, plasticizers, stabilizers, lubricants, and flame-retardants are used to produce robust plastics. Most of the chemicals are not bound chemically with polymer but some flame retardants are bound chemically with polymer. The most dangerous types of additives are phthalates, lead compounds used as heat stabilizers, and flame retardants (Campanale et al., 2020). There has been a substantial rise in the relative risk of liver angiosarcoma among employees at PVC or vinyl chloride monomer manufacturing facilities (Ward et al., 2001). One study found that long-term exposure to styrene oxide and inhalation can cause cancer in Dawley rats (McConnell et al., 1994).

Microplastics themselves may not be harmful, but chemicals attached to them can affect the health of marine animals, terrestrial animals, and humans. Plasticizers used for flexibility in plastic use phthalic acid esters such as diisodecyl phthalate (DIDP), benzyl butyl phthalate (BBP), di-(2-ethylhexyl) phthalate (DEHP) etc., which can be carcinogenic (Benjamin et al., 2015). Bisphenol A (BPA) use in plastic can cause hormone imbalance, and almost 80 diseases, including testicular cancer, obesity, and reproductive abnormalities, are thought to be caused by the hormone-disrupting properties of BPA, which is a chemical used in plastic and plasticizers that has been linked to exposure to endocrine-disrupting chemicals (Biemann et al., 2021) (Figure 4.3).

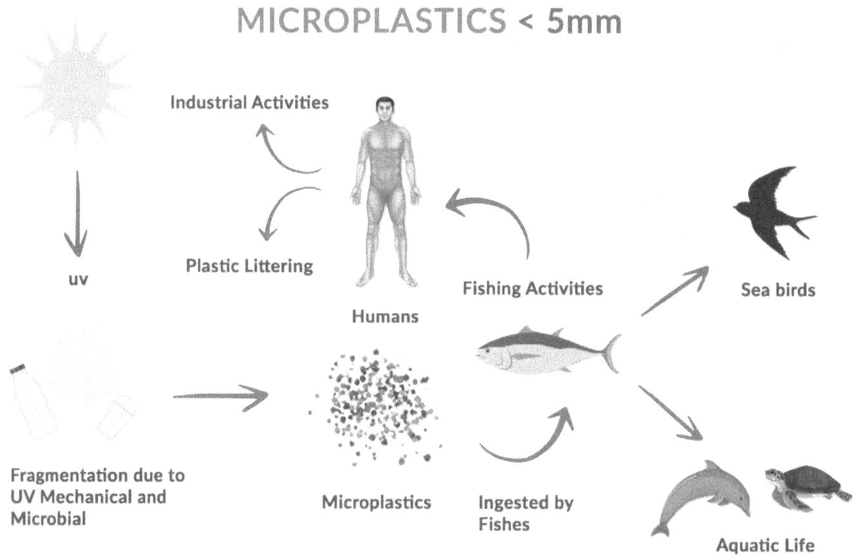

FIGURE 4.3 Effect of microplastics on the health of humans and other animals.

3.1. Microplastics in Aquatic Animals

Microplastics are rapidly spread by wind and water because of the small size of the particles. As a result, the particles can be found in living things as well as in polar ice, soil, water, and even in the air (Fackelmann & Sommer, 2019). According to a review conducted across 24 countries, out of 198 species, 60% of fish ingested microplastics (Sequeira et al., 2020). For instance, *Chlorella pyrenoidosa* microalgae have experienced alterations in morphology, reduced growth, and dropped photosynthetic activity as a result of microplastics (polystyrene) (Mao et al., 2018). Microplastics exhibit neurotoxicity, according to experiments conducted on juvenile European sea bass (*Dicentrarchus labrax*) (Barboza et al., 2018). The existence of trophic transfer in aquatic systems has now been supported by some evidence. In this case, *Pleuronectes platessa*, European sea bass, and *Ammodytes tobianus*, a sand eel and prey species, both had microplastics in their stomach, but there was no evidence of accumulation (Welden et al., 2018). Fish may suffer negative health effects if chemicals on plastic material's surface accumulate. About 150 species of fish have found to be injected with plastic. Around, 3.7 to 17.7 particles per individual were found in five different species of shellfish from the Persian Gulf, including gastropods and bivalves (Naji et al., 2018). Fish will ingest microplastics more frequently as a result of the increasing amount of microplastics that are present in marine environments each year (Figure 4.4).

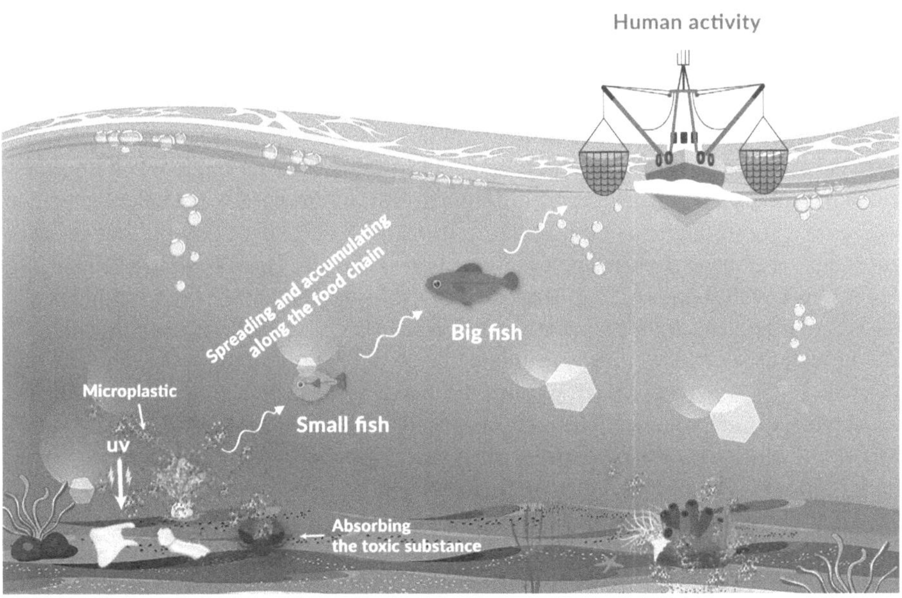

FIGURE 4.4 Tropic transfer of microplastics in sea animal.

3.2. Microplastics in Terrestrial Animals

Despite the significance of terrestrial ecosystems for human exposure, the majority of studies have focused on the marine environment and aquatic life. Initially moving microplastics from (0.87 ± 1.9 particles g^{-1}) earthworm casts to (129.8 ± 82.3 particles g^{-1}) chicken feces, was observed (Huerta et al., 2017). In times of famine, animals may even purposefully consume plastic foreign material.

Human activity is the main means of production of terrestrial microplastics, microplastics from sewage, industrial wastewater, and wastewater irrigation, which enter into soil and crops and eventually into land animals. Microplastics are also present in the air and can be inhaled, directly entering the animal body. Plastic ingested by earthworms gets converted into secondary microplastic and by excrement enters into soil and can be spread and fed by other soil animals. Additionally, birds can transport microplastics over long distances (Rhodes, 2018).

4. MICROPLASTICS AND HUMAN HEALTH

Microplastics can be classified as primary and secondary, according to their sources. Abrasives for cosmetic use, drug delivery systems, engineering and industrial practices such as air blasting are a few examples through which primary microplastics are specifically produced. Once they enter wastewater, these microplastics, which are typically challenging to get removed using waste disposal techniques, will eventually accumulate in the surroundings. Larger secondary microplastics are produced when plastics are repeatedly fragmented into smaller pieces by numerous, complex environmental factors like temperature, waves, wind, and UV light.

Only recently, experts have started to think about possible consequences on human health. The presence of microplastic in the human placenta was first observed by Ragusa and coworkers (2021); four placentas contained 12 microplastic fragments with spherical or irregular shapes, measuring 5 to 10 μm in size. Microplastics can enter the human body primarily through seafood, drinking water, and beer. It is also present in some fruits and vegetables, such as apples, broccoli, and carrots. There have been found 174 (kg/kg) fibers and 9 (kg/kg) fragments on average for honey, and 217 (kg/kg) fibers and 32 (kg/kg) fragments on average for sugar (Karbalaei et al., 2018). The contamination of fruit as well as vegetables results from the plants' uptake of microplastic-contaminated water through their roots. Microplastics are present in most crops, including lettuce, pears, radishes, and potatoes. The highest concentrations of microplastic particles, however, are found in apples and carrots. Microplastics eaten by fish, sea birds, and other marine animals can enter into the human food chain. It is anticipated that the buildup of these microplastics in aquatic organisms will have detrimental effects on human health, including changes in growth and development, neurotoxic reactions, and metabolic disorders (Deng et al., 2017).

Human beings also absorb microplastics themselves. According to research, in the feces of individuals from Europe, Russia, and Japan, there were minute plastic particles.

TABLE 4.1 Different diseases associated with microplastic in humans.

DISEASES	*REFERENCE*
Inflammatory bowel disease (IBD)	(Yan et al., 2021)
Antibiotic resistance microbe	(Kaur et al., 2022)
Neurotoxicity	(Barboza et al., 2020)
Metabolic syndrome	(Ragusa et al., 2022)
Genotoxicity	(Çobanoğlu et al., 2021)
Uterine arterioles reduction	(Hu et al., 2021)
Inflammation	(Prata et al., 2020)
Oxidative stress	(Kim et al., 2021)
Endothelial dysfunction	(Lee et al., 2021)
Chronic disease	(Leslie et al., 2020)

Some amounts of microplastics can also stay in our body and can enter tissue. As studied in rats, microplastics were found in their intestines, kidneys, heart, and stomach. Small plastics were also found in the brains of fish. Oxidative stress, DNA damage, and inflammation can occur due to exposure to microplastics. According to research, microplastics themselves may not be harmful but microorganisms present on the surface of microplastics can cause harm, for instance, *Vibrio cholerae* was observed on the surface of microplastics (Sangkham et al., 2022). Microplastics can also affect human health through inhalation and contamination present in food and water (Kinigopoulou et al., 2022).

Microplastics can act as carriers for a variety of toxins, such as persistent contaminants and additives from industrial production processes by solubilizing in water. These toxins pose a significant risk to people's health (Li et al., 2018) (Table 4.1).

The International Agency for Research on Cancer has determined that vinyl chloride, a monomer used to make PVC, has been found to be carcinogenic for humans (Tickner et al., 2001). Some plastics, including PU, PVC, styrene polymers (like PS), and epoxy resin, can produce dangerous monomers, which can be mutagenic/cancerous (Lithner et al., 2011). Therefore, mitigation of microplastic is necessary. There are two main approaches for the degradation of plastic: microplastic thermoregulation, and biodegradation. Biodegradation of microplastic occurs with the help of extracellular enzymes or bio-catalyst (Thakur et al., 2023).

5. BIO-CATALYST AND ITS EFFECT ON MICROPLASTICS

Removal of microplastics through conventional wastewater treatment systems is not very efficient. Also, the chemical method of plastic degradation can further lead to harmful chemical waste and can cause environmental hazards. To substitute the chemical/

conventional method of microplastic degradation, the biological methods represent important and efficient ways for plastic degradation. Henceforth in this chapter, we discuss the application of bio catalyst in removal of microplastics.

5.1. Enzymes

Enzymes represents a novel alternative for biodegradation of microplastics. Among the different class of the enzymes, the hydrolases are the ones that can further open ways for the bioremediation of the microplastics. Hydrolases and oxidoreductases are the two main classes of enzymes involved in the biodegradation of plastic. They include laccases, peroxidases, cutinase, lipase, and PETase. Blue copper oxidases include the class of enzymes known as laccases (EC 1.10.3.2), which are monomeric multicopper glycoproteins. *Rhus vernicifera,* a species of plant, was the first plant in which laccases were found in 1883. The breakdown of lignin and the elimination of the poisonous phenols that are created during this process both involve laccases (Janusz et al., 2020).

Heme-containing oxidoreductases known as peroxidases (EC 1.11.1) catalyze the oxidation of inorganic and organic compounds as well as the reduction of hydrogen peroxide. Peroxidases can be categorized as lignin peroxidases (EC 1.11.1.14), versatile peroxidases (EC 1.11.1.16), or dye decolorizing peroxidases (EC 1.11.1.19) depends on the type of reducing substrate they use. They can be non-specific and have a high redox potential (Dhagat et al., 2022). Apart from the plant species, microbes also represent important group possessing diverse enzymatic potential for biodegradation of plastics.

An important group of microbial species representatives are *Phanerochaete tremellosa, Trametes villosa, B. subtilis,* and *Ceriporiopsis subvermispora,* which produce the majority of manganese peroxidases (MnPs EC 1.11.1.13). They are a group of extracellular glycoproteins with heme prosthetic properties that are primarily tasked with removing lignin. They produce Mn (III) complexes by catalyzing the stabilizing ligands that cause the peroxide-dependent oxidation of Mn (II) to Mn (III), which can then carry out the oxidation of the organic substrate (Aitken et al., 1990).

Another group of enzymes are Esterases. Esterases are represented by the four classes based on its reactivity. Among the group of esterase enzymes cutinases are extracellular serine esterases that are split into two bacterial and two fungal subfamilies (E.C. 3.1.1.74). The primary structure of these subfamilies varies depending on whether they are prokaryotic or eukaryotic. Cutinases have a central sheet made up of five parallel strands and are folded in half. On either side of the central sheet are two or three helices. Their active site is exposed and contains the catalytic triad Ser-His-Asp/Glu. As an exception is *Trichoderma reesei* cutinase having a covered active site that resembles lipase. Here, the histidine's imidazole eliminates a proton from the serine hydroxyl group, while the serine oxygen engages in a nucleophilic attack on the substrate's acyl carbonyl carbon and transacylation follows next. These enzymes have numerous industrial uses due to esterification, transesterification, and catalyze hydrolysis reactions (Cambou et al., 1984).

Extracellular triacylglycerol acyl hydrolases called lipases (EC 3.1.1.3) can hydrolyze insoluble tri-, di-, and monoglyceride substrates, which are converted into free fatty acids as well as glycerol by ester bonds. Moreover, applications for fungal lipases in the industry include the food, textile, and manufacturing sectors, as well as the creation of detergents, cosmetics, and medications (Houde et al., 2004).

5.2. Microbial Degradation of Plastics

Microorganisms can degrade plastic in two ways, either by aerobic biodegradation or via- anaerobic biodegradation. In aerobic biodegradation, the large compounds are broken into small compounds by using oxygen as a terminal electron acceptor. While in anaerobic biodegradation, bacteria uses iron, sulfate, carbon dioxide, manganese, and nitrate as a terminal electron acceptor to break down large organic compounds into small compounds. Microorganisms first secret an extracellular enzyme, which then adheres to the surface of a polymer and cleaves the polymer (Reineke et al., 2001). Some important bacteria, i.e., *Delftia sp., Achromobacter xylosoxidans, Brevibacillus borstelensis, Comamonas sp., Stenotrophomonas sp., Bacillus sp. YP1, Bacillus weihenstephanensis, Bacillus subtiis, Pseudomonas aeruginosa, Pseudomasnas putida, Pseudomonas syringae,* and *Ideonella sakaiensis,* are capable of biodegradation of plastics (Walter et al., 2022). Fungi that have shown biodegradation are *Aspergillus flavus, Acremonium kiliense, Aspergillus oryzae, Alternaria dauci, Alternaria solani, Aspergillus glaucus, Aspergillus Niger, Chrysonilia setophila, Fusarium oxysporium, Penicillium funiculosum*, etc. (Zhang et al., 2020).

Recycling is the most obvious way to cut back on plastic waste. Nevertheless, globally, only less than 10% of all plastic had been sent to recycling. Synthetic plastics are necessary to our modern way of life, but their overuse puts the environment and public health at serious risk. Therefore, mitigation is necessary, Microorganisms can degrade plastic with the complex physio-chemical transformation of polymer into monomer (Muthuraj et al., 2021). Given the prevalence of plastics in various ecosystems and the enormous diversity of microorganisms and their metabolic processes, microbial communities in various habitats have probably developed skills for utilizing and decomposing plastic. In addition to their roles in cell regulation and function, enzymes also play a role in the degradation of anthropogenic pollutants, such as microplastics. In some cases, removal efficiency values of more than 90% are reached, according to the few studies that have looked at the evolution of microplastics in WWTPs (Menéndez-Manjón et al., 2022). Recently, *Ideonella sakaiensis* 201-F6 was identified and demonstrated to grow on PET films with low crystallinity (Palm, 2019). Exploiting extremophile microorganisms for use in bioremediation is another intriguing area of research (Kaushik et al., 2021). Additionally, it is conceivable to introduce mutations into bacterial genes that increase their enzymes' capacity for biodegradation (Santero & Díaz, 2020).

Microplastics can be deteriorated via biological ways in the environment using other degradative mechanisms:

5.2.1. Hydrolytic degradation

Breakdown of the polymer occurs when a bond of the polymer interacts with a water molecule and breaks down into smaller segments (Mondal & Martin, 2012).

5.2.2. Photodegradation

Breakdown occurs when exposed to photons, especially wavelengths similar to sunlight, such as ultraviolet light, visible light, and infrared radiation, a photodegradable molecule degrades. Other types of electromagnetic radiation can also cause photo degradation (Yousif & Haddad, 2013).

5.2.3. Thermal degradation

Breakdown occurs when the polymer changes its properties under the influence of increased temperature (Ashter et al., 2016).

5.2.4. Biodegradation

The environment's organic chemicals undergo natural mineralization, disintegration into simpler compounds, and redistribution through elemental cycles like the carbon, nitrogen, and sulfur cycles (Das et al., 2012). It is the process that modifies the surface of the material to change the chemical, physical, and mechanical properties of plastics.

6. COMMON PLASTIC AND THEIR DEGRADATION

Polyester: Polyesters are polymers in which the component monomers are joined together by ester linkages. Esterases, the enzymes that break down the many different kinds of esters found in nature, are found in a wide variety of microorganisms. Typically, ester linkages hydrolyze quickly (Edlund et al., 2003).

Polyhydroxybutyrate (PHB): PHB and other polyhydroxyalkanoate (PHA) polymers are metabolized by a wide variety of microorganisms. On the basis of this, several bacterial genes of PHA depolymerases have been isolated. Basically, the four regions—the signal sequence, a catalytic domain with the lipase box, a substrate-binding domain that acts as an adsorption site for polymer substrates, and a domain connecting the catalytic and substrate-binding domains—build the protein sequences of the serine hydrolases known as PHA depolymerases (Poirier et al., 1995).

Polycaprolactone (PCL): PCL is a synthetic polyester that is readily degraded by microorganisms; however, little is known about PCL depolymerases. It was discovered that a PCL trimer shares a chemical structure with two cutin monomers and studies have shown the degradation of PCL by some cutinase-like enzyme (Caruso et al., 2015).

Polyvinyl chloride: PVC, a well-known synthetic polymer, is frequently used in rigid or plasticized materials. Pollution caused by used PVC-based plastics, however, is a serious problem. When cellulose and PVC were available as sources of carbon, *Trichocladium* species and *Chaetomium* species grew more quickly. This suggests that cellulose and PVC are metabolized simultaneously. It has also been found that *Bacillus flexus* and *Pseudomonas citronellolis* biodegrade PVC film, with additives showing higher depolymerizing activity than the PVC polymer chains (Chiellini et al., 2013).

Polystyrene: It is the most widely used type of plastic produced globally, and is primarily used to package foods and single-use items. Like other synthetic plastics, polystyrene is widely used because it has good mechanical properties and is reasonably priced. Polystyrene is frequently used for foam packaging, food containers, building materials (insulation), cassette boxes, CDs, disposable cups, plates, and cutlery. In 2013, production of polystyrene reached approximately 21 million tons worldwide. High molecular weight and strong hydrophobic properties in polystyrene make these polymers extremely stable and highly resistant to biodegradation. Without any pretreatment, the degradation of atactic PS by *Curvularia* species was examined (Mohanan et al., 2020).

Low-density polyethylene (LDPE): LDPE (69.13%) is the most common type of plastic waste found in landfills in the form of plastic bags. The majority of LDPE, which is amorphous, short-branched, and contains 10-30 CH3 per 1,000 carbon atoms, is made up of one or more comonomers, including 1-butene, 1-hexene, and 1-octene. A few comparative studies on the biodegradability of various pre-treated polyethylene materials by *Rhodococcus rhodochrous* (one of the most efficient bacteria), including HDPE, LDPE, and linear low-density polyethylene (LLDPE) films of different thickness for PE biodegradation (Jeon et al., 2014).

Nylon: High molecular weight nylon-66 membrane was observed to be significantly degraded by lignin-degrading white rot fungi grown in ligninolytic conditions with little ammonium tartrate or glucose (Shimao et al., 2001).

The enigma of using microbes to degrade PE is that PE has a high molecular weight, which restricts the variety of probable enzymatic reactions. In their enzymatic systems, microorganisms typically use substrates with 10–50 carbons. Consequently, a decrease in PE's molecular weight is required for its degradation by microbes. Furthermore, molecules must be transported through cell membranes with a lower molecular weight (Restrepo-Flórez et al., 2014). Due to the widespread use of polyethylene terephthalate (PET) in the manufacture of disposable bottles and other products, a significant amount of PET waste has accumulated in the environment (Sinha et al., 2010) and PETase will breakdown PET (polyethylene terephthalate) into BHET, MHET, and TPA (Djapovic et al., 2021).

PET decomposition into BHET, MHET, and TPA acylation and deacylation, the two stages of the reaction mechanism, happen in a single, associative, coordinated, and asynchronous step for each. Ser131 attacks the substrate by nucleophilically transferring a proton to His208, which causes a tetrahedral transition state and the release of MHET after the ester bond is broken. The acylated Ser131 intermediate is attacked by the deprotonation of an active site water molecule by His208, leading to the loss of the intermediate's bond to the substrate. After that, the water proton is moved from Ser131 to His208, forming MHET and regenerating the enzyme (Jerves et al., 2021) (Figure 4.5).

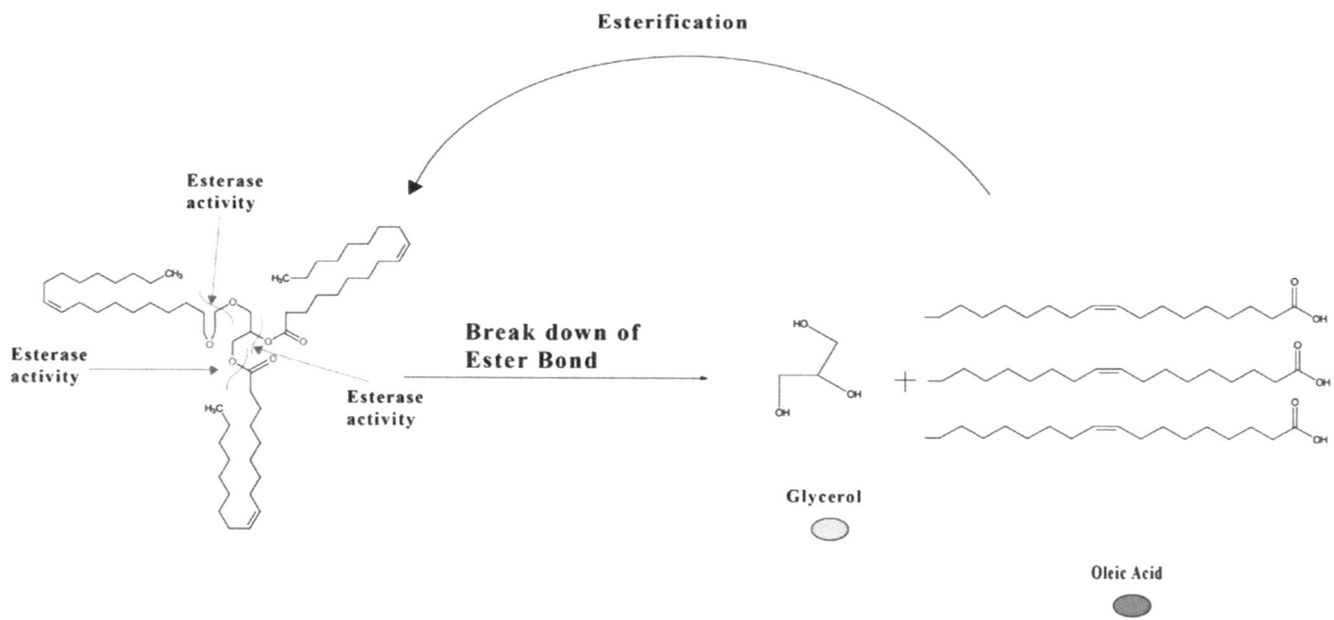

FIGURE 4.5 Esterase activity leading to the degradation of PET (polyethylene terephthalate).
Figure created by author.

7. LIMITATION OF BIODEGRADATION

The main limitation of several biodegradable plastics is the release of methane when these get decomposed in landfills. As a consequence, plenty of methane is produced annually. Because methane is 84-fold more stronger than carbon dioxide and absorbs heat more quickly, it can speed up climate change (Chidambarampadmavathy et al., 2017). Biodegradable plastics cannot solve the issue because the ocean's waters are too cold for them to break down. These plastics will consequently either float on the water or disintegrate into minute pieces that are harmful to marine life. Therefore, using this kind of plastic won't be able to solve every pollution problem (Urbanek et al., 2018). On the other hand, production of bioplastic is 20 to 50% more expensive than that of traditional plastic. However, with more advanced technology and easier access to resources, this cost can be significantly reduced (Zeller et al., 2013). It brought up concerns about the toxic residue leftover from decomposition. While the inner part shouldn't be easily available for biodegradation, only the surface is influenced by it (Isangedighi et al., 2020).

8. FUTURE ASPECTS

It is challenging to gather degradation products from the soil or other ecosystems, in contrast to the aquatic environment. Thus, the development of an extraction method for degradation products has to get considerably more attention (Du et al., 2021). In order to protect oneself from the COVID-19 epidemic, more plastic products (such as facemasks and gloves) were consumed, which eventually exacerbated plastic pollution (Rivas et al., 2022). However, given their potential long-term effects, significant research is being done to comprehend the environmental fate of microplastics and any potential remediation methods (Anik et al., 2021). Because microplastics are widely used in the environment, novel removal methods have been developed, one of which involves immobilizing the enzymes responsible for breaking down microplastics (Tang et al., 2022). Furthermore, due to the lack of a suitable or particular enzyme, a single microbial species cannot break down all types of plastic, therefore synergies between various microbial species must be investigated to improve plastics biodegradation (Ali et al., 2021).

9. CONCLUSION

According to prediction analyses, it takes 320 years for a typical marine environment to transform a 1 mm plastic particle into a plastic nanoparticle (Koelmans et al. 2015).

Research on a self-degradable polymeric substance, also known as bioplastics, is rising (Pham et al., 2021). Finding a new biodegradable material is a great way to address the threat of plastic pollution on a global scale, but we cannot ignore the fact that the quantity of microplastics that are discarded is quite significant and is still increasing (Arpia et al., 2021). Although biodegradation of microplastics is effective, pre-treatment of microplastics is required for better degradation. Plastic degradation could be accelerated by adding natural polymers to it (Singh & Sharma, 2008). Several microbial enzymes that break down plastic have been discovered thus far, and they represent interesting biocatalyst candidates for the depolymerization of plastic. However, naturally occurring plastic-degrading enzymes are not well adapted for synthetic plastic degradation in commercial applications due to weak thermostability and limited catalytic activity. Moreover, no single technology will be able to offer a comprehensive solution due to the diversity and complexity of the synthetic polymers.

REFERENCES

Aitken, M. D., & Irvine, R. L. (1990). Characterization of reactions catalyzed by manganese peroxidase from Phanerochaete chrysosporium. *Archives of Biochemistry and Biophysics, 276*(2), 405–414.

Ali, S. S., Elsamahy, T., Koutra, E., Kornaros, M., El-Sheekh, M., Abdelkarim, E. A., & Sun, J. (2021). Degradation of conventional plastic wastes in the environment: A review on current status of knowledge and future perspectives of disposal. *Science of the Total Environment, 771*, 144719.

Anik, A. H., Hossain, S., Alam, M., Sultan, M. B., Hasnine, M. T., & Rahman, M. M. (2021). Microplastics pollution: A comprehensive review on the sources, fates, effects, and potential remediation. *Environmental Nanotechnology Monitoring & Management, 16*, 100530.

Arpia, A. A., Chen, W. H., Ubando, A. T., Naqvi, S. R., & Culaba, A. B. (2021). Microplastic degradation as a sustainable concurrent approach for producing biofuel and obliterating hazardous environmental effects: A state-of-the-art review. *Journal of Hazardous Materials, 418*, 126381.

Ashter, S.A. (2016). Types of biodegradable polymers. In S. A. Ashter (Ed.), *Introduction to bioplastics engineering* (pp. 81–151). Amsterdam: William Andrew.https://doi.org/10.1016/B978-0-323-39396-6.00005-1

Atanasova, N., Stoitsova, S., Paunova-Krasteva, T., & Kambourova, M. (2021). Plastic degradation by extremophilic bacteria. *International Journal of Molecular Sciences, 22*(11), 5610.

Bank, M. S., & Hansson, S. V. (2019). The plastic cycle: A novel and holistic paradigm for the anthropocene. *Environmental Science & Technology, 53*(13), 7177–7179.

Barboza, L. G. A., Lopes, C., Oliveira, P., Bessa, F., Otero, V., Henriques, B., … & Guilhermino, L. (2020). Microplastics in wild fish from North East Atlantic Ocean and its potential for causing neurotoxic effects, lipid oxidative damage, and human health risks associated with ingestion exposure. *Science of the Total Environment, 717*, 134625.

Barboza, L. G. A., Vethaak, A. D., Lavorante, B. R., Lundebye, A. K., & Guilhermino, L. (2018). Marine microplastic debris: An emerging issue for food security, food safety and human health. *Marine Pollution Bulletin, 133*, 336–348.

Barrett, J., Chase, Z., Zhang, J., Holl, M. M. B., Willis, K., Williams, A., ... & Wilcox, C. (2020). Microplastic pollution in deep-sea sediments from the Great Australian Bight. *Frontiers in Marine Science*, 808.

Benjamin, S., Pradeep, S., Josh, M. S., Kumar, S., & Masai, E. (2015). A monograph on the remediation of hazardous phthalates. *Journal of Hazardous Materials*, *298*, 58–72.

Biemann, R., Blüher, M., & Isermann, B. (2021). Exposure to endocrine-disrupting compounds such as phthalates and bisphenol A is associated with an increased risk for obesity. *Best Practice & Research Clinical Endocrinology & Metabolism*, *35*(5), 101546.

Boucher, J., & Friot, D. (2017). *Primary microplastics in the oceans: A global evaluation of sources* (Vol. 10). Gland, Switzerland: Iucn.

Cambou, B., & Klibanov, A. M. (1984). Comparison of different strategies for the lipase-catalyzed preparative resolution of racemic acids and alcohols: Asymmetric hydrolysis, esterification, and transesterification. *Biotechnology and Bioengineering*, *26*(12), 1449–1454.

Campanale, C., Massarelli, C., Savino, I., Locaputo, V., & Uricchio, V. F. (2020). A detailed review study on potential effects of microplastics and additives of concern on human health. *International Journal of Environmental Research and Public Health*, *17*(4), 1212.

Carmen, S. (2021). Microbial capability for the degradation of chemical additives present in petroleum-based plastic products: A review on current status and perspectives. *Journal of Hazardous Materials*, *402*, 123534.

Caruso, G. (2015). Plastic degrading microorganisms as a tool for bioremediation of plastic contamination in aquatic environments. *J Pollut Eff Cont*, *3*(3), 1–2.

Castillo-Carvajal, L. C., Sanz-Martín, J. L., & Barragán-Huerta, B. E. (2014). Biodegradation of organic pollutants in saline wastewater by halophilic microorganisms: A review. *Environmental Science and Pollution Research*, *21*, 9578–9588.

Chidambarampadmavathy, K., Karthikeyan, O. P., & Heimann, K. (2017). Sustainable bio-plastic production through landfill methane recycling. *Renewable and Sustainable Energy Reviews*, *71*, 555–562.

Chiellini, F., Ferri, M., Morelli, A., Dipaola, L., & Latini, G. (2013). Perspectives on alternatives to phthalate plasticized poly (vinyl chloride) in medical devices applications. *Progress in Polymer Science*, *38*(7), 1067-1088.

Çobanoğlu, H., Belivermiş, M., Sıkdokur, E., Kılıç, Ö., & Çayır, A. (2021). Genotoxic and cytotoxic effects of polyethylene microplastics on human peripheral blood lymphocytes. *Chemosphere*, *272*, 129805.

Cózar, A., Echevarría, F., González-Gordillo, J. I., Irigoien, X., Úbeda, B., Hernández-León, S., Palma, ... & Duarte, C. M. (2014). Plastic debris in the open ocean. *Proceedings of the National Academy of Sciences*, *111*(28), 10239–10244.

Das, G., Bordoloi, N. K., Rai, S. K., Mukherjee, A. K., & Karak, N. (2012). Biodegradable and biocompatible epoxidized vegetable oil modified thermostable poly (vinyl chloride): Thermal and performance characteristics post biodegradation with Pseudomonas aeruginosa and Achromobacter sp. *Journal of Hazardous Materials*, *209*, 434–442.

Deng, Y., Zhang, Y., Lemos, B., & Ren, H. (2017). Tissue accumulation of microplastics in mice and biomarker responses suggest widespread health risks of exposure. *Scientific Reports*, *7*(1), 46687.

Dhagat, S., & Jujjavarapu, S. E. (2022). Utility of lignin-modifying enzymes: a green technology for organic compound mycodegradation. *Journal of Chemical Technology & Biotechnology*, *97*(2), 343–358.

Di Donato, P., Buono, A., Poli, A., Finore, I., Abbamondi, G. R., Nicolaus, B., & Lama, L. (2018). Exploring marine environments for the identification of extremophiles and their enzymes for sustainable and green bioprocesses. *Sustainability*, *11*(1), 149.

Djapovic, M., Milivojevic, D., Ilic-Tomic, T., Lješević, M., Nikolaivits, E., Topakas, E., & Nikodinovic-Runic, J. (2021). Synthesis and characterization of polyethylene terephthalate (PET) precursors and potential degradation products: Toxicity study and application in discovery of novel PETases. *Chemosphere*, *275*, 130005.

Du, H., Xie, Y., & Wang, J. (2021). Microplastic degradation methods and corresponding degradation mechanism: research status and future perspectives. *Journal of Hazardous Materials*, *418*, 126377.

Edlund, U., & Albertsson, A. C. (2003). Polyesters based on diacid monomers. *Advanced Drug Delivery Reviews*, *55*(4), 585–609.

Espinosa, C., Esteban, M. Á., & Cuesta, A. (2016). Microplastics in aquatic environments and their toxicological implications for fish. *Toxicology-New Aspects to This Scientific Conundrum*, 113.

Fackelmann, G., & Sommer, S. (2019). Microplastics and the gut microbiome: How chronically exposed species may suffer from gut dysbiosis. *Marine Pollution Bulletin*, *143*, 193–203.

Forster, N. A., Tighe, M. K., & Wilson, S. C. (2020). Microplastics in soils of wilderness areas: what is the significance of outdoor clothing and footwear?. *Geoderma*, *378*, 114612.

Geyer, R., Jambeck, J. R., & Law, K. L. (2017). Production, use, and fate of all plastics ever made. *Science Advances, 3*(7), e1700782.

Hahladakis, J. N., Velis, C. A., Weber, R., Iacovidou, E., & Purnell, P. (2018). An overview of chemical additives present in plastics: Migration, release, fate and environmental impact during their use, disposal and recycling. *Journal of Hazardous Materials*, *344*, 179–199.

Houde, A., Kademi, A., & Leblanc, D. (2004). Lipases and their industrial applications: An overview. *Applied Biochemistry and Biotechnology*, *118*, 155–170.

Hu, J., Qin, X., Zhang, J., Zhu, Y., Zeng, W., Lin, Y., & Liu, X. (2021). Polystyrene microplastics disturb maternal-fetal immune balance and cause reproductive toxicity in pregnant mice. *Reproductive Toxicology*, *106*, 42-50.

Huerta Lwanga, E., Mendoza Vega, J., Ku Quej, V., Chi, J. D. L. A., Sanchez del Cid, L., Chi, C., ... & Geissen, V. (2017). Field evidence for transfer of plastic debris along a terrestrial food chain. *Scientific Reports*, *7*(1), 14071.

Isangedighi, I. A., David, G. S., & Obot, O. I. (2020). Plastic waste in the aquatic environment: impacts and management. In *Analysis of nanoplastics and microplastics in food* (pp. 15–43). CRC Press.

Janusz, G., Pawlik, A., Świderska-Burek, U., Polak, J., Sulej, J., Jarosz-Wilkołazka, A., & Paszczyński, A. (2020). Laccase properties, physiological functions, and evolution. *International Journal of Molecular Sciences*, *21*(3), 966.

Jeon, H. J., & Kim, M. N. (2014). Degradation of linear low density polyethylene (LLDPE) exposed to UV-irradiation. *European polymer journal*, *52*, 146–153.

Jerves, C., Neves, R. P., Ramos, M. J., da Silva, S., & Fernandes, P. A. (2021). Reaction mechanism of the PET degrading enzyme PETase studied with DFT/MM molecular dynamics simulations. *ACS Catalysis*, *11*(18), 11626–11638

Jiménez-Skrzypek, G., Hernández-Sánchez, C., Ortega-Zamora, C., González-Sálamo, J., González-Curbelo, M. Á., & Hernández-Borges, J. (2021). Microplastic-adsorbed organic contaminants: analytical methods and occurrence. *TrAC Trends in Analytical Chemistry*, *136*, 116186.

Karbalaei, S., Hanachi, P., Walker, T. R., & Cole, M. (2018). Occurrence, sources, human health impacts and mitigation of microplastic pollution. *Environmental Science and Pollution Research*, *25*, 36046–36063.

Kaur, K., Reddy, S., Barathe, P., Oak, U., Shriram, V., Kharat, S. S., ... & Kumar, V. (2022). Microplastic-associated pathogens and antimicrobial resistance in environment. *Chemosphere*, *291*, 133005.

Kaushik, S., Alatawi, A., Djiwanti, S. R., Pande, A., Skotti, E., & Soni, V. (2021). Potential of extremophiles for bioremediation. *Microbial rejuvenation of polluted environment* (Vol. 1, pp. 293–328). Springer.

Kim, J. H., Yu, Y. B., & Choi, J. H. (2021). Toxic effects on bioaccumulation, hematological parameters, oxidative stress, immune responses and neurotoxicity in fish exposed to microplastics: A review. *Journal of Hazardous Materials*, *413*, 125423.

Koelmans, A. A., Besseling, E., & Shim, W. J. (2015). Nanoplastics in the aquatic environment. Critical review. *Marine Anthropogenic Litter*, 325–340.

Lee, H. S., Amarakoon, D., Wei, C. I., Choi, K. Y., Smolensky, D., & Lee, S. H. (2021). Adverse effect of polystyrene microplastics (PS-MPs) on tube formation and viability of human umbilical vein endothelial cells. *Food and Chemical Toxicology*, *154*, 112356.

Leslie, H. A., & Depledge, M. H. (2020). Where is the evidence that human exposure to microplastics is safe?. *Environment International*, *142*, 105807.

Li, J., Liu, H., & Chen, J. P. (2018). Microplastics in freshwater systems: A review on occurrence, environmental effects, and methods for microplastics detection. *Water Research*, *137*, 362–374.

Li, Y., Sun, Y., Li, J., Tang, R., Miu, Y., & Ma, X. (2021). Research on the influence of microplastics on marine life. In IOP Conference Series: Earth and Environmental Science, Vol. 631, No.1 (p. 012006). IOP Publishing.

Lithner, D., Larsson, Å., & Dave, G. (2011). Environmental and health hazard ranking and assessment of plastic polymers based on chemical composition. *Science of the Total Environment*, *409*(18), 3309–3324.

Malankowska, M., Echaide-Gorriz, C., & Coronas, J. (2021). Microplastics in marine environment: a review on sources, classification, and potential remediation by membrane technology. *Environmental Science: Water Research & Technology*, *7*(2), 243–258.

Mao, Y., Ai, H., Chen, Y., Zhang, Z., Zeng, P., Kang, L., Li, ... & Li, H. (2018). Phytoplankton response to polystyrene microplastics: perspective from an entire growth period. *Chemosphere*, *208*, 59–68.

McConnell, E. E., & Swenberg, J. A. (1994). Review of styrene and styrene oxide long-term animal studies. *Critical Reviews in Toxicology*, *24*(sup1), S49–S55.

Meikle, J. L. (1995). *American plastic: A cultural history*. Rutgers University Press.

Menéndez-Manjón, A., Martínez-Díez, R., Sol, D., Laca, A., Laca, A., Rancaño, A., & Díaz, M. (2022). Long-term occurrence and fate of microplastics in WWTPs: A case study in Southwest Europe. *Applied Sciences*, *12*(4), 2133.

Mohanan, N., Montazer, Z., Sharma, P. K., & Levin, D. B. (2020). Microbial and enzymatic degradation of synthetic plastics. *Frontiers in Microbiology*, *11*, 580709.

Mondal, S., & Martin, D. (2012). Hydrolytic degradation of segmented polyurethane copolymers for biomedical applications. *Polymer Degradation and Stability*, *97*(8), 1553–1561.

Moore, C. J., Moore, S. L., Weisberg, S. B., Lattin, G. L., & Zellers, A. F. (2002). A comparison of neustonic plastic and zooplankton abundance in southern California's coastal waters. *Marine Pollution Bulletin*, *44*(10), 1035–1038.

Morris, P. J. (2005). *Polymer pioneers: A popular history of the science and technology of large molecules* (No. 5). Chemical Heritage Foundation.

Mukherjee, A. G., Wanjari, U. R., Bradu, P., Patil, M., Biswas, A., Murali, R., ... & Gopalakrishnan, A. V. (2022). Elimination of microplastics from the aquatic milieu: A dream to achieve. *Chemosphere*, 135232.

Muthuraj, R., Valerio, O., & Mekonnen, T. H. (2021). Recent developments in short-and medium-chain-length Polyhydroxyalkanoates: Production, properties, and applications. *International Journal of Biological Macromolecules*, *187*, 422–440.

Naji, A., Nuri, M., & Vethaak, A. D. (2018). Microplastics contamination in molluscs from the northern part of the Persian Gulf. *Environmental Pollution*, *235*, 113–120.

Othman, A. R., Hasan, H. A., Muhamad, M. H., Ismail, N. I., & Abdullah, S. R. S. (2021). Microbial degradation of microplastics by enzymatic processes: A review. *Environmental Chemistry Letters*, *19*, 3057–3073.

Palm, G. J., Reisky, L., Böttcher, D., Müller, H., Michels, E. A., Walczak, M. C., … & Weber, G. (2019). Structure of the plastic-degrading Ideonella sakaiensis MHETase bound to a substrate. *Nature Communications*, *10*(1), 1717.

Papari, S., Bamdad, H., & Berruti, F. (2021). Pyrolytic conversion of plastic waste to value-added products and fuels: A review. *Materials*, *14*(10), 2586.

Parolini, M., Antonioli, D., Borgogno, F., Gibellino, M. C., Fresta, J., Albonico, C., … & Cavallo, R. (2021). Microplastic contamination in snow from Western Italian Alps. *International Journal of Environmental Research and Public Health*, *18*(2), 768.

Pham, T. H., Do, H. T., Thi, L. A. P., Singh, P., Raizada, P., Wu, J. C. S., & Nguyen, V. H. (2021). Global challenges in microplastics: from fundamental understanding to advanced degradations toward sustainable strategies. *Chemosphere*, *267*, 129275.

Poirier, Y., Nawrath, C., & Somerville, C. (1995). Production of polyhydroxyalkanoates, a family of biodegradable plastics and elastomers, in bacteria and plants. *Nature Biotechnology*, *13*(2), 142–150.

Prata, J. C., da Costa, J. P., Lopes, I., Duarte, A. C., & Rocha-Santos, T. (2020). Environmental exposure to microplastics: An overview on possible human health effects. *Science of the Total Environment*, *702*, 134455.

Ragusa, A., Matta, M., Cristiano, L., Matassa, R., Battaglione, E., Svelato, A., & Nottola, S. A. (2022). Deeply in plasticenta: Presence of microplastics in the intracellular compartment of human placentas. *International Journal of Environmental Research and Public Health*, *19*(18), 11593.

Reineke, W. (2001). Aerobic and anaerobic biodegradation potentials of microorganisms. *Biodegradation and Persistence*, 1–161.

Restrepo-Flórez, J. M., Bassi, A., & Thompson, M. R. (2014). Microbial degradation and deterioration of polyethylene: A review. *International Biodeterioration & Biodegradation*, *88*, 83–90.

Rhodes, C. J. (2018). Plastic pollution and potential solutions. *Science progress*, *101*(3), 207–260.

Rivas, M. L., Albion, I., Bernal, B., Handcock, R. N., Heatwole, S. J., Parrott, M. L., ... & Deschaseaux, E. (2022). The plastic pandemic: COVID-19 has accelerated plastic pollution, but there is a cure. *Science of the Total Environment*, *847*, 157555.

Santero, E., & Díaz, E. (2020). Genetics of biodegradation and bioremediation. *Genes*, *11*(4), 441.

Shimao, M. (2001). Biodegradation of plastics. *Current Opinion in Biotechnology*, *12*(3), 242–247.

Singh, B., & Sharma, N. (2008). Mechanistic implications of plastic degradation. *Polymer Degradation and Stability*, *93*(3), 561–584.

Sinha, V., Patel, M. R., & Patel, J. V. (2010). PET waste management by chemical recycling: A review. *Journal of Polymers and the Environment*, *18*(1), 8–25.

Tagg, A. S., Sapp, M., Harrison, J. P., Sinclair, C. J., Bradley, E., Ju-Nam, Y., & Ojeda, J. J. (2020). Microplastic monitoring at different stages in a wastewater treatment plant using reflectance micro-FTIR imaging. *Frontiers in Environmental Science*, *8*, 145.

Tang, K. H. D., Lock, S. S. M., Yap, P. S., Cheah, K. W., Chan, Y. H., Yiin, C. L., … & Chai, Y. H. (2022). Immobilized enzyme/microorganism complexes for degradation of microplastics: A review of recent advances, feasibility and future prospects. *Science of the Total Environment*, 154868

Thakur, B., Singh, J., Singh, J., Angmo, D., & Vig, A. P. (2023). Biodegradation of different types of microplastics: Molecular mechanism and degradation efficiency. *Science of the Total Environment, 877*, 162912.

Thompson, R. C., Olsen, Y., Mitchell, R. P., Davis, A., Rowland, S. J., John, A. W., … & Russell, A. E. (2004). Lost at sea: where is all the plastic?. *Science, 304*(5672), 838–838.

Tickner, J. A., Schettler, T., Guidotti, T., McCally, M., & Rossi, M. (2001). Health risks posed by use of Di-2-ethylhexyl phthalate (DEHP) in PVC medical devices: A critical review. *American Journal of Industrial Medicine, 39*(1), 100–111.

Tong, H., Jiang, Q., Hu, X., & Zhong, X. (2020). Occurrence and identification of microplastics in tap water from China. *Chemosphere, 252*, 126493.

Urbanek, A. K., Rymowicz, W., & Mirończuk, A. M. (2018). Degradation of plastics and plastic-degrading bacteria in cold marine habitats. *Applied Microbiology and Biotechnology, 102*, 7669–7678.

Usman, S., Abdull Razis, A. F., Shaari, K., Amal, M. N. A., Saad, M. Z., Mat Isa, N., & Nazarudin, M. F. (2021). Polystyrene microplastics exposure: an insight into multiple organ histological alterations, oxidative stress and neurotoxicity in javanese medaka fish (*Oryzias javanicus* Bleeker, 1854). *International Journal of Environmental Research and Public Health, 18*(18), 9449.

Walter, A., Sopracolle, L., Mutschlechner, M., Spruck, M., & Griesbeck, C. (2022). Biodegradation of different PET variants from food containers by Ideonella sakaiensis. *Archives of Microbiology, 204*(12), 711.

Wang, F., Wang, Q., Adams, C. A., Sun, Y., & Zhang, S. (2022). Effects of microplastics on soil properties: Current knowledge and future perspectives. *Journal of Hazardous Materials, 424*, 127531.

Ward, E., Boffetta, P., Andersen, A., Colin, D., Comba, P., Deddens, J. A., … & Simonato, L. (2001). Update of the follow-up of mortality and cancer incidence among European workers employed in the vinyl chloride industry. *Epidemiology*, 710–718.

Welden, N. A., Abylkhani, B., & Howarth, L. M. (2018). The effects of trophic transfer and environmental factors on microplastic uptake by plaice, Pleuronectes plastessa, and spider crab, Maja squinado. *Environmental Pollution, 239*, 351–358.

Yan, Z., Liu, Y., Zhang, T., Zhang, F., Ren, H., & Zhang, Y. (2021). Analysis of microplastics in human feces reveals a correlation between fecal microplastics and inflammatory bowel disease status. *Environmental Science & Technology, 56*(1), 414–421.

Yousif, E., & Haddad, R. (2013). Photodegradation and photostabilization of polymers, especially polystyrene. *SpringerPlus, 2*(1), 1–32.

Zarfl, C., Fleet, D., Fries, E., Galgani, F., Gerdts, G., Hanke, G., & Matthies, M. (2011). Microplastics in oceans. *Marine Pollution Bulletin, 62*, 1589–1591.

Zeller, M. A., Hunt, R., Jones, A., & Sharma, S. (2013). Bioplastics and their thermoplastic blends from Spirulina and Chlorella microalgae. *Journal of Applied Polymer Science, 130*(5), 3263–3275.

Zhang, J., Gao, D., Li, Q., Zhao, Y., Li, L., Lin, H., … & Zhao, Y. (2020). Biodegradation of polyethylene microplastic particles by the fungus *Aspergillus flavus* from the guts of wax moth *Galleria mellonella*. *Science of the Total Environment, 704*, 135931.

5

Recent Biotechnological Interventions for Sustainable Treatment of Industrial Wastewater Following Membrane-based Methods

Simranjeet Singh, Harry Kaur, Hailemariam Assefa, Daljeet Singh Dhanjal, Jaskaran Dhiman, Joginder Singh and Praveen C. Ramamurthy

1. INTRODUCTION

The accelerating population is leading to rapid growth in the agricultural and industrial sectors, consequently increasing demand for water globally. Rivers, groundwater, and

DOI: 10.1201/9781003496014-5

lakes are the primary water sources of irrigation, industry, and households (Al et al., 2023). Furthermore, the discharge of waste from diverse sectors such as industry, agriculture, households and municipalities has significantly contributed to the deterioration of both the quality and quantity of drinkable water (Bolan et al., 2024). Anthropogenic activities, such as industrial waste disposal, sewage leaks, oil spills, heavy metals, deforestation, pesticides, etc., are the major contributors to water pollution (Bashir et al., 2020).

Industrial wastewater contains elevated levels of heavy metals, organic compounds, hazardous chemicals, and other pollutants, which presents unique challenges in treatment due to its intricate composition and fluctuating contaminant concentrations (Nasr, 2022; Sathya et al., 2022). The industries utilise approximately 33% of the freshwater and its emissions contain diverse pollutants (Holt, 2000). According to the AQUASTAT database, 3,928 cubic kilometres of freshwater are removed annually worldwide. Of this amount, 22% (equivalent to 865 cubic kilometres) is utilised by the industrial sector and its effluents are causing irreversible deterioration of the ecosystem (https://www.fao.org/aquastat/en/overview/methodology/water-use; https://ourworldindata.org/water-use-stress). The untreated release of wastewater into the sewage system ultimately contaminates the underground water and other natural resources causing environmental pollution (Lin et al., 2022). This further result in significant degradation of crop quality and potential contamination of the food chain (Singh et al., 2023). Additionally, water pollution leads to the spread of several waterborne diseases such as diarrhoea, giardiasis, typhoid, cholera, hepatitis, jaundice, and cancer (Shayo et al., 2023). Because of this concern, several nations have established logical regulations for determining the permissible levels of contaminants that can be safely absorbed by certain water bodies, such as rivers and lakes (Chidiac et al., 2023). Several programmes, such as the total maximum daily load (TMDL) under the US Clean Water Act, Integrated Pollution Prevention and Control (IPPC) in Europe, and the Central Pollution Control Board (CPCB) in India, determine the maximum amount of pollutants that can be carried and the standards for their discharge. These programmes establish minimum acceptable standards (MINAS) for the release of municipal and industrial wastes (Rajaram & Das, 2008). Additionally, multiple treatment facilities are being established that employ chemical, electrochemical, biological, and physical methods to produce drinkable water (Hand & Cusick, 2021; Koul et al., 2022). Industrial developers have adopted sustainable technology for cleaner production, reduced water use, and less pollution (Boretti & Rosa, 2019). Hence, proper wastewater treatment before releasing it into water bodies is crucial for water quality and quantity optimization amid economic expansion and limited clean water availability (Wang & Choi, 2019).

2. MEMBRANE TECHNOLOGIES FOR WASTEWATER TREATMENT

Despite the longstanding existence of membrane technology, the diverse and intricate characteristics of wastewater allow for further enhancements in terms of efficiency, spatial demands, energy consumption, permeate quality, and technical expertise prerequisites (Ezugbe & Rathilal, 2020). Continual modifications are being made to membrane modules to prevent membrane fouling, a significant obstacle in membrane processes. The ongoing development of the wastewater treatment plants includes the integration of multiple membrane processes with other technologies like coagulation or adsorption in a hybrid manner (Du et al., 2020). Numerous enhancements have been implemented to optimise the performance of membranes, rendering them more suitable for a wide range of applications (Ahmad et al., 2020).

Membranes can be categorised as either isotropic or anisotropic based on their characteristic properties. Isotropic membranes possess uniform surface and in the case of microporous materials, their permeation fluxes are significantly higher compared to nonporous (dense) ones. Isotropic microporous membranes have extensive applications in microfiltration membranes (Baker, 2023). Anisotropic membranes, in contrast, exhibit non-uniformity and consist of distinct layers with varying architectures and compositions (Sagle & Freeman, 2004). They are specifically utilised in reverse osmosis (RO) procedures (Yu et al., 2021).

Membranes are categorised as either organic or inorganic based on their composition. Synthetic organic polymers are used to create organic membranes (Radu et al., 2023). Typically, membranes used in pressure-driven separation procedures are constructed from synthetic organic polymers. The materials encompass polyethylene (PE), polytetrafluorethylene (PTFE), polypropylene, and cellulose acetate, among other options (Aliyu et al., 2018). Inorganic membranes are composed of materials such as ceramics, metals, zeolites, or silica (Kayvani Fard et al., 2018). These substances find extensive use in many industrial applications such as hydrogen separation, ultrafiltration, and microfiltration (Lin, 2019). The movement of wastewater across membranes is determined by many driving forces; there are processes that are driven by pressure, as well as those that are non-pressure driven (Jhaveri & Murthy, 2016). Pressure-driven membrane technologies utilise hydraulic pressure to accomplish separation (Bruggen et al., 2003). These processes can be classified into four primary categories. The four types of filtrations are microfiltration (MF), ultrafiltration (UF), nanofiltration (NF), and reverse osmosis (RO) (Charcosset, 2016). RO, one of the pressure-driven membrane processes, is widely recognised for its exceptional efficacy in the separation of minute particles like bacteria, as well as monovalent ions (Birniwa et al., 2024; Bruggen et al., 2003). Reverse osmosis generates a hydrostatic pressure that is sufficiently powerful to surpass the inherent osmotic pressure of the feed (Blandin et al., 2016). In the entirety of the procedure, water molecules are assimilated onto the surface of the membrane. These molecules permeate the membrane material and subsequently desorb at the permeate side of the membrane for collection (Wang et al., 2023). Undoubtedly, pressure driven membrane techniques have made water reclamation from wastewater a viable choice. Nevertheless, the energy demands persist because of the pressure (Verma et al., 2021).

The properties of the membrane have a significant impact on the overall efficiency of the separation process and extensive research has been conducted to allow substances to pass through, reject contaminants, and prevent fouling (Ahmed et al., 2023; Tang et al., 2009). Currently, the use of the altered membranes for large-scale purification of process water is not feasible due to economic and durability constraints. Moreover, most of the investigations simply focused on enhancing a solitary quality (Mansi et al., 2023).

For instance, the antifouling properties of the membrane are enhanced by its surface coating, but this improvement is counteracted by the increased thickness, which leads to a higher resistance to mass transfer and a decrease in water flux (Shahkaramipour et al., 2017). Similarly, although grafting nanomaterials onto the membrane surface significantly enhances its overall performance, the release of these nanomaterials into the environment is a significant environmental issue (Nain et al., 2022). Furthermore, it is necessary to streamline the membrane modification techniques that have proven to be effective in laboratory settings to make them more applicable for industrial-scale implementation (Shoshaa et al., 2023). Future study should conduct a thorough investigation to conclusively establish the efficacy of membrane technologies to the existing technologies in the practical treatment of industrial water (Ma et al., 2024).

3. MICROBIAL CONSORTIA FOR MEMBRANE BIOREACTORS (MBRS)

With the alarming rise of pollutants in the environment, it is imperative to detoxify the pollutants and safeguard the natural resources. The adoption of membrane bioreactors (MBRs) has demonstrated remarkable advantages in remediation of several pollutants (Figure 5.1). The MBR method is utilized for treating the activated sludge process in both municipal and industrial wastewater treatment sectors (Friha et al., 2014; Iorhemen et al., 2016). This technique has garnered recognition due to its hybrid nature, integrating biological and physical separation principles through the use of a membrane in a single setup (Chang et al., 2002; Tay et al., 2007). The entire system requires minimal space, facilitating a smooth separation process, eliminating the need for secondary clarifiers, and allowing for the automated removal of sludge (Mutamim et al., 2013). Moreover, this technique generates superior effluent quality, increased volumetric loading rates, minimal sludge formation, reduced hydraulic retention times (HRT) and efficient removal of nutrients and pollutants (Judd, 2008; Meng et al., 2009; Mutamim et al., 2013). The microbiology, biotechnology, bioprocess and environmental engineering serve as fundamental components in fabricating effective microbial consortia for membrane bioreactors (MBRs). A significant factor that needs attention while creation of MBRs is the addition of a membrane anti-fouling agent. The membrane can witness biofouling due to the release of microbial products or polymeric contents by the microbes (Kunacheva et al., 2020). The cellular processes such as substrate metabolism releases cell content causing membrane fouling. The fouling may also be caused by the heterogenous solid particles in the form of sludge flocs, colloids, etc., clogging the pores of the membranes (Meng et al., 2009; Tay et al., 2007). The solution to this lies in the inclusion of biological entities, which can degrade the polysaccharides or biopolymers in the sludge and thereby prevent membrane fouling (Xu et al., 2019). Previous studies have reported that use of enzyme such as lyase from *Sphingomonas* sp. can prevent the biofilm formation of *Pseudomonas aeruginosa* (Blanco-Cabra et al., 2020). Another study has reported that anti-fouling microbes such as *P. curvatum* HO1 are capable of breaking uronic acid and hence preventing the filtration blockage (Okamura

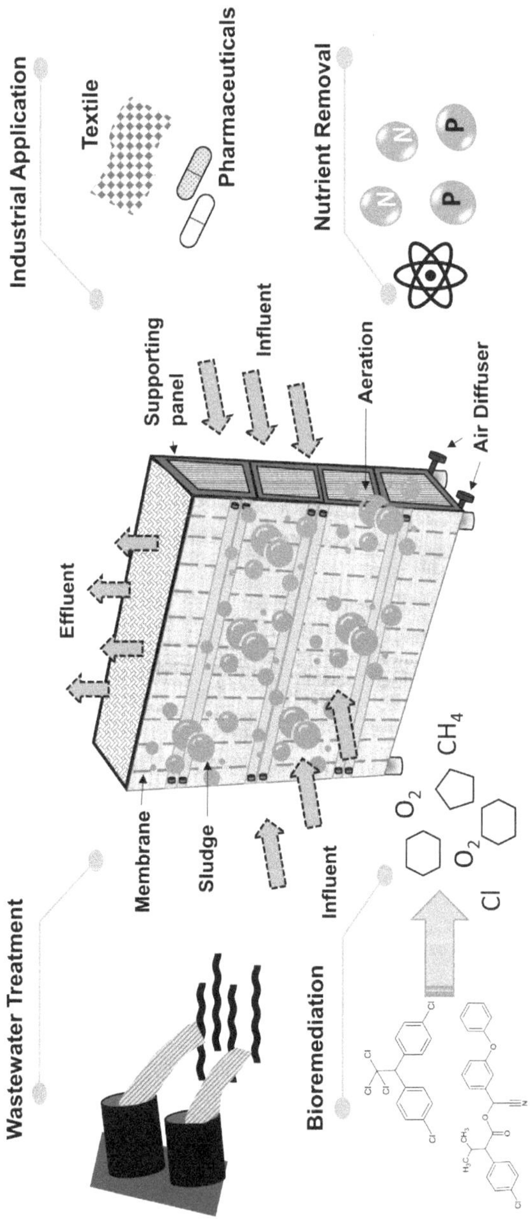

FIGURE 5.1 Schematic representation of membrane bioreactors (MBRs) and its application in diverse sectors.

et al., 2010). On the contrary, the use of enzymes and isolated microbes requires ideal conditions, high maintenance and they have low stability issues (Qi et al., 2023). To address these limitations, the microbial consortia have eventually replaced enzymes in the MREs owing to its diverse metabolic process, synergistic effects and their capability to adapt to the environmental stress (Singh et al., 2019; Tshikantwa et al., 2018; Zhang & Wang, 2016). The utilization of more than one microorganism (microbial consortia) has earned escalating recognition due to the enhanced outcome. The microbial consortia have proven to be impressive in remediating environmental contaminants efficiently and are thus strategically employed in various industrial applications (Bhatia et al., 2018). One distinctive characteristic of microbial consortia is their ability to undergo reactions employing diverse mechanisms to decontaminate large biomass. The biological communication among microbes makes them capable of addressing a wide range of waste efficiently (Kouzuma et al., 2015; Waghmode et al., 2011) (Figure 5.1).

3.1. Selection and Optimization of Microbial Communities

Organisms can exert positive, negative, or neutral effects on each other when collectively employed in consortia. The selection of microorganisms within consortia is determined by the available substrate and operating parameters. These microbes may originate from diverse taxonomic backgrounds, and achieving the right combination ensures a sustained, harmonious coexistence with the intended outcomes (Tshikantwa et al., 2018). Microbial consortia can either be naturally occurring or artificially assembled depending on the targeted site of action. While natural consortia may exhibit inhibitory or neutral effects under specific circumstances, constructed or synthetic consortia are meticulously designed to enhance the reaction process and maintain stability even in challenging conditions (Jeong & Kim, 2019). Numerous aspects, including manipulating the composition of microorganisms, optimizing aeration, and adjusting steering mechanisms, can be customized in membrane bioreactors to enhance membrane stability and lifespan. Numerous studies have underscored the importance of employing synthetic microbial consortia to improve the stability, and overall efficiency of water treatment processes (Lawson et al., 2019). One such recent study reports that the degradation of alginate was enriched by introducing *Bacteroides* into the consortia (Geng et al., 2021). Monitoring of several factors such as aeration, membrane fouling mitigation, nutrient removal process, pH and other parameters needs to be optimized to achieve stable treatment performance by the microbial consortia (Xiao et al., 2019).

3.2. Enhanced Nutrient Removal and Pollutant Degradation

To facilitate the removal of nutrients, different combinations of pre-anoxic, anerobic and anaerobic conditions in the bioreactors play a pivotal role in degrading the organic pollutant (Pellegrin, 2015; Xiao et al., 2014). The pre-anoxic environment encourages the bacteria to perform denitrification process and reduces the nitrogen in the effluent. Whereas, in the anaerobic condition, the removal of phosphorus with the aid of specific phosphorus accumulating bacteria occurs. In different conditions, the microbes undergo biological processes such as nitrification, denitrification and phosphorus removal to ensure low nutrient content in the effluent (Xiao et al., 2019). Moreover, the high sludge content contains higher nitrifying bacteria and hence be advantageous in removal of nitrogen (Pellegrin, 2015). Recent studies have reported the removal of pollutants of emerging concern, which are microplastics using MBR, with 99.9% efficiency (Lares et al., 2018; Talvitie et al., 2017). Thus, the MBR based method has proved to be advantageous over the conventional secondary sedimentation approach for microplastic removal (Xiao et al., 2019).

Contrarily to the conventional approaches (including radiation treatment and advanced oxidation processes), MBRs have also shown promising results in removing the pathogenic organisms (De Luca et al., 2013; Harb & Hong, 2017; Xiao et al., 2019). The mechanism behind the excellent pathogen removal efficiency of the membrane has been investigated, and four mechanisms have been accepted so far by the researchers (Chaudhry et al., 2015; Xagoraraki et al., 2014). The proposed mechanisms include: pathogen attachment to liquor solids, retention of pathogen by the biocake, removal of pathogen by backwashed membrane, and inactivation process. Moreover, the MBR-based method prevents the occurrence of mutations or formation of germ resistance due to its smooth physical processing steps. Thus, this method proves to be preferable in disinfection as compared to other conventional approaches including radiation treatment and advanced oxidation processes.

3.3. Case Studies for Successful Applications of Microbial Consortia-Based Membrane Bioreactors

There are several reports in the utilization of microbial consortia-based membrane bioreactors that have been successfully employed in treatment of different pollutants. There are diverse types of bioreactors that have been reported to treat various industrial waste. Some of them includes simple membrane bioreactor, enhanced membrane bioreactor, aerated membrane bioreactor, dynamic membrane bioreactor and membrane distillation bioreactor (Meena et al., 2022). The microbial consortia of methanogens including *Methanosaeta, Methanomethylovorans* and *Methanosarcina* have been utilized in the anaerobic fluidized membrane bioreactor (AFMBR) to detoxify the industrial wastewater (Cheng et al., 2018). Another study has demonstrated the detoxification of phenolic contaminants in the wastewater with the help of *Micrococcus luteus, Pseudomonas* and *Rhizobium* microbial consortia (Su et al., 2019). In the treatment of pharmaceutical wastewater, the microbial consortia including Thermoplasmata, Firmicutes, Bacteroidetes, and Methanobacteria have effectively employed (Chen et al., 2014). These instances collectively highlight numerous instances of industrial wastewater treatment through the strategic use of microbial consortia-based membrane bioreactors. The resultant treated wastewater is transformed into water suitable for drinking, supporting aquatic biota, and serving agricultural purposes (Meena et al., 2022).

The microbial consortia have been utilized in remediation of several toxic pollutants. For instance, *Brevibacillus* sps. and *Aneurinibacillus* sp have effectively remediated pollutants such as polyethylene and polypropylene (Skariyachan et al., 2018). For remediation of petroleum pollutants, the microbes *Bacillus pumilus* and *B. cereus* have reported by (Patowary et al., 2016). The microbial combination of *Pseudomonas putida/ Shewanella oneidensis* and *Galactomyces geotrichum/ Brevibacillus laterosporus* have been reported to remediate azo dyes (Wang et al., 2014). For the remediation of polychlorinated biphenyls, researchers have reported the use of microbial consortia including *Pseudomonas sp.* and *Debaryomyces maramus* (Chen et al., 2015). The aromatic hydrocarbons have been reported to be remediated by the aid of *Burkholderia multivorans*, *Pandoraea sp.* and *Hyphomicrobium facile* microorganism combination (Ozaki et al., 2006). Moreover, the combination of *Staphylococcus sp.* and *Pseudomonas sp* have been employed for wastewater treatment (Zhang et al., 2007). For the degradation of 17β estradiol, the consortia of *Rhodococcus equi* and *Comamonas testosterone* have been utilized by the researchers in the past (Wang et al., 2019). Thus, for the remediation of several pollutants and for the wastewater treatment the microbial consortia have shown promising results in the past. Thus, the microbial consortia-based membrane bioreactors have gained significant recognition worldwide for wastewater treatment.

4. ENZYME IMMOBILIZED MEMBRANES

The enzyme immobilized membrane is adversely used in food processing, wastewater treatment plants and pharmaceutical industry. Owing to the continuous reaction, enhanced stability, its reusability and reduced product inhibition, this technique is widely adopted (Naseer et al., 2020; Zdarta et al., 2018). The enzyme immobilized members speedily convert the toxic substrate to product with minimal expenditure and with high sustainability (Zeuner et al., 2014). The porous property of the membrane makes the separation process smooth by increasing the area of attachment and area of reaction of the enzyme (Wenten et al., 2020).

4.1. Immobilization Techniques and Strategies

The membrane shelters higher number of enzyme entities with reduction of membrane fouling. The enzymes are immobilized in a way that it resembles the process of membrane fouling thereby preventing the fouling phenomena to occur (Luo et al., 2013, 2014). There are mainly four methods employed for the effective immobilization of enzymes on/in membranes (Figure 5.2).

4.1.1. Physical method

The first immobilization method includes enzyme adsorption which resembles the fouling phenomena of soil adsorption (Sassolas et al., 2012). This process occurs

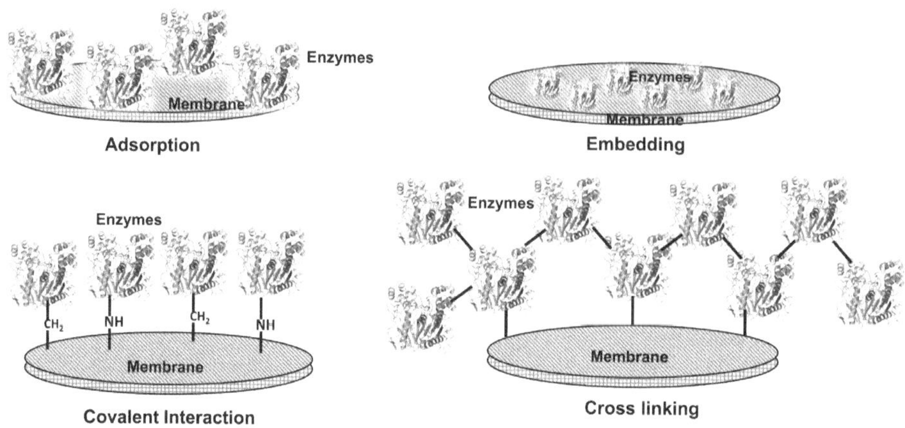

FIGURE 5.2 Representation of four different methods employed for the effective immobilization of enzymes on/in membranes. Green coloured enzyme represents any enzyme that needs to be immobilized.

either by dipping the membrane in the enzyme solution or by following the ultrafiltration cycle (Taheran et al., 2017; Zeuner et al., 2014). There are several disadvantages of this method which includes easy desorption, low loading capacity of enzyme on membranes. The second well-adopted method is enzyme embedding, it resembles the process of fouling caused by pore blockage. This method addressed the limitations of the adsorption method by overcoming low loading limitation and preventing the pore blockage.

The integration of both the methods, i.e., adsorption and embedding provides higher advantage with significantly increased catalytic activity. The advantages of employing both immobilization methods ensured higher recovery activity, catalytic efficiency, stability and loading capacity (Cen et al., 2019). One such example is the use of polymeric composite membrane consisting of two layers, each with unique properties. In normal mode skin layer, the pore diameter was small and hence the adsorption method was employed for enzyme immobilization. In reverse mode support layer, the pores were larger and hence the embedding method was employed (Luo et al., 2013). It was observed that the reverse mode layer exhibited better results and are hence widely adopted. There are several reports on utilization of such physical immobilization methods. The enzyme glutamate dehydrogenase was immobilized in the reverse layer of the reactors, it ensured higher catalytic efficiency with heightened stability (Luo et al., 2013). Another biocatalytic membrane was reported with immobilization of glucose oxidase and catalase enzymes (Morthensen et al., 2017). The enzymes were initially adsorbed by reverse ultrafiltration method and eventually were embedded into the polydopamine membrane by normal mode thus ensuring enzyme stability. Another method of enzyme immobilization adopted by the researchers is the sandwich method (Luo et al., 2014; Morthensen et al., 2017). In one such study, alcohol dehydrogenase was immobilized by reverse ultrafiltration method and was transferred with an additional support layer on another layer. This configuration of three layers in the form of sandwich exhibited higher biotransformation property, hydrophilic nature and higher compatibility (Luo et al., 2014).

4.1.2. Chemical method

The chemical method employed is immobilization by covalent interaction, it resembles the process of surface modification fouling. In this method the enzyme interacts with the functional group present on the membrane surface or in the pores. The chemical entities are activated by treating the membrane by several method including plasma, γ-irradiation, wet chemical and UV-induced method (Jochems et al., 2011). Some of them are listed in Table 5.1.

The different functional groups that participate in such immobilization techniques include amide, epoxy, hydroxyl, sulfhydryl, amino, azido, hydrazide, nitrene, dichlorophenol, carboxyl, imidazole, metals, etc., (Cen et al., 2019; Jochems et al., 2011). The modification of membrane with amino, carboxyl or aldehyde groups encouraged the enzyme immobilization process via bond formation (Meller et al., 2017). In case of the amino group modification, the 1,6-hexamethylenediamine, 1,12-diaminododecane and 3-aminopropyltriethoxysilane are mostly used to modify the membrane (Cen et al., 2019; Sulaiman et al., 2015). The amino group activated membrane interacts with the carboxyl

TABLE 5.1 Chemical methods employed for immobilization of different enzymes on membranes.

METHOD EMPLOYED	*ENZYME UTILIZED*	*FUNCTIONAL GROUP*	*REFERENCES*
UV-induced	Lipase	Epoxy	(Cen et al., 2019)
	Invertase	Epoxy	(Bayramoğlu et al., 2003), (Cen et al., 2019)
	α-Amylase	Epoxy	(Bayramoğlu et al., 2004)
	Urease	Epoxy	(Jochems et al., 2011)
Plasma	Glucose isomerase	Hydroxyl	(Jochems et al., 2011)
	Lipase	poly (α-allyl glucoside)	(Deng et al., 2005)
	Lipase	poly (γ-stearyl l-glutamate)	(Deng et al., 2004)
	Acetylcholinesterase	Amino	(Alp et al., 2000)
γ-irradiation	Lipase	Amino	(Abrol et al., 2007)
Wet chemical	Horseradish peroxidase	Polyaniline	(Piletsky et al., 2003)
	Cellulase	Amide	(Jochems et al., 2011)
	Urease	Epoxy	(Godjevargova & Gabrovska, 2003; Tyagi et al., 2009)
	Glucose oxidase	Polyaniline	(Bayramoğlu et al., 2010)
	β-Galactosidase	Amino	(Grano et al., 2004)

TABLE 5.2 Different enzymes employed for pollutant degradation.

ENZYME	*METHOD FOR APPLICATION OF ENZYME TO MEMBRANES USED IN WASTEWATER TREATMENT*	*POLLUTANT*	*REFERENCES*
Laccase	Adsorption	Phenolic compounds	(Liu et al., 2012)
Peroxidase	Cross-linking	Textile dyes	(Darwesh et al., 2019)
Phosphotriesterase	Covalent binding	Organophosphate pesticides	(Gao et al., 2014)
Laccase	Entrapment	Bisphenol A	(Olajuyigbe et al., 2019)
Horseradish peroxidase	Encapsulation	Pentachlorophenol	(Niu et al., 2013)
Tyrosinase	Adsorption	Phenol	(Seetharam & Saville, 2003)
Horseradish peroxidase	Covalent binding	Dye	(Bilal, Iqbal et al., 2016)
Metallohydrolases	Cross-linking	Paraoxon ethyl	(Pedroso et al., 2021)
Peroxidase	Covalent binding	Carbamazepine	(Calza et al., 2016)
Horseradish peroxidase	Entrapment	Sulfamethoxazole	(Zdarta et al., 2021)
PETase	Cross-linking	Polyethylene terephthalate (PET)	(Schwaminger et al., 2021)
Chloroperoxidase	Covalent binding	Azo dye	(Cen et al., 2019)
Lipase	Adsorption	Oils	(Hoq et al., 1985)
Lipase	Entrapment	Ethyl alcohol and lactic acid	(Ugur Nigiz, 2016)
Tyrosinase	Cross-linking	Phenol	(Abdollahi et al., 2018)
Manganese peroxidase	Cross-linking	Dye removal	(Bilal, Asgher et al., 2016)
Horseradish peroxidase	Entrapment	Textile dyes	(Bilal et al., 2017)
Tyrosinase	Cross-linking	Wastewater	(Wu et al., 2017)
Laccase	Cross-linking	Methyl orange, cibacron blue, and reactive black 5	(Berradi et al., 2019)
Horseradish peroxidase	Cross-linking	Anthraquinone dye acid violet	(Sekuljica et al., 2016)

group of the protein or another mediator to generate biocatalytic membrane. There are several methods that are utilized to modify the membrane and allow the covalent interaction of protein with the membrane.

The another chemical method is cross linking, which is associated with the phenomena of bio cake formation causing resistance in the passage of the solution (Luo et al., 2013). In this method, the enzyme is either directly linked with each other or it is linked via another entity (example BSA) with the membrane (Sassolas et al., 2012). This method is beneficial over others because of its higher enzyme loading capacity (Palai & Bhattacharya, 2013). Examples of different enzymes employed for pollutant degradation are depicted in the Table 5.2.

4.2. Examples of Applications in Different Industries

The enzyme immobilized membranes have diverse applications in the industrial sectors. Different classes of enzymes are employed in different industries including food, textile, medical, water treatment, etc. (Maghraby et al., 2023). The enzyme proteases are used in the degradation of proteins and play significant role in the food industry, pharmaceutical, leather, detergent and textile industry (Bai et al., 2012; Moslemi et al., 2018). In textile industries, enzymes such as cellulase, keratinase, protease and laccase are immobilized by different methods to modify fibres, enhance the quality of fibre and in decolorization of textile effluents (Heikinheimo et al., 2000; Soares et al., 2011). Enzymes such as lipases and amylases are utilized in the detergent industries for cleaning efficiency and assuring oil and starch stain removal and to improve the colour as well as the quality of the fabric (Mine et al., 2003; Sharma et al., 2001). Enzymes such as horseradish peroxidases are immobilized on graphene oxide to treat wastewater efficiently and remove the phenolic contaminants from the water (Vineh et al., 2018; Sellami et al., 2021). Other enzymes employed in the wastewater treatment include laccases and tyrosinases (Abdollahi et al., 2018; Seetharam & Saville, 2003; Wu et al., 2017).

5. BIOSORPTION MEMBRANES

The biosorption membrane is based on the principle of sorption process. The living or non-living cells are used to remediate the pollutants by adopting bioaccumulation or passive biosorption method (Wenten et al., 2020). In the case of bioaccumulation, the pollutant adheres to the cells, enters the cells and undergoes cellular metabolism and ultimately releases non-toxic compounds (Bilal et al., 2018; Gavrilescu, 2004). Whereas, in the passive biosorption method, the physical and chemical method of sorption occurs which includes precipitation, ion exchange, chelation, interaction and complexation (Asgher, 2012; Gadd, 2009). The sorption with the aid of non-living biomass is preferable over the bioaccumulation method, because of the regeneration and reusable property of the sorbent (Aksu, 2005; Farooq et al., 2010). The biosorbents that are employed

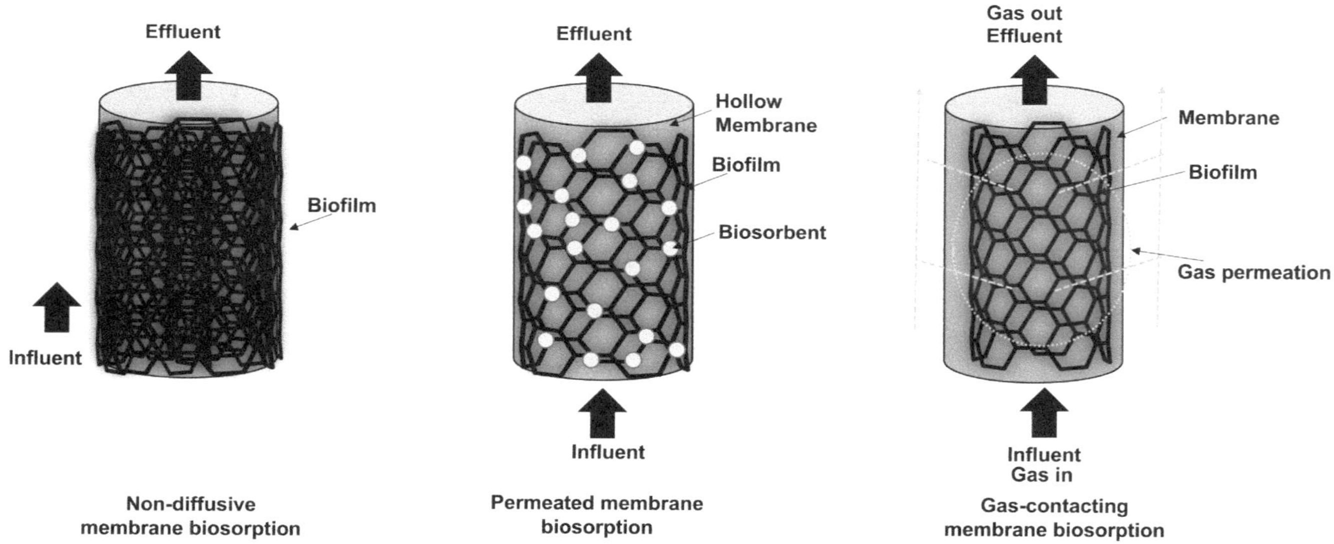

FIGURE 5.3 Schematic representation of three types of membrane biosorption configurations.

for remediation of various pollutants include algal, bacterial and fungal sources. Besides these, the activated sludge, chitinous biomass and biological waste materials are also utilized as sorbents (Dadwal & Mishra, 2017; Davis et al., 2003). The biosorbents are advantageous over conventional sorbents because of low cost, reusability, easy availability and higher efficiency. The pollutants such as metals may be easily recovered in the sorbate when biosorbents are employed in the wastewater treatment process (Davis et al., 2000; Volesky, 2001; Wenten et al., 2020). The cell surface contains several functional groups, these in turn interacts with specific pollutants chemically and via sorption phenomena, it serves as the remediating tool (Aksu, 2005; Das, 2012; Dhankhar & Hooda, 2011; Khoo & Ting, 2001). In the reactor, the biosorption membrane can remediate the pollutants in the influent either in batch or in continuous fashion. The biological entities are adhered to the membranes using several immobilization techniques that include physical and chemical immobilization methods. With the continuous passage of the pollutants when the sorbate gets heavily accumulated, the centrifugation or filtration method is then used to regenerate the sorbent (Aksu, 2005; Khoo & Ting, 2001; Ting & Sun, 2000). The biosorption is strategically integrated with the membrane separation process to ensure higher proficiency and maximal mass transfer kinetics thus ensuring efficient treatment of the wastewater (Qiu et al., 2016; Qiu & Ting, 2014; Raghavan et al., 2018; Zhang et al., 2017). This membrane bioreactor ensures higher surface area, pollutant specific treatment, higher remediation kinetics and greater separation of effluent and the biomass (Beolchini et al., 2008; Koltuniewicz et al., 2004). The deposition of biosorbents on the membrane increases the interaction of the sorbate with the biomass and hence results in removal of the sorbate with higher efficiency.

The membrane biosorption is of three types based on the nature of the membrane (Figure 5.3). The types include non-diffusive, permeated and gas contacting (Wenten et al., 2020). The diffusive membrane biosorption includes the biofilm formation on the membrane, and it filters the pollutants via this sorption membrane which is the biofilm. The pollutants interact with the biological entities of the biofilm and hence do not allow the pollutants to diffuse through membrane. This membrane behaves as filter which effectively separates the sorbate from the effluent, allowing the release of contaminant free water via the membrane. The second type is the permeated biosorption membrane. This membrane is selective in nature and allows specific pollutants to penetrate the membrane and hence to help in extracting them in particular. The specific pollutants get absorbed by the biosorbents which are placed in the other side of the membrane. In the case of gas contacting type of membrane, the membrane is porous in nature. The gas pollutants get absorbed by the membrane biofilm, and it may even facilitate oxygenation of the biosorbents (Himma et al., 2019; Siagian et al., 2019).

5.1. Utilization of Biopolymers and Microorganisms for Adsorption

The polymers that are mainly employed are biodegradable and are non-toxic in nature. The advantage of biopolymers includes presence of several chemical functionalities, flexibility, specific affinity and cost-effective approach. Several polymers that are mostly used as membranes for enzyme immobilization are poly(D,L-lactide-co-glycolide)

(PLGA), poly(D,L-lactide) (PLA), polyvinyl alcohol (PVA), cellulose electrospun fibres, Polyhydroxyalkanoates (PHA), chitosan and alginate-based polymers (Giovagnoli et al., 2004; Hosseini et al., 2022; Shinde et al., 2018; Yuan et al., 2023). For instance, agarose-based biopolymers with superior properties and manipulative pore size with the aid of agarose concentration and the crosslinking agent have been widely used (Zdarta et al., 2018; Zucca et al., 2016). The enzyme lipase when immobilized on alginate-based biopolymer exhibited greater stability and catalytic efficiency (İspirli et al., 2017). For wastewater treatment, the biopolymer from pulp and paper waste mix was employed to immobilize biosorbent and detoxify pharmaceutical pollutants and toxic dyes (Bankeeree et al., 2016). The protein-based membranes and the polypeptide chains as additives are widely used for ensuring rapid degradation process with higher efficiency (Winartasaputra et al., 1982).

Several investigations have reported the utilization of different microorganism in the remediation of diverse classes of pollutants. The pollutants get adsorbed by the biosorbents with the aid of different physical and chemical processes (Escudero et al., 2019). For instance, *Deinococcus radiodurans* have been reported to successfully remediate Cd (II) and Pb (II) ions by the process of chelation (Jaafar et al., 2015). Moreover, *Cystoseira crinitophylla* has been reported to remove Cu (II) ions via ion exchange method (Christoforidis et al., 2015). Another organism, *Rhodococcus pyridinivorans XB* have been reported to degrade 98% of phthalic acid into non-toxic products (Zhao et al., 2018).

5.2. Removal of Heavy Metals and Other Contaminants

With the aid of non-diffusive membrane biosorption method, the removal of several toxic metals has been reported. For instance, the *Arthrobacter sp.* has been reported for the removal of metals such as copper by non-diffusive membrane type (Veglio et al., 1999, 2000). In this case, the biosorbent possesses carboxylic and aminic groups, which interacted with the metal ions and hence helped in its removal. Other microorganism reported for cooper remediation are *Sphaerotilus natans* and *Saccharomyces cerevisiae* (Koltuniewicz & Bezak, 2002; Pagnanelli et al., 2003). They have been reported to remove copper independently via the non-diffusive membrane biosorption. For the removal of another toxic metal, cadmium, *Arthrobacter sp.* is used as the biosorbent in the membrane (Pagnanelli et al., 2003). Some other toxic metals such as Pb, Cu, Cd are effectively removed from the solution with the aid of *Sphaerotilus natans* acting as the biosorbent in the membrane (Pagnanelli et al., 2004). Another microorganism that have been reported as metal biosorbent is *Sargassum glaucescens*. A study reports the removal of metals such as Ni and Co by using non-diffusive membrane biosorption method (Esmaeili & Aghababai, 2015).

In case of permeated membrane biosorption method, several toxic pollutants have been removed from wastewater. For instance, by employing silicone rubber membrane, several toxic pollutants have been removed by several microorganism acting as the biosorbent. HAB halogenated aromatic bacteria have been reported to remove pollutants such as 3-Chloronitrobenzene, Aniline, 4-Chloroaniline and 2,3-Dichloroaniline from wastewater by employing silicone rubber tubes (Brookes & Livingston, 1994; Wenten et al., 2020). The previous studies have reported that the pollutants such as toluene and dicholoromethane can be removed by the microbial culture acting as the biosorbent on the silicon rubber membrane (Jorge & Livingston, 2000). By employing anion exchange membrane, the pollutants such as nitrate and bromate have been removed with the permeated membrane biosorption configuration (Fonseca et al., 2000; Velizarov et al., 2008).

The gas contacting membrane biosorption configuration is employed to remove specific pollutants with the help of a gas stream. The spirally wound silicone rubber acted as the gas contacting membrane, and the *Xanthobacter autotrophicus* GJ10 was employed as the biosorbent on the other side of the membrane. This set up has been reported to remove 1,2-Dichloroethane with 91% efficiency (Freitas Santos et al., 1995). Another PDMS/PSf-ZrO_2 membrane has been reported to remove Dimethyl sulfide in the presence of *Hyphomicrobium* VS as biosorbent with 74% to 99% removal efficiency (De et al., 2002). By employing polypropylene membrane as the gas coating membrane, pollutants such as propene and ammoniacal–N has been removed by using *Xanthobacter* Py2 as the biosorbent (De, 2003; Sun et al., 2015). Silicone membrane has been used to remove pollutants such as chlorophenols, phenol, xylene and acetate by utilizing *Vibrio natriegens* or halophilic soil microorganisms as the biosorbents in the other side of the membrane (Wenten et al., 2020).

5.3. Regeneration and Reusability of Biosorption Membranes

Biosorption membranes play a pivotal role in remediating several contaminants including dyes, phenolic contents, heavy metals and several other pollutants sustainably. This membrane is utilized in industrial applications and in wastewater treatment. The regeneration techniques include chemical regeneration, thermal regeneration and biological regeneration (Wenten et al., 2020). In chemical regeneration and thermal method, the desorption step is conducted by exposing the membrane to chemical and heat, respectively. In biological regeneration method, the membrane is regenerated using microbes and the biofilm is recreated. This method is also termed as green regeneration method.

To maintain the biological efficiency of the biosorbent used and increase its longevity, the membrane monitoring and appropriate regeneration steps play a significant role. The biosorbent thickness plays a significant role in remediation of the pollutants (Koltuniewicz & Bezak, 2002). The thicker biosorbent layer limits the permeation of the influent and also fail to generate pollutant free effluent (Liao et al., 2018). Thus, the diffusion limit and biofilm formation have to be designed scientifically at regularly intervals to obtain better results. The use of biosorption membrane has proved to immensely advantageous over conventional wastewater treatment process (Wenten et al., 2020).

6. CONCLUSION

Membrane technology has emerged as a promising and powerful tool for wastewater treatment. It offers several advantages over conventional methods and can remove a wide range of contaminants, including organic matter, heavy metals, microorganisms and emerging pollutants. Membrane systems can be compact and modular, making them suitable for space-constrained locations. Compared to some traditional methods, certain membrane processes require less energy for operation. Membrane technology can be used to treat various types of wastewater from different industries. Membrane processes can be integrated effectively with other wastewater treatment technologies, such as biological treatment, to achieve optimal results. However, challenges remain in optimizing membrane technology for broader application. The accumulation of contaminants on the membrane surface can hinder its performance and require regular cleaning or replacement. Implementing membrane systems can involve significant upfront costs. Although lower than some methods, certain pressure-driven membrane processes still require considerable energy for operation. Ongoing research and development efforts are focused on addressing these challenges. Advancements in membrane materials, process optimization and the integration of innovative technologies like bioreactors and enzyme-immobilized membranes hold promise for enhancing the efficiency, sustainability, and affordability of membrane-based wastewater treatment. Membrane technology has the potential to play a crucial role in addressing the global water crisis by providing a sustainable and effective solution for wastewater treatment. As the technology continues to evolve, it is expected to become even more widely adopted, contributing to a cleaner and more sustainable future.

ACKNOWLEDGEMENT

Dr. Simranjeet Singh would like to acknowledge Regional Center for Biotechnology, Faridabad, Haryana, for Research Associateship (DBT-RA), funding under award letter No DBT-RA/2022/July/N/2044 dated January 12, 2023, and thanks to Ministry of Education (MoE) for their support under the grant MoE-STARS/STARS-2/2023-0714, dated September 26, 2023.

REFERENCES

Abdollahi, K., Yazdani, F., Panahi, R., & Mokhtarani, B. (2018). Biotransformation of phenol in synthetic wastewater using the functionalized magnetic nano-biocatalyst particles carrying tyrosinase. *3 Biotech*, *8*(10), 419. https://doi.org/10.1007/s13205-018-1445-2

Abrol, K., Qazi, G. N., & Ghosh, A. K. (2007). Characterization of an anion-exchange porous polypropylene hollow fiber membrane for immobilization of ABL lipase. *Journal of Biotechnology*, *128*(4), 838–848. https://doi.org/10.1016/j.jbiotec.2006.12.031

Ahmed, M. A., Amin, S., & Mohamed, A. A. (2023). Fouling in reverse osmosis membranes: Monitoring, characterization, mitigation strategies and future directions. *Heliyon*, *9*(4), e14908. https://doi.org/10.1016/j.heliyon.2023.e14908

Ahmad, N. A., Goh, P. S., Yogarathinam, L. T., Zulhairun, A. K., & Ismail, A. F. (2020). Current advances in membrane technologies for produced water desalination. *Desalination, 493*, 114643. https://doi.org/10.1016/j.desal.2020.114643

Aksu, Z. (2005). Application of biosorption for the removal of organic pollutants: A review. *Process Biochemistry, 40*(3–4), 997–1026. https://doi.org/10.1016/j.procbio.2004.04.008

Al Hamedi, F. H., Kandhan, K., Liu, Y., Ren, M., Jaleel, A., & Alyafei, M. A. (2023). Wastewater irrigation: A promising way for future sustainable agriculture and food security in the United Arab Emirates. *Water, 15*(12), 2284. https://doi.org/10.3390/w15122284

Aliyu, U. M., Rathilal, S., & Isa, Y. M. (2018). Membrane desalination technologies in water treatment: A review. *Water Practice & Technology, 13*(4), 738–752.

Alp, B., Mutlu, S., & Mutlu, M. (2000). Glow-discharge-treated cellulose acetate (CA) membrane for a high linearity single-layer glucose electrode in the food industry. *Food Research International, 33*(2), 107–112. https://doi.org/10.1016/S0963-9969(00)00013-2

Asgher, M. (2012). Biosorption of reactive dyes: A review. *Water, Air, & Soil Pollution, 223*(5), 2417–2435. https://doi.org/10.1007/s11270-011-1034-z

Bai, Y., Huang, H., Meng, K., Shi, P., Yang, P., Luo, H., … & Yao, B. (2012). Identification of an acidic α-amylase from Alicyclobacillus sp. A4 and assessment of its application in the starch industry. *Food Chemistry, 131*(4), 1473–1478. https://doi.org/10.1016/j.foodchem.2011.10.036

Baker, R. W. (2023). *Membrane technology and applications*. John Wiley & Sons.

Bankeeree, W., Prasongsuk, S., Imai, T., Lotrakul, P., & Punnapayak, H. (2016). A novel xylan-polyvinyl alcohol hydrogel bead with laccase entrapment for decolorization of reactive black 5. *BioResources, 11*(3), 6984–7000. https://doi.org/10.15376/biores.11.3.6984-7000

Bashir, I., Lone, F. A., Bhat, R. A., Mir, S. A., Dar, Z. A., & Dar, S. A. (2020). Concerns and threats of contamination on aquatic ecosystems. In K. R. Hakeem, R. A. Bhat, & H. Qadri (Eds.), *Bioremediation and biotechnology: Sustainable approaches to pollution degradation* (pp. 1–26). Springer International Publishing. https://doi.org/10.1007/978-3-030-35691-0_1

Bayramoğlu, G., Akgöl, S., Bulut, A., Denizli, A., & Yakup Arıca, M. (2003). Covalent immobilisation of invertase onto a reactive film composed of 2-hydroxyethyl methacrylate and glycidyl methacrylate: Properties and application in a continuous flow system. *Biochemical Engineering Journal, 14*(2), 117–126. https://doi.org/10.1016/S1369-703X(02)00170-5

Bayramoğlu, G., Metin, A. Ü., Altıntas, B., & Arıca, M. Y. (2010). Reversible immobilization of glucose oxidase on polyaniline grafted polyacrylonitrile conductive composite membrane. *Bioresource Technology, 101*(18), 6881–6887. https://doi.org/10.1016/j.biortech.2010.04.025

Bayramoğlu, G., Yilmaz, M., & Arica, M. Y. (2004). Immobilization of a thermostable α-amylase onto reactive membranes: Kinetics characterization and application to continuous starch hydrolysis. *Food Chemistry, 84*(4), 591–599. https://doi.org/10.1016/S0308-8146(03)00283-8

Beolchini, F., Pagnanelli, F., Toro, L., & Veglio, F. (2008). A methodological approach for dynamic modelling of membrane integrated biosorption: A case study on Sphaerotilus natans as biosorbent. *International Journal of Environment and Pollution, 34*(1/2/3/4), 325. https://doi.org/10.1504/IJEP.2008.020801

Berradi, M., Hsissou, R., Khudhair, M., Assouag, M., Cherkaoui, O., El Bachiri, A., & El Harfi, A. (2019). Textile finishing dyes and their impact on aquatic environs. *Heliyon, 5*(11), e02711. https://doi.org/10.1016/j.heliyon.2019.e02711

Besharati Vineh, M., Saboury, A. A., Poostchi, A. A., Rashidi, A. M., & Parivar, K. (2018). Stability and activity improvement of horseradish peroxidase by covalent immobilization on functionalized reduced graphene oxide and biodegradation of high phenol concentration.

International Journal of Biological Macromolecules, *106*, 1314–1322. https://doi.org/10.1016/j.ijbiomac.2017.08.133

Bhatia, S. K., Bhatia, R. K., Choi, Y.-K., Kan, E., Kim, Y.-G., & Yang, Y.-H. (2018). Biotechnological potential of microbial consortia and future perspectives. *Critical Reviews in Biotechnology*, *38*(8), 1209–1229. https://doi.org/10.1080/07388551.2018.1471445

Bilal, M., Asgher, M., Iqbal, M., Hu, H., & Zhang, X. (2016). Chitosan beads immobilized manganese peroxidase catalytic potential for detoxification and decolorization of textile effluent. *International Journal of Biological Macromolecules*, *89*, 181–189. https://doi.org/10.1016/j.ijbiomac.2016.04.075

Bilal, M., Iqbal, H. M. N., Hu, H., Wang, W., & Zhang, X. (2017). Enhanced bio-catalytic performance and dye degradation potential of chitosan-encapsulated horseradish peroxidase in a packed bed reactor system. *Science of The Total Environment*, *575*, 1352–1360. https://doi.org/10.1016/j.scitotenv.2016.09.215

Bilal, M., Iqbal, H. M. N., Hussain Shah, S. Z., Hu, H., Wang, W., & Zhang, X. (2016). Horseradish peroxidase-assisted approach to decolorize and detoxify dye pollutants in a packed bed bioreactor. *Journal of Environmental Management*, *183*, 836–842. https://doi.org/10.1016/j.jenvman.2016.09.040

Bilal, M., Rasheed, T., Sosa-Hernández, J., Raza, A., Nabeel, F., & Iqbal, H. (2018). Biosorption: An interplay between marine algae and potentially toxic elements—a review. *Marine Drugs*, *16*(2), 65. https://doi.org/10.3390/md16020065

Birniwa, A. H., Habibu, S., Abdullahi, S. S., Mohammad, R. E. A., Hussaini, A., Magaji, H., … & Jagaba, A. H. (2024). Membrane technologies for heavy metals removal from water and wastewater: A mini review. *Case Studies in Chemical and Environmental Engineering*, *9*, 100538. https://doi.org/10.1016/j.cscee.2023.100538

Blanco-Cabra, N., Paetzold, B., Ferrar, T., Mazzolini, R., Torrents, E., Serrano, L., & LLuch-Senar, M. (2020). Characterization of different alginate lyases for dissolving Pseudomonas aeruginosa biofilms. *Scientific Reports*, *10*(1), 9390. https://doi.org/10.1038/s41598-020-66293-2

Blandin, G., Verliefde, A. R., Comas, J., Rodriguez-Roda, I., & Le-Clech, P. (2016). Efficiently combining water reuse and desalination through forward osmosis—Reverse osmosis (FO-RO) hybrids: A critical review. *Membranes*, *6*(3), 37.

Bolan, S., Padhye, L. P., Jasemizad, T., Govarthanan, M., Karmegam, N., Wijesekara, H., … & Bolan, N. (2024). Impacts of climate change on the fate of contaminants through extreme weather events. *Science of the Total Environment*, *909*, 168388. https://doi.org/10.1016/j.scitotenv.2023.168388

Boretti, A., & Rosa, L. (2019). Reassessing the projections of the World Water Development Report. *NPJ Clean Water*, *2*(1), 1–6. https://doi.org/10.1038/s41545-019-0039-9

Brookes, P. R., & Livingston, A. G. (1994). Biological detoxification of a 3-chloronitrobenzene manufacture wastewater in an extractive membrane bioreactor. *Water Research*, *28*(6), 1347–1354. https://doi.org/10.1016/0043-1354(94)90301-8

Calza, P., Zacchigna, D., & Laurenti, E. (2016). Degradation of orange dyes and carbamazepine by soybean peroxidase immobilized on silica monoliths and titanium dioxide. *Environmental Science and Pollution Research*, *23*(23), 23742–23749. https://doi.org/10.1007/s11356-016-7399-1

Cen, Y., Liu, Y., Xue, Y., & Zheng, Y. (2019). Immobilization of enzymes in/on membranes and their applications. *Advanced Synthesis & Catalysis*, *361*(24), 5500–5515. https://doi.org/10.1002/adsc.201900439

Chang, I.-S., Le Clech, P., Jefferson, B., & Judd, S. (2002). Membrane fouling in membrane bioreactors for wastewater treatment. *Journal of Environmental Engineering*, *128*(11), 1018–1029. https://doi.org/10.1061/(ASCE)0733-9372(2002)128:11(1018)

Charcosset, C. (2016). Ultrafiltration, microfiltration, nanofiltration and reverse osmosis in integrated membrane processes. *Integrated Membrane Systems and Processes*, 1–22.

Chaudhry, R. M., Nelson, K. L., & Drewes, J. E. (2015). Mechanisms of pathogenic virus removal in a full-scale membrane bioreactor. *Environmental Science & Technology*, *49*(5), 2815–2822. https://doi.org/10.1021/es505332n

Chen, F., Hao, S., Qu, J., Ma, J., & Zhang, S. (2015). Enhanced biodegradation of polychlorinated biphenyls by defined bacteria-yeast consortium. *Annals of Microbiology*, *65*(4), 1847–1854. https://doi.org/10.1007/s13213-014-1023-8

Chen, Z., Wang, Y., Li, K., & Zhou, H. (2014). Effects of increasing organic loading rate on performance and microbial community shift of an up-flow anaerobic sludge blanket reactor treating diluted pharmaceutical wastewater. *Journal of Bioscience and Bioengineering*, *118*(3), 284–288. https://doi.org/10.1016/j.jbiosc.2014.02.027

Cheng, H.-H., Whang, L.-M., Yi, T.-F., Liu, C.-P., Lin, T.-F., & Yeh, M.-S. (2018). Pilot study of cold-rolling wastewater treatment using single-stage anaerobic fluidized membrane bioreactor. *Bioresource Technology*, *263*, 418–424. https://doi.org/10.1016/j.biortech.2018.04.124

Chidiac, S., El Najjar, P., Ouaini, N., El Rayess, Y., & El Azzi, D. (2023). A comprehensive review of water quality indices (WQIs): History, models, attempts and perspectives. *Reviews in Environmental Science and Bio/Technology*, *22*(2), 349–395. https://doi.org/10.1007/s11157-023-09650-7

Christoforidis, A. K., Orfanidis, S., Papageorgiou, S. K., Lazaridou, A. N., Favvas, E. P., & Mitropoulos, A. Ch. (2015). Study of Cu(II) removal by Cystoseira crinitophylla biomass in batch and continuous flow biosorption. *Chemical Engineering Journal*, *277*, 334–340. https://doi.org/10.1016/j.cej.2015.04.138

Dadwal, A., & Mishra, V. (2017). Review on biosorption of arsenic from contaminated water. *CLEAN – Soil, Air, Water*, *45*(7), 1600364. https://doi.org/10.1002/clen.201600364

Darwesh, O. M., Matter, I. A., & Eida, M. F. (2019). Development of peroxidase enzyme immobilized magnetic nanoparticles for bioremediation of textile wastewater dye. *Journal of Environmental Chemical Engineering*, *7*(1), 102805. https://doi.org/10.1016/j.jece.2018.11.049

Das, N. (2012). Remediation of radionuclide pollutants through biosorption – an overview. *CLEAN – Soil, Air, Water*, *40*(1), 16–23. https://doi.org/10.1002/clen.201000522

Davis, T. A., Volesky, B., & Mucci, A. (2003). A review of the biochemistry of heavy metal biosorption by brown algae. *Water Research*, *37*(18), 4311–4330. https://doi.org/10.1016/S0043-1354(03)00293-8

Davis, T. A., Volesky, B., & Vieira, R. H. S. F. (2000). Sargassum seaweed as biosorbent for heavy metals. *Water Research*, *34*(17), 4270–4278. https://doi.org/10.1016/S0043-1354(00)00177-9

De Bo, I. (2003). Investigation of the permeability and selectivity of gases and volatile organic compounds for polydimethylsiloxane membranes. *Journal of Membrane Science*, *215*(1–2), 303–319. https://doi.org/10.1016/S0376-7388(03)00024-3

De Bo, I., Van Langenhove, H., & Heyman, J. (2002). Removal of dimethyl sulfide from waste air in a membrane bioreactor. *Desalination*, *148*(1–3), 281–287. https://doi.org/10.1016/S0011-9164(02)00715-4

De Luca, G., Sacchetti, R., Leoni, E., & Zanetti, F. (2013). Removal of indicator bacteriophages from municipal wastewater by a full-scale membrane bioreactor and a conventional activated sludge process: Implications to water reuse. *Bioresource Technology*, *129*, 526–531. https://doi.org/10.1016/j.biortech.2012.11.113

Deng, H.-T., Xu, Z.-K., Dai, Z.-W., Wu, J., & Seta, P. (2005). Immobilization of Candida rugosa lipase on polypropylene microfiltration membrane modified by glycopolymer: Hydrolysis of olive oil in biphasic bioreactor. *Enzyme and Microbial Technology*, *36*(7), 996–1002. https://doi.org/10.1016/j.enzmictec.2005.01.025

Deng, H.-T., Xu, Z.-K., Wu, J., Ye, P., Liu, Z.-M., & Seta, P. (2004). A comparative study on lipase immobilized polypropylene microfiltration membranes modified by sugar-containing polymer and polypeptide. *Journal of Molecular Catalysis B: Enzymatic*, *28*(2–3), 95–100. https://doi.org/10.1016/j.molcatb.2004.01.004

Dhankhar, R., & Hooda, A. (2011). Fungal biosorption – an alternative to meet the challenges of heavy metal pollution in aqueous solutions. *Environmental Technology*, *32*(5), 467–491. https://doi.org/10.1080/09593330.2011.572922

Du, X., Shi, Y., Jegatheesan, V., & Haq, I. U. (2020). A review on the mechanism, impacts and control methods of membrane fouling in MBR system. *Membranes*, *10*(2). https://doi.org/10.3390/membranes10020024

Escudero, L. B., Quintas, P. Y., Wuilloud, R. G., & Dotto, G. L. (2019). Recent advances on elemental biosorption. *Environmental Chemistry Letters*, *17*(1), 409–427. https://doi.org/10.1007/s10311-018-0816-6

Esmaeili, A., & Aghababai B. A. (2015). Biosorption of nickel and cobalt from plant effluent by Sargassum glaucescens nanoparticles at new membrane reactor. *International Journal of Environmental Science and Technology*, *12*(6), 2055–2064. https://doi.org/10.1007/s13762-014-0744-3

Farooq, U., Kozinski, J. A., Khan, M. A., & Athar, M. (2010). Biosorption of heavy metal ions using wheat based biosorbents – A review of the recent literature. *Bioresource Technology*, *101*(14), 5043–5053. https://doi.org/10.1016/j.biortech.2010.02.030

Fonseca, A. D., Crespo, J. G., Almeida, J. S., & Reis, M. A. (2000). Drinking water denitrification using a novel ion-exchange membrane bioreactor. *Environmental Science & Technology*, *34*(8), 1557–1562. https://doi.org/10.1021/es9910762

Freitas Santos, L. M. D., Hömmerich, U., & Livingston, A. G. (1995). Dichloroethane removal from gas streams by an extractive membrane bioreactor. *Biotechnology Progress*, *11*(2), 194–201. https://doi.org/10.1021/bp00032a011

Friha, I., Karray, F., Feki, F., Jlaiel, L., & Sayadi, S. (2014). Treatment of cosmetic industry wastewater by submerged membrane bioreactor with consideration of microbial community dynamics. *International Biodeterioration & Biodegradation*, *88*, 125–133. https://doi.org/10.1016/j.ibiod.2013.12.015

Gadd, G. M. (2009). Biosorption: Critical review of scientific rationale, environmental importance and significance for pollution treatment. *Journal of Chemical Technology & Biotechnology*, *84*(1), 13–28. https://doi.org/10.1002/jctb.1999

Gao, Y., Truong, Y. B., Cacioli, P., Butler, P., & Kyratzis, I. L. (2014). Bioremediation of pesticide contaminated water using an organophosphate degrading enzyme immobilized on nonwoven polyester textiles. *Enzyme and Microbial Technology*, *54*, 38–44. https://doi.org/10.1016/j.enzmictec.2013.10.001

Gavrilescu, M. (2004). Removal of heavy metals from the environment by biosorption. *Engineering in Life Sciences*, *4*(3), 219–232. https://doi.org/10.1002/elsc.200420026

Geng, Z.-Q., Qian, D.-K., Hu, Z.-Y., Wang, S., Yan, Y., Van Loosdrecht, M. C. M., … & Zhang, F. (2021). Identification of extracellular key enzyme and intracellular metabolic pathway in alginate-degrading consortia via an integrated metaproteomic/metagenomic analysis. *Environmental Science & Technology*, *55*(24), 16636–16645. https://doi.org/10.1021/acs.est.1c05289

Giovagnoli, S., Blasi, P., Ricci, M., & Rossi, C. (2004). Biodegradable microspheres as carriers for native superoxide dismutase and catalase delivery. *AAPS PharmSciTech*, *5*(4), 1–9. https://doi.org/10.1208/pt050451

Godjevargova, T., & Gabrovska, K. (2003). Immobilization of urease onto chemically modified acrylonitrile copolymer membranes. *Journal of Biotechnology*, *103*(2), 107–111. https://doi.org/10.1016/S0168-1656(03)00107-X

Grano, V., Diano, N., Rossi, S., Portaccio, M., Attanasio, A., Cermola, M., … & Mita, D. G. (2004). *Production of Low-Lactose Milk by Means of Nonisothermal Bioreactors, 20*, 1393–1401.

Hand, S., & Cusick, R. D. (2021). Electrochemical disinfection in water and wastewater treatment: identifying impacts of water quality and operating conditions on performance. *Environmental Science & Technology*, *55*(6), 3470–3482. https://doi.org/10.1021/acs.est.0c06254

Harb, M., & Hong, P.-Y. (2017). Molecular-based detection of potentially pathogenic bacteria in membrane bioreactor (MBR) systems treating municipal wastewater: A case study. *Environmental Science and Pollution Research*, *24*(6), 5370–5380. https://doi.org/10.1007/s11356-016-8211-y

Heikinheimo, L., Buchert, J., Miettinen-Oinonen, A., & Suominen, P. (2000). Treating denim fabrics with trichoderma reesei cellulases. *Textile Research Journal*, *70*(11), 969–973. https://doi.org/10.1177/004051750007001106

Himma, N. F., Wardani, A. K., Prasetya, N., Aryanti, P. T. P., & Wenten, I. G. (2019). Recent progress and challenges in membrane-based O_2 /N_2 separation. *Reviews in Chemical Engineering*, *35*(5), 591–625. https://doi.org/10.1515/revce-2017-0094

Holt, M. S. (2000). Sources of chemical contaminants and routes into the freshwater environment. *Food and Chemical Toxicology*, *38*, S21–S27. https://doi.org/10.1016/S0278-6915(99)00136-2

Hoq, M. M., Koike, M., Yamane, T., & Shimizu, S. (1985). Continuous hydrolysis of olive oil by lipase in microporous hydrophobic hollow fiber bioreactor. *Agricultural and Biological Chemistry*, *49*(11), 3171–3178. https://doi.org/10.1080/00021369.1985.10867235

Hosseini, A., Ramezani, S., Tabibiazar, M., Mohammadi, M., Golchinfar, Z., Mahmoudzadeh, M., & Jahanban-Esfahlan, A. (2022). Immobilization of α-amylase in ethylcellulose electrospun fibers using emulsion-electrospinning method. *Carbohydrate Polymers*, *278*, 118919. https://doi.org/10.1016/j.carbpol.2021.118919

Iorhemen, O., Hamza, R., & Tay, J. (2016). Membrane bioreactor (MBR) technology for wastewater treatment and reclamation: Membrane fouling. *Membranes*, *6*(2), 33. https://doi.org/10.3390/membranes6020033

İspirli Doğaç, Y., Deveci, İ., Mercimek, B., & Teke, M. (2017). A comparative study for lipase immobilization onto alginate based composite electrospun nanofibers with effective and enhanced stability. *International Journal of Biological Macromolecules*, *96*, 302–311. https://doi.org/10.1016/j.ijbiomac.2016.11.120

Jaafar, R., Al-sulami, A., Al Taee, A., Aldoghachi, F., & Napes, S. (2015). Biosorption and bioaccumulation of some heavy metals by deinococcus radiodurans isolated from soil in Basra Governorate, Iraq. *Journal of Biotechnology & Biomaterials*, *05*(02). https://doi.org/10.4172/2155-952X.1000190

Jeong, S.-Y., & Kim, T. G. (2019). Development of a novel methanotrophic process with the helper micro-organism *Hyphomicrobium* sp. NM3. *Journal of Applied Microbiology*, *126*(2), 534–544. https://doi.org/10.1111/jam.14140

Jhaveri, J. H., & Murthy, Z. (2016). A comprehensive review on anti-fouling nanocomposite membranes for pressure driven membrane separation processes. *Desalination*, *379*, 137–154.

Jochems, P., Satyawali, Y., Diels, L., & Dejonghe, W. (2011). Enzyme immobilization on/in polymeric membranes: Status, challenges and perspectives in biocatalytic membrane reactors (BMRs). *Green Chemistry*, *13*(7), 1609. https://doi.org/10.1039/c1gc15178a

Jorge, R. M. F., & Livingston, A. G. (2000). Biological treatment of an alternating source of organic compounds in a single tube extractive membrane bioreactor. *Journal of Chemical Technology & Biotechnology*, *75*(12), 1174–1182. https://doi.org/10.1002/1097-4660(200012)75:12<1174::AID-JCTB339>3.0.CO;2-2

Judd, S. (2008). The status of membrane bioreactor technology. *Trends in Biotechnology*, *26*(2), 109–116. https://doi.org/10.1016/j.tibtech.2007.11.005

Kayvani Fard, A., McKay, G., Buekenhoudt, A., Al Sulaiti, H., Motmans, F., Khraisheh, M., & Atieh, M. (2018). Inorganic membranes: Preparation and application for water treatment and desalination. *Materials*, *11*(1), 74.

Khoo, K.-M., & Ting, Y.-P. (2001). Biosorption of gold by immobilized fungal biomass. *Biochemical Engineering Journal*, *8*(1), 51–59. https://doi.org/10.1016/S1369-703X(00)00134-0

Koltuniewicz, A. B., & Bezak, K. (2002). *Engineering of Membrane Biosorption, 144*, 219–226.

Koltuniewicz, A. B., Witek, A., & Bezak, K. (2004). Efficiency of membrane-sorption integrated processes. *Journal of Membrane Science*, *239*(1), 129–141. https://doi.org/10.1016/j.memsci.2004.02.037

Koul, B., Bhat, N., Abubakar, M., Mishra, M., Arukha, A. P., & Yadav, D. (2022). Application of natural coagulants in water treatment: A sustainable alternative to chemicals. *Water*, *14*(22), 3751. https://doi.org/10.3390/w14223751

Kouzuma, A., Kato, S., & Watanabe, K. (2015). Microbial interspecies interactions: Recent findings in syntrophic consortia. *Frontiers in Microbiology*, *6*. https://doi.org/10.3389/fmicb.2015.00477

Kunacheva, C., Soh, Y. N. A., & Stuckey, D. C. (2020). Identification of soluble microbial products (SMPs) from the fermentation and methanogenic phases of anaerobic digestion. *Science of the Total Environment*, *698*, 134177. https://doi.org/10.1016/j.scitotenv.2019.134177

Lares, M., Ncibi, M. C., Sillanpää, M., & Sillanpää, M. (2018). Occurrence, identification and removal of microplastic particles and fibers in conventional activated sludge process and advanced MBR technology. *Water Research*, *133*, 236–246. https://doi.org/10.1016/j.watres.2018.01.049

Lawson, C. E., Harcombe, W. R., Hatzenpichler, R., Lindemann, S. R., Löffler, F. E., O'Malley, M. A., … & McMahon, K. D. (2019). Common principles and best practices for engineering microbiomes. *Nature Reviews Microbiology*, *17*(12), 725–741. https://doi.org/10.1038/s41579-019-0255-9

Liao, Y., Goh, S., Tian, M., Wang, R., & Fane, A. G. (2018). Design, development and evaluation of nanofibrous composite membranes with opposing membrane wetting properties for extractive membrane bioreactors. *Journal of Membrane Science*, *551*, 55–65. https://doi.org/10.1016/j.memsci.2018.01.029

Lin, L., Yang, H., & Xu, X. (2022). Effects of water pollution on human health and disease heterogeneity: A review. *Frontiers in Environmental Science*, *10*. https://doi.org/10.3389/fenvs.2022.880246

Lin, Y. S. (2019). Inorganic membranes for process intensification: Challenges and perspective. *Industrial & Engineering Chemistry Research*, *58*(15), 5787–5796. https://doi.org/10.1021/acs.iecr.8b04539

Liu, Y., Zeng, Z., Zeng, G., Tang, L., Pang, Y., Li, Z., … & Xie, G. (2012). Immobilization of laccase on magnetic bimodal mesoporous carbon and the application in the removal of phenolic compounds. *Bioresource Technology*, *115*, 21–26. https://doi.org/10.1016/j.biortech.2011.11.015

Luo, J., Meyer, A. S., Jonsson, G., & Pinelo, M. (2013). Fouling-induced enzyme immobilization for membrane reactors. *Bioresource Technology*, *147*, 260–268. https://doi.org/10.1016/j.biortech.2013.08.019

Luo, J., Meyer, A. S., Jonsson, G., & Pinelo, M. (2014). Enzyme immobilization by fouling in ultrafiltration membranes: Impact of membrane configuration and type on flux behavior and biocatalytic conversion efficacy. *Biochemical Engineering Journal*, *83*, 79–89. https://doi.org/10.1016/j.bej.2013.12.007

Ma, R., Li, J., Zeng, P., Duan, L., Dong, J., Ma, Y., & Yang, L. (2024). The application of membrane separation technology in the pharmaceutical industry. *Membranes*, *14*(1), 24.

Maghraby, Y. R., El-Shabasy, R. M., Ibrahim, A. H., & Azzazy, H. M. E.-S. (2023). Enzyme immobilization technologies and industrial applications. *ACS Omega*, *8*(6), 5184–5196. https://doi.org/10.1021/acsomega.2c07560

Mansi, A. E., El-Marsafy, S. M., Elhenawy, Y., & Bassyouni, M. (2023). Assessing the potential and limitations of membrane-based technologies for the treatment of oilfield produced water. *Alexandria Engineering Journal*, *68*, 787–815. https://doi.org/10.1016/j.aej.2022.12.013

Meena, M., Yadav, G., Sonigra, P., & Shah, M. P. (2022). A comprehensive review on application of bioreactor for industrial wastewater treatment. *Letters in Applied Microbiology*, *74*(2), 131–158. https://doi.org/10.1111/lam.13557

Meller, K., Szumski, M., & Buszewski, B. (2017). Microfluidic reactors with immobilized enzymes—Characterization, dividing, perspectives. *Sensors and Actuators B: Chemical*, *244*, 84–106. https://doi.org/10.1016/j.snb.2016.12.021

Meng, F., Chae, S.-R., Drews, A., Kraume, M., Shin, H.-S., & Yang, F. (2009). Recent advances in membrane bioreactors (MBRs): Membrane fouling and membrane material. *Water Research*, *43*(6), 1489–1512. https://doi.org/10.1016/j.watres.2008.12.044

Mine, Y., Fukunaga, K., Itoh, K., Yoshimoto, M., Nakao, K., & Sugimura, Y. (2003). Enhanced enzyme activity and enantioselectivity of lipases in organic solvents by crown ethers and cyclodextrins. *Journal of Bioscience and Bioengineering*, *95*(5), 441–447. https://doi.org/10.1016/S1389-1723(03)80042-7

Morthensen, S. T., Meyer, A. S., Jørgensen, H., & Pinelo, M. (2017). Significance of membrane bioreactor design on the biocatalytic performance of glucose oxidase and catalase: Free vs. immobilized enzyme systems. *Biochemical Engineering Journal*, *117*, 41–47. https://doi.org/10.1016/j.bej.2016.09.015

Moslemi, M., Homaei, A., & Toiserkani, H. (2018). Aspartic acid introduce the functional amine groups on the surface of superparamagnetic Fe(OH)3@Fe3O4 nanoparticles for efficient immobilization of Penaeus vannamei protease. *Bioprocess and Biosystems Engineering*, *41*(6), 749–756. https://doi.org/10.1007/s00449-018-1908-1

Mutamim, N. S. A., Noor, Z. Z., Hassan, M. A. A., Yuniarto, A., & Olsson, G. (2013). Membrane bioreactor: Applications and limitations in treating high strength industrial wastewater. *Chemical Engineering Journal*, *225*, 109–119. https://doi.org/10.1016/j.cej.2013.02.131

Nain, A., Sangili, A., Hu, S.-R., Chen, C.-H., Chen, Y.-L., & Chang, H.-T. (2022). Recent progress in nanomaterial-functionalized membranes for removal of pollutants. *iScience*, *25*(7), 104616. https://doi.org/10.1016/j.isci.2022.104616

Naseer, S., Ouyang, J., Chen, X., Pu, S., Guo, Y., Zhang, X., … & Yang, C. (2020). Immobilization of β-glucosidase by self-catalysis and compared to crosslinking with glutaraldehyde. *International Journal of Biological Macromolecules*, *154*, 1490–1495. https://doi.org/10.1016/j.ijbiomac.2019.11.030

Nasr, M. (2022). Chapter 2—Aquatic pollution and wastewater treatment system. In V. C. Pandey (Ed.), *Algae and aquatic macrophytes in cities* (pp. 23–37). Elsevier. https://doi.org/10.1016/B978-0-12-824270-4.00006-7

Niu, J., Xu, J., Dai, Y., Xu, J., Guo, H., Sun, K., & Liu, R. (2013). Immobilization of horseradish peroxidase by electrospun fibrous membranes for adsorption and degradation of pentachlorophenol in water. *Journal of Hazardous Materials*, *246–247*, 119–125. https://doi.org/10.1016/j.jhazmat.2012.12.023

Obotey Ezugbe, E., & Rathilal, S. (2020). Membrane technologies in wastewater treatment: A review. *Membranes*, *10*(5), 89. https://doi.org/10.3390/membranes10050089

Okamura, D., Mori, Y., Hashimoto, T., & Hori, K. (2010). Effects of microbial degradation of biofoulants on microfiltration membrane performance in a membrane bioreactor. *Environmental Science & Technology*, *44*(22), 8644–8648. https://doi.org/10.1021/es102321m

Olajuyigbe, F. M., Adetuyi, O. Y., & Fatokun, C. O. (2019). Characterization of free and immobilized laccase from Cyberlindnera fabianii and application in degradation of bisphenol A. *International Journal of Biological Macromolecules*, *125*, 856–864. https://doi.org/10.1016/j.ijbiomac.2018.12.106

Ozaki, S., Kishimoto, N., & Fujita, T. (2006). Isolation and phylogenetic characterization of microbial consortia able to degrade aromatic hydrocarbons at high rates. *Microbes and Environments*, *21*(1), 44–52. https://doi.org/10.1264/jsme2.21.44

Pagnanelli, F., Beolchini, F., Di Biase, A., & Vegliò, F. (2003). Effect of equilibrium models in the simulation of heavy metal biosorption in single and two-stage UF/MF membrane reactor systems. *Biochemical Engineering Journal*, *15*(1), 27–35. https://doi.org/10.1016/S1369-703X(02)00179-1

Pagnanelli, F., Beolchini, F., Di Biase, A., & Vegliò, F. (2004). Biosorption of binary heavy metal systems onto Sphaerotilus natans cells confined in an UF/MF membrane reactor: Dynamic simulations by different Langmuir-type competitive models. *Water Research*, *38*(4), 1055–1061. https://doi.org/10.1016/j.watres.2003.10.031

Pagnanelli, F., Beolchini, F., Esposito, A., Toro, L., & Vegliò, F. (2003). Mechanistic modeling of heavy metal biosorption in batch and membrane reactor systems. *Hydrometallurgy*, *71*(1–2), 201–208. https://doi.org/10.1016/S0304-386X(03)00157-9

Palai, T., & Bhattacharya, P. K. (2013). Kinetics of lactose conversion to galacto-oligosaccharides by β-galactosidase immobilized on PVDF membrane. *Journal of Bioscience and Bioengineering*, *115*(6), 668–673. https://doi.org/10.1016/j.jbiosc.2012.12.014

Patowary, K., Patowary, R., Kalita, M. C., & Deka, S. (2016). Development of an efficient bacterial consortium for the potential remediation of hydrocarbons from contaminated sites. *Frontiers in Microbiology*, *7*. https://doi.org/10.3389/fmicb.2016.01092

Pedroso, M. M., Hine, D., Hahn, S., Chmielewicz, W. M., Diegel, J., Gahan, L., & Schenk, G. (2021). Pesticide degradation by immobilised metalloenzymes provides an attractive avenue for bioremediation. *EFB Bioeconomy Journal*, *1*, 100015. https://doi.org/10.1016/j.bioeco.2021.100015

Pellegrin, M.-L. (2015). Application of membrane bioreactor processes for achieving low effluent nutrient concentrations. *Water Intelligence Online*, *14*. https://doi.org/10.2166/9781780406756

Piletsky, S., Piletska, E., Bossi, A., Turner, N., & Turner, A. (2003). Surface functionalization of porous polypropylene membranes with polyaniline for protein immobilization. *Biotechnology and Bioengineering*, *82*(1), 86–92. https://doi.org/10.1002/bit.10544

Qi, J., Zhou, Q., Huang, D., Yu, Z., & Meng, F. (2023). Construction of synthetic anti-fouling consortia: Fouling control effects and polysaccharide degradation mechanisms. *Microbial Cell Factories*, *22*(1), 230. https://doi.org/10.1186/s12934-023-02235-7

Qiu, G., & Ting, Y.-P. (2014). Short-term fouling propensity and flux behavior in an osmotic membrane bioreactor for wastewater treatment. *Desalination*, *332*(1), 91–99. https://doi.org/10.1016/j.desal.2013.11.010

Qiu, G., Zhang, S., Srinivasa Raghavan, D. S., Das, S., & Ting, Y.-P. (2016). The potential of hybrid forward osmosis membrane bioreactor (FOMBR) processes in achieving high throughput treatment of municipal wastewater with enhanced phosphorus recovery. *Water Research*, *105*, 370–382. https://doi.org/10.1016/j.watres.2016.09.017

Radu, E. R., Voicu, S. I., & Thakur, V. K. (2023). Polymeric membranes for biomedical applications. *Polymers*, *15*(3), 619. https://doi.org/10.3390/polym15030619

Rajaram, T., & Das, A. (2008). Water pollution by industrial effluents in India: Discharge scenarios and case for participatory ecosystem specific local regulation. *Futures*, *40*(1), 56–69. https://doi.org/10.1016/j.futures.2007.06.002

Sagle, A., & Freeman, B. (2004). Fundamentals of membranes for water treatment. *The Future of Desalination in Texas*, *2*(363), 137.

Sassolas, A., Blum, L. J., & Leca-Bouvier, B. D. (2012). Immobilization strategies to develop enzymatic biosensors. *Biotechnology Advances*, *30*(3), 489–511. https://doi.org/10.1016/j.biotechadv.2011.09.003

Sathya, K., Nagarajan, K., Carlin Geor Malar, G., Rajalakshmi, S., & Raja Lakshmi, P. (2022). A comprehensive review on comparison among effluent treatment methods and modern methods of treatment of industrial wastewater effluent from different sources. *Applied Water Science*, *12*(4), 70. https://doi.org/10.1007/s13201-022-01594-7

Schwaminger, S. P., Fehn, S., Steegmüller, T., Rauwolf, S., Löwe, H., Pflüger-Grau, K., & Berensmeier, S. (2021). Immobilization of PETase enzymes on magnetic iron oxide nanoparticles for the decomposition of microplastic PET. *Nanoscale Advances*, *3*(15), 4395–4399. https://doi.org/10.1039/D1NA00243K

Seetharam, G. B., & Saville, B. A. (2003). Degradation of phenol using tyrosinase immobilized on siliceous supports. *Water Research*, *37*(2), 436–440. https://doi.org/10.1016/S0043-1354(02)00290-7

Sekuljica, N., Prlainovic, N., Jovanovic, J., Stefanovic, A., Grbavcic, S., Mijin, D., & Knezevic-Jugovic, Z. (2016). Immobilization of horseradish peroxidase onto kaolin by glutaraldehyde method and its application in decolorization of anthraquinone dye. *Hemijska Industrija*, *70*(2), 217–224. https://doi.org/10.2298/HEMIND150220028S

Sellami, K., Couvert, A., Nasrallah, N., Maachi, R., Tandjaoui, N., Abouseoud, M., & Amrane, A. (2021). Bio-based and cost effective method for phenolic compounds removal using cross-linked enzyme aggregates. *Journal of Hazardous Materials*, *403*, 124021. https://doi.org/10.1016/j.jhazmat.2020.124021

Shahkaramipour, N., Tran, T. N., Ramanan, S., & Lin, H. (2017). Membranes with surface-enhanced antifouling properties for water purification. *Membranes*, *7*(1), 13.

Sharma, R., Chisti, Y., & Banerjee, U. C. (2001). Production, purification, characterization, and applications of lipases. *Biotechnology Advances*, *19*(8), 627–662. https://doi.org/10.1016/S0734 9750(01)00086-6

Shayo, G. M., Elimbinzi, E., Shao, G. N., & Fabian, C. (2023). Severity of waterborne diseases in developing countries and the effectiveness of ceramic filters for improving water quality. *Bulletin of the National Research Centre*, *47*(1), 113. https://doi.org/10.1186/s42269-023-01088-9

Shinde, P., Musameh, M., Gao, Y., Robinson, A. J., & Kyratzis, I. (Louis). (2018). Immobilization and stabilization of alcohol dehydrogenase on polyvinyl alcohol fibre. *Biotechnology Reports*, *19*, e00260. https://doi.org/10.1016/j.btre.2018.e00260

Shoshaa, R., Ashfaq, M. Y., & Al-Ghouti, M. A. (2023). Recent developments in ultrafiltration membrane technology for the removal of potentially toxic elements, and enhanced anti-fouling performance: A review. *Environmental Technology & Innovation*, *31*, 103162. https://doi.org/10.1016/j.eti.2023.103162

Siagian, U. W. R., Raksajati, A., Himma, N. F., Khoiruddin, K., & Wenten, I. G. (2019). Membrane-based carbon capture technologies: Membrane gas separation vs. membrane contactor. *Journal of Natural Gas Science and Engineering*, *67*, 172–195. https://doi.org/10.1016/j.jngse.2019.04.008

Singh, B. J., Chakraborty, A., & Sehgal, R. (2023). A systematic review of industrial wastewater management: Evaluating challenges and enablers. *Journal of Environmental Management, 348*, 119230. https://doi.org/10.1016/j.jenvman.2023.119230

Singh, R., Ryu, J., & Kim, S. W. (2019). Microbial consortia including methanotrophs: Some benefits of living together. *Journal of Microbiology, 57*(11), 939–952. https://doi.org/10.1007/s12275-019-9328-8

Skariyachan, S., Patil, A. A., Shankar, A., Manjunath, M., Bachappanavar, N., & Kiran, S. (2018). Enhanced polymer degradation of polyethylene and polypropylene by novel thermophilic consortia of Brevibacillus sps. And Aneurinibacillus sp. Screened from waste management landfills and sewage treatment plants. *Polymer Degradation and Stability, 149*, 52–68. https://doi.org/10.1016/j.polymdegradstab.2018.01.018

Soares, J. C., Moreira, P. R., Queiroga, A. C., Morgado, J., Malcata, F. X., & Pintado, M. E. (2011). Application of immobilized enzyme technologies for the textile industry: A review. *Biocatalysis and Biotransformation, 29*(6), 223–237. https://doi.org/10.3109/10242422.2011.635301

Srinivasa Raghavan, D. S., Qiu, G., & Ting, Y.-P. (2018). Fate and removal of selected antibiotics in an osmotic membrane bioreactor. *Chemical Engineering Journal, 334*, 198–205. https://doi.org/10.1016/j.cej.2017.10.026

Su, X., Wang, Y., Xue, B., Hashmi, M. Z., Lin, H., Chen, J., … & Sun, F. (2019). Impact of resuscitation promoting factor (Rpf) in membrane bioreactor treating high-saline phenolic wastewater: Performance robustness and Rpf-responsive bacterial populations. *Chemical Engineering Journal, 357*, 715–723. https://doi.org/10.1016/j.cej.2018.09.197

Sulaiman, S., Mokhtar, M. N., Naim, M. N., Baharuddin, A. S., & Sulaiman, A. (2015). A review: Potential usage of cellulose nanofibers (CNF) for enzyme immobilization via covalent interactions. *Applied Biochemistry and Biotechnology, 175*(4), 1817–1842. https://doi.org/10.1007/s12010-014-1417-x

Sun, L., Wang, Z., Wei, X., Li, P., Zhang, H., Li, M., … & Wang, S. (2015). Enhanced biological nitrogen and phosphorus removal using sequencing batch membrane-aerated biofilm reactor. *Chemical Engineering Science, 135*, 559–565. https://doi.org/10.1016/j.ces.2015.07.033

Taheran, M., Naghdi, M., Brar, S. K., Knystautas, E. J., Verma, M., & Surampalli, R. Y. (2017). Degradation of chlortetracycline using immobilized laccase on Polyacrylonitrile-biochar composite nanofibrous membrane. *Science of the Total Environment, 605–606*, 315–321. https://doi.org/10.1016/j.scitotenv.2017.06.185

Talvitie, J., Mikola, A., Koistinen, A., & Setälä, O. (2017). Solutions to microplastic pollution – Removal of microplastics from wastewater effluent with advanced wastewater treatment technologies. *Water Research, 123*, 401–407. https://doi.org/10.1016/j.watres.2017.07.005

Tang, C. Y., Kwon, Y.-N., & Leckie, J. O. (2009). Effect of membrane chemistry and coating layer on physiochemical properties of thin film composite polyamide RO and NF membranes: I. FTIR and XPS characterization of polyamide and coating layer chemistry. *Desalination, 242*(1–3), 149–167.

Tay, J. H., Yang, P., Zhuang, W. Q., Tay, S. T. L., & Pan, Z. H. (2007). Reactor performance and membrane filtration in aerobic granular sludge membrane bioreactor. *Journal of Membrane Science, 304*(1–2), 24–32. https://doi.org/10.1016/j.memsci.2007.05.028

Ting, Y. P., & Sun, G. (2000). Comparative study on polyvinyl alcohol and alginate for cell immobilization in biosorption. *Water Science and Technology, 42*(5–6), 85–90. https://doi.org/10.2166/wst.2000.0499

Tshikantwa, T. S., Ullah, M. W., He, F., & Yang, G. (2018). Current trends and potential applications of microbial interactions for human welfare. *Frontiers in Microbiology, 9*, 1156. https://doi.org/10.3389/fmicb.2018.01156

Tyagi, C., Tomar, L. K., & Singh, H. (2009). Surface modification of cellulose filter paper by glycidyl methacrylate grafting for biomolecule immobilization: Influence of grafting parameters and urease immobilization. *Journal of Applied Polymer Science*, *111*(3), 1381–1390. https://doi.org/10.1002/app.28933

Ugur Nigiz, F. (2016). Rhizomucor miehei lipase-immobilized sodium alginate membrane preparation and usage in a pervaporation biocatalytic membrane reactor. *Chemical and Biochemical Engineering Quarterly Journal*, *30*(3), 381–391. https://doi.org/10.15255/CABEQ.2016.2345

Van Der Bruggen, B., Vandecasteele, C., Van Gestel, T., Doyen, W., & Leysen, R. (2003). A review of pressure-driven membrane processes in wastewater treatment and drinking water production. *Environmental Progress*, *22*(1), 46–56. https://doi.org/10.1002/ep.670220116

Veglio, F., Beolchini, F., & Barba, D. (2000). Experimental study and simulation on the biosorption of copper(ii) in membrane reactors: A preliminary study. *Industrial & Engineering Chemistry Research*, *39*(7), 2480–2484. https://doi.org/10.1021/ie990786d

Veglio, F., Beolchini, F., Quaresima, R., & Toro, L. (1999). Biosorption of toxic metals by immobilised biomass and UF/MF membrane reactor. In *Process Metallurgy* (Vol. 9, pp. 463–472). Elsevier. https://doi.org/10.1016/S1572-4409(99)80135-7

Velizarov, S., Matos, C., Oehmen, A., Serra, S., Reis, M., & Crespo, J. (2008). Removal of inorganic charged micropollutants from drinking water supplies by hybrid ion exchange membrane processes. *Desalination*, *223*(1–3), 85–90. https://doi.org/10.1016/j.desal.2007.01.217

Verma, B., Balomajumder, C., Sabapathy, M., & Gumfekar, S. P. (2021). Pressure-driven membrane process: A review of advanced technique for heavy metals remediation. *Processes*, *9*(5), 752.

Volesky, B. (2001). Detoxification of metal-bearing effluents: Biosorption for the next century. *Hydrometallurgy*, *59*(2–3), 203–216. https://doi.org/10.1016/S0304-386X(00)00160-2

Waghmode, T. R., Kurade, M. B., Khandare, R. V., & Govindwar, S. P. (2011). A sequential aerobic/microaerophilic decolorization of sulfonated mono azo dye Golden Yellow HER by microbial consortium GG-BL. *International Biodeterioration & Biodegradation*, *65*(7), 1024–1034. https://doi.org/10.1016/j.ibiod.2011.08.002

Wang, L., He, J., Heiranian, M., Fan, H., Song, L., Li, Y., & Elimelech, M. (2023). Water transport in reverse osmosis membranes is governed by pore flow, not a solution-diffusion mechanism. *Science Advances*, *9*(15), eadf8488.

Wang, N., & Choi, Y. (2019). Challenges for sustainable water use in the urban industry of Korea based on the global non-radial directional distance function model. *Sustainability*, *11*(14), 3895. https://doi.org/10.3390/su11143895

Wang, V. B., Chua, S.-L., Cai, Z., Sivakumar, K., Zhang, Q., Kjelleberg, S., … & Yang, L. (2014). A stable synergistic microbial consortium for simultaneous azo dye removal and bioelectricity generation. *Bioresource Technology*, *155*, 71–76. https://doi.org/10.1016/j.biortech.2013.12.078

Wang, Y., Shao, H., Zhu, S., Tian, K., Qiu, Q., & Huo, H. (2019). Degradation of 17β-estradiol and products by a mixed culture of *Rhodococcus equi* DSSKP-R-001 and *Comamonas testosteroni* QYY20150409. *Biotechnology & Biotechnological Equipment*, *33*(1), 268–277. https://doi.org/10.1080/13102818.2019.1568913

Wenten, I. G., Khoiruddin, K., Harimawan, A., Ting, Y. P., & Boopathy, R. (2020). Membrane biosorption: Recent advances and challenges. *Current Pollution Reports*, *6*(2), 152–172. https://doi.org/10.1007/s40726-020-00145-5

Winartasaputra, Handani., Kuan, S. S., & Guilbault, G. G. (1982). Amperometric enzymic determination of triglycerides in serum. *Analytical Chemistry*, *54*(12), 1987–1990. https://doi.org/10.1021/ac00249a017

Wu, Q., Xu, Z., Duan, Y., Zhu, Y., Ou, M., & Xu, X. (2017). Immobilization of tyrosinase on polyacrylonitrile beads: Biodegradation of phenol from aqueous solution and the relevant cytotoxicity assessment. *RSC Advances*, *7*(45), 28114–28123. https://doi.org/10.1039/C7RA03174B

Xagoraraki, I., Yin, Z., & Svambayev, Z. (2014). Fate of viruses in water systems. *Journal of Environmental Engineering*, *140*(7), 04014020. https://doi.org/10.1061/(ASCE)EE.1943-7870.0000827

Xiao, K., Liang, S., Wang, X., Chen, C., & Huang, X. (2019). Current state and challenges of full-scale membrane bioreactor applications: A critical review. *Bioresource Technology*, *271*, 473–481. https://doi.org/10.1016/j.biortech.2018.09.061

Xiao, K., Xu, Y., Liang, S., Lei, T., Sun, J., Wen, X., … & Huang, X. (2014). Engineering application of membrane bioreactor for wastewater treatment in China: Current state and future prospect. *Frontiers of Environmental Science & Engineering*, *8*(6), 805–819. https://doi.org/10.1007/s11783-014-0756-8

Xu, R., Yu, Z., Zhang, S., & Meng, F. (2019). Bacterial assembly in the bio-cake of membrane bioreactors: Stochastic vs. deterministic processes. *Water Research*, *157*, 535–545. https://doi.org/10.1016/j.watres.2019.03.093

Yu, J., Gu, K., Yang, B., Wang, K., Zhou, Y., & Gao, C. (2021). The permeability and selectivity of the polyamide reverse osmosis membrane were significantly enhanced by PhSiCl(3). *Membranes*, *11*(2), 142. https://doi.org/10.3390/membranes11020142

Yuan, Y., Shen, J., & Salmon, S. (2023). Developing enzyme immobilization with fibrous membranes: Longevity and characterization considerations. *Membranes*, *13*(5), 532. https://doi.org/10.3390/membranes13050532

Zdarta, J., Degórska, O., Jankowska, K., Rybarczyk, A., Piasecki, A., Ciesielczyk, F., & Jesionowski, T. (2021). Removal of persistent sulfamethoxazole and carbamazepine from water by horseradish peroxidase encapsulated into poly(vinyl chloride) electrospun fibers. *International Journal of Molecular Sciences*, *23*(1), 272. https://doi.org/10.3390/ijms23010272

Zdarta, J., Meyer, A. S., Jesionowski, T., & Pinelo, M. (2018). Developments in support materials for immobilization of oxidoreductases: A comprehensive review. *Advances in Colloid and Interface Science*, *258*, 1–20. https://doi.org/10.1016/j.cis.2018.07.004

Zeuner, B., Luo, J., Nyffenegger, C., Aumala, V., Mikkelsen, J. D., & Meyer, A. S. (2014). Optimizing the biocatalytic productivity of an engineered sialidase from Trypanosoma rangeli for 3′-sialyllactose production. *Enzyme and Microbial Technology*, *55*, 85–93. https://doi.org/10.1016/j.enzmictec.2013.12.009

Zhang, H., & Wang, X. (2016). Modular co-culture engineering, a new approach for metabolic engineering. *Metabolic Engineering*, *37*, 114–121. https://doi.org/10.1016/j.ymben.2016.05.007

Zhang, W., Tang, B., & Bin, L. (2017). Research progress in biofilm-membrane bioreactor: A critical review. *Industrial & Engineering Chemistry Research*, *56*(24), 6900–6909. https://doi.org/10.1021/acs.iecr.7b00794

Zhang, Z., Lin, B., Xia, S., Wang, X., & Yang, A. (2007). Production and application of a novel bioflocculant by multiple-microorganism consortia using brewery wastewater as carbon source. *Journal of Environmental Sciences*, *19*(6), 667–673. https://doi.org/10.1016/S1001-0742(07)60112-0

Zhao, G., Rui, K., Dou, S. X., & Sun, W. (2018). Heterostructures for electrochemical hydrogen evolution reaction: A review. *Advanced Functional Materials*, *28*(43), 1803291. https://doi.org/10.1002/adfm.201803291

Zucca, P., Fernandez-Lafuente, R., & Sanjust, E. (2016). Agarose and its derivatives as supports for enzyme immobilization. *Molecules*, *21*(11), 1577. https://doi.org/10.3390/molecules21111577

Recent Advances in Nanotechnology and Nanofiltration Methods for Industrial Wastewater Treatment

6

Bhairav Prasad, Palki Sahib Kaur, Tripti Malik, Prerna Rana and Abhinandan Sharma

1. INTRODUCTION

Clean and safe water is the basic need of humans, animals, plants and other forms of life for their harmonious development and development of a sustained ecosystem. Many natural and anthropogenic activities, including mining, volcanic eruption, acid drainage, rapid industrialization, rise in population and agriculture practices, use water and generate a huge amount of wastewater. The wastewater generated by such practices contain many toxic and hazardous chemicals, and upon exposure to natural systems, i.e., soil, water bodies, river, groundwater and sea, contaminate them deleteriously and impact their toxicity to the consumer. Wastewater contains many toxic compounds such as carcinogens, mutagens, hydrocarbons, heavy metals and pathogenic microorganisms. Every year, millions of people die worldwide due to illness acquired through the consumption of wastewater (Jain et al., 2021). Therefore, it is our fundamental duty to treat and detoxify this wastewater before it is discharged into the natural ecosystem. Many traditional methods have been well-established and applied for the treatment of wastewater. Conventional wastewater methods, including, adsorption, coagulation,

DOI: 10.1201/9781003496014-6

flocculation, chlorination, ion-exchange, filtration, aerobic and anaerobic, have been routinely employed. However, it has also been observed that the conventional methods are not efficient in removing many toxicants such as nitrates, phosphorous and heavy metals significantly, plus these methods are time consuming, laborious rendering their restricted use under certain conditions (Rajasulochana & Preethy, 2016; Zinicovscaia, 2016). Recently, more emphasis has been given on nanomaterials and nanotechnology-based methods for wastewater treatments. Nanotechnology fabricates and manipulates materials at the molecular or atomic level to synthesized molecules with new dimension, structure, and with superior optical, magnetic, electrical, mechanical and conductive properties which potentially improve wastewater treatment above conventional methods (Khan et al., 2019). Additionally, nanomaterials-based methods facilitate a high surface-to-volume ratio, superior reactivity, sensitivity, adsorption capacity and ease to synthesize, which make it appropriate for wastewater reclamation. In this chapter, we will discuss the nanotechnology-based wastewater treatments methods such as nanofiltration, nanozymes, adsorption photocatalysis, disinfection and sensing technology (Jain et al., 2021). We will also highlight the production cost and risk associated with application of nanomaterials in wastewater treatment.

2. SOURCES OF WASTEWATER WATER AND ITS COMPOSITION

Wastewater is generated from diverse sources. Natural sources include acid drainage and volcanic eruption. Anthropogenic activities, which are the major producers of wastewater, include mining, urbanization, rapid industrialization and pharmaceutical industries, domestic and municipal wastewater. The characteristics and composition of wastewater mainly depend on the type of origin. The composition of typical municipal wastewater is illustrated in Table 6.1.

2.1. Pharmaceutical Industry

The pharmaceutical industry is also a major contributor of aromatics and organic compounds that are considered as emerging pollutants (Samal et al., 2022; Aliste et al., 2022). The key ingredients found in pharmaceutical wastewater are hormone stimulants, β-blockers, antibiotics, painkillers, analgesics, anti-inflammatory drugs, blood–lipid regulators, cytostatic drugs, antidepressants and other drugs (Samal et al., 2022; Polianciuc et al., 2020). These pollutants are resistant to physical, chemical and light treatments; not degraded by conventional methods; and remain present in the ecosystem. For a long time thereafter they contaminate surface and groundwater (Frascaroli et al., 2021). The persistent addition of these pharmaceutical compounds in aquatic ecosystems results in the emergence of antibiotic resistance and cause serious threats to aquatic life and eventually human health through trophic transfer (Yang et al., 2021).

TABLE 6.1 Composition of typical municipal wastewater.

COMPONENTS	*HEAVY METALS*	*EMERGING POLLUTANTS*	*MICROORGANISMS*
Nitrogen	Lead (Pb)	Personal care products	Bacteria
Phosphorous	Arsenic (As)	Microplastics	Protozoa
Organics	Mercury (Hg)	Pharmaceutical products	Fungi
Suspended solids	Cadmium (Cd)	Antibiotics	Virus
Dissolved solids	Nickle (Ni)	Microfibers	Nematodes
Oil and grease	Chromium (Cr)	Pesticides	Microalgae

2.2. Food Processing and Dairy Industry

Food processing and dairy industries are the imperative sector worldwide which generates huge amounts of wastewater and requires large amounts of water for daily operation. Many industries, such as fruit and vegetable industries, beverages and dairy industries, along with other edible products and their byproducts, generate harmful wastes and wastewater. A significant amount of wastewater is generated from many activities such as cleaning, cooling and production, operation rich in organic matter, high BOD, COD, TSS, total nitrogen and phosphorous. Sometimes wastewater may contain pathogenic bacteria and other toxic compounds, and is discharged in natural water bodies without proper treatment, which results in severe environmental and health hazards (Rani et al., 2024). The wastewater produced from the slaughter and meat industry is rich in organic matter. Their composition also depends on the frequency of slaughtering, type of meat being processed, the disinfectant used to maintain hygiene and the size of the plant (Keškić et al., 2024).

2.3. Domestic and Municipal Wastewater

Domestic wastewater is principally the water generated from residential settlements and flow out from kitchens, bathrooms and toilets. Eighty percent of the domestic wastewater worldwide is discharged to natural water bodies without adequate treatment. The composition of domestic wastewater mainly includes household chemicals, disinfectants, heavy metals, oil and grease, nutrients, bio-waste, detergents, bathing, soaps, medicinal residues and soluble organic matter (Aggrwal et al., 2022). Municipal wastewater is the water generated from cities and towns that primarily consists of organic, inorganic and suspended solids. These organic substances include lignin, carbohydrates, fats, soaps, oils, detergents, proteins and other synthetic organic chemicals. The inorganic contaminants present in municipal wastewater includes inorganic salts, their ions and heavy metals, which may have harmful effects on plants, animals, humans and the overall environment (Pereira et al., 2014: Meena et al., 2018).

2.4. Agricultural Practices

Agriculture practices directly and indirectly affect the quality of water and other aquatic ecosystems. The enormous application of pesticides, fertilizers and other sediments from point and non-point sources contaminate water bodies through surface runoff. The rapid increase in population has extended pressure on the agriculture industry to produce high yields, requiring them to apply more fertilizers and pesticides (Zahoor & Mustaq, 2023). Water pollution due to agriculture is a worldwide problem and equally affects developed and developing countries. Chemical-based fertilizers and pesticides leach from farmland to nearby lakes, rivers and ponds, polluting them and making these water bodies nutritionally rich, thereby causing eutrophication. The surface runoff from farmland may percolate and can cause pollution of groundwater (Babu et al., 2022). In developing countries, agriculture is the backbone and many farmers directly depend on agriculture as the prime source of income. Farmers in these countries cannot implement a sustainable agriculture practice due to lack of resources and technology. The traditional techniques of agriculture cause widespread pollution of water bodies and degrade the nearby ecosystem, causing health hazards for communities that rely on them (Khalil et al., 2022).

3. CONVENTIONAL WASTEWATER TREATMENT METHODS

Many conventional methods have been employed for a many years to treat wastewater in both developed and developing countries globally. The wastewater generated from various sources has different levels of toxicants and pose a serious threat to the environment if left untreated. Therefore, various conventional methods including primary, secondary and tertiary or advanced treatment methods are employed to tackle the various pollutants and contaminants present in wastewater. Different types of conventional methods used for wastewater treatments are illustrated in Figure 6.1.

Preliminary treatment of wastewater mainly includes screening of floatable material and grit removal from effluents, thereby preparing it for the further treatment process. The main objective of this treatment is to remove or reduce the size of pollutants, which are either suspended or floatable. These materials include floatable plastics and pieces of wood, paper, garbage, cloths, etc. Preliminary treatment also facilitates the removal of organic matter such as fecal matter, sand, gravel, metals and glass. It also removes excess amounts of oil and grease suspended in wastewater (Sonune & Ghate, 2004).

Primary wastewater treatments, also known as conditioning of the wastewater, includes sedimentation, floatation and neutralization of the wastewater that significantly reduces the organic load including oil, grease and lighter floatable materials present on the surface of the wastewater (Ramalho, 2012). Secondary wastewater is a biological treatment process including an activated sludge process, trickling filter, aerobic and anaerobic treatment, wastewater lagoons and oxidation ponds. The microbial consortia decompose the organic matter in the wastewater and significantly reduce the amount of

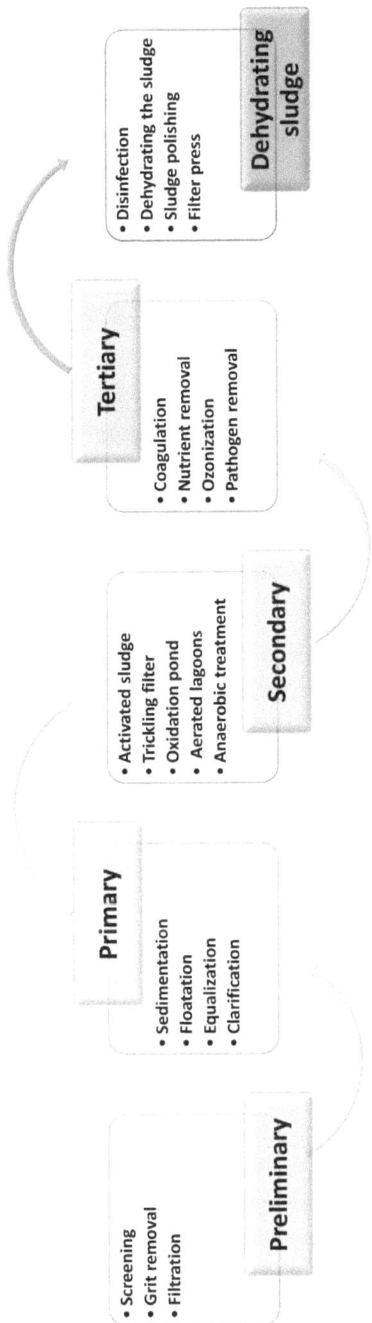

FIGURE 6.1 Conventional methods used for wastewater treatments.

BOD and COD (Ramalho, 2012). Tertiary or advanced wastewater treatments mainly comprise activated carbon adsorption, reverse osmosis, electro- dialysis and ion-exchange, which remove the non-biodegradable pollutants, followed by the removal of nutrients from the wastewater, mainly nitrogen and phosphorus. Tertiary treatment also includes the disinfection of the wastewater through ozonization and chlorination to kill the pathogenic microbes before being discharged into natural reservoirs (Zagklis & Bampos, 2022).

4. CURRENT SCENARIO: NANOTECHNOLOGY FOR WASTEWATER TREATMENT

Due to rapid industrialization, increase in human population, urbanization, agriculture practice, mining and domestic practices, huge amounts of wastewater are released, which not only pollute natural water sources but are also hazardous to living beings. A large population of the world still does not have clean and safe water for their daily needs, and many die every year due to water borne diseases, despite the treatment of wastewater by many conventional methods. Recently, special attention has been given to nanotechnology for the treatment of wastewater with higher efficiency and quality. Nanotechnology implies the fabrication of matter to atomic or molecular level to produce new configuration, devices and systems that possess superior mechanical, electric, optical, magnetic and conductive properties. Nanomaterials have special characteristics, such as high reactivity and sensitivity, high surface-to-volume ratio, high adsorption capacity and ease of functionality, and hence are suitable for wastewater treatment. Various nanotechnology-based approaches for wastewater treatment are adsorption, sorption, photocatalysis, nanofiltration, sensing and disinfection (Jain et al., 2021).

4.1. Nanofiltration

Nanofiltration (NF) is a membrane-pressure driven filtration technique that uses nanometer-sized pores through which particles lesser than 1–10 nm can pass through. It falls between ultra-filtration (UF) and reverse osmosis (RO), driven by both convective and diffusive mechanisms in addition to trans-membrane pressure difference during filtration. The presence of membrane charges also plays a crucial role in the filtration process; NF is considered one of the important methods and is employed for water and wastewater treatment in different industrial applications (Nagy, 2018; Abdel-Fatah, 2018). NF has become very popular recently in the field of wastewater treatment, due to its flexible nature, ease of production and economical issues. There are two types of NF membranes commonly used: ceramic and polymeric membrane. Ceramic membranes have higher mechanical, chemical and thermal stabilities (Parham et al., 2013), while polymeric membranes display a high fouling rate and low chemical resistance (Mostafavi et al., 2009). Applications of NF for the treatment of wastewater generated

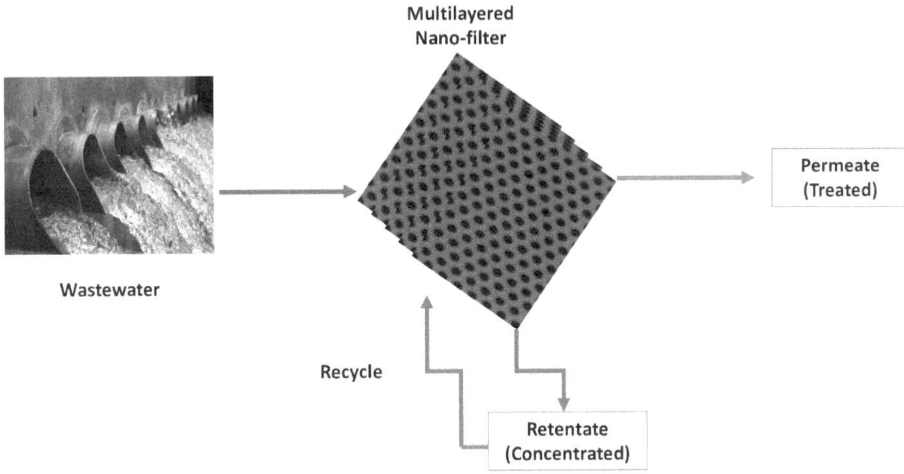

FIGURE 6.2 Mechanism of nanofiltration.

from various sources include mining, food, pharmaceutical, textile, oil and gas, paper and pulp industries and agriculture activities (Ahmad et al., 2023). NF membranes made up of benzoate ferroxane NPs have been reported to remove dye (Moradi et al., 2020). Another NF membrane consists of PVC (poly vinyl alcohol) coated with TiO_2 NPs which has been used to treat paper and pulp mill wastewater. It was also capable of removing dissolved pollutants and microorganisms that also enhanced anti-fouling properties (Khalid et al., 2020). NF blended with gold particles potentially separates phosphate from wastewater, thereby reducing the chances of eutrophication in water bodies (Shalaby et al., 2020). A Ti3C2Tx–MXene lamellar nanofiltration membrane has been found to remove certain dyes from wastewater by modulating an external electrostatic field, particularly negative voltage applied for the removal of cationic dye, increasing sieving potential of the membrane but decreasing water flux (Li et al., 2022). The efficiencies of a nanofilter composite membrane (NFCM), graphene oxide (GO) and cellulose nanocrystal (CNC) membrane were evaluated for the removal of heavy metals Cd^{2+}, Co^{2+}, Cr^{3+}, Pb^{2+}, Ni^{2+} and methylene blue dye from industrial wastewater (Sheikh et al., 2023). The basic mechanism of nanofiltration is illustrated in Figure 6.2.

4.2. Nanozymes

Nanozymes, or nanomaterial-based enzymes resembling natural proteinaceous enzymes in terms of catalytic activities, have higher stability and are easily produced, conveniently transported and stored. To date more than 50 nanozymes have been artificially produced that can mimic natural enzymes such as catalase, peroxide, dismutase, phosphatase, laccase, protease, glucose oxidase, sulfite oxidase and cysteine oxidase (Wang et al., 2019). Catalytic degradation of wastewater is another very important method to reduce the toxicity in wastewater. Some microbial enzymes have been reported to reduce the pollution load in industrial effluents. Recently, nano-material-based enzymes fabricated artificially are termed as nanozymes. The nanozymes mimic natural enzymes and can replace the conventional enzymes due to their superior and similar catalytic activity. Nanozymes retain their functionality in harsh conditions and resist denaturation and inactivation. Additionally these are easily-generated, cost effective, reusable, show less impact on the environment and are easy to prepare. The physiochemical properties of nanozymes aid its application in wastewater treatment. The physical adsorption of nanozymes not only removes the pollutants but also promotes catalytic degradation of the binding pollutants, and also enhances the efficiency of nanozymes (Hong et al., 2022). The catalytic activity of peroxide nanozymes has already been explored for the catalytic degradation of dye rhodamine B (Ibarbia et al., 2020), xylenol orange (Jain et al., 2019) and methylene blue (Huang et al., 2020) in industrial wastewater. Similarly, Cu- nanoparticles have been reported as haloperoxide nanozymes which prevent the signaling and biofilm formation (Hu et al., 2018; Wang et al., 2019). Cerium dioxide (CeO_2) nanoparticles mimic phosphatase activity and catalyze the degradation of paraoxon, an organophosphate pesticide (Trenque et al. 2019). In another study, the laccase nanozymes play a remarkable role in catalyzing the degradation of phenol pollutants (Wang et al., 2019). Fe_3O_4@Cu/C and Fe_3O_4@CuO composites mimic peroxidase-like activity used to degrade methylene blue dye with hydrogen peroxide oxidant and are successfully used in industrial wastewater treatment (Huang et al., 2020). It has also been observed that nanozymes based approaches are potentially used to remove emerging pollutants, such as personnel care products, pharmaceuticals, microplastics, disinfectants, dye and endocrine disrupting chemicals from groundwater, surface water and landfill leachates (Martínez-Zamudio et al., 2023; Panday et al., 2022). Currently a number of organic pollutants, including rhodamine B, methylene blue, Phenol, aniline and xylenol orange, were remarkably removed up to nanomolar concentration by the application of nanozymes, and are employed for environmental monitoring and treatment (He & Liang, 2023). Fe_3O_4 nanoparticles mimic peroxidase like activity and act like the natural horse radish peroxidase enzyme used for the oxidation of chromogenic substance, including TMB (3,3′,5,5′-tetramethylbenzidine), OPD (*o*-phenylenediamine dihydrochloride) and ABTS (2,2′-azino-bis (3-ethylbenzothiazoline-6-sulfonic acid) in the presence of H_2O_2 (Gao et al., 2013). Some nanozymes play bi-functional roles and catalyze reaction including oxidase, peroxidase, catalase, and superoxide dismutase-like activities, and are employed for environmental monitoring and treatment. In one study, it has been observed that gold nanoparticles (AuNPs) mimic oxidase and cobalt doped magnetic iron oxide nanoparticle (Co-MIONzyme) exhibit peroxidase-like activities; amalgamated in inorganic-inorganic bi-nanozymes cascade for the catalytic degradation of dye methylene blue (Mansur et al., 2022). Fe_3O_4@Cu/GMP bio-mimetic, exhibiting laccase-like activities catalyzes the oxidation of o-phenylenediamine (OPD) and remove phenolic compounds and naphthol. The Fe_3O_4@Cu/GMP bio-mimetic enzyme has been determined to be more robust than natural laccase enzyme, and can be recycled for at least five years while retaining functionality up to 70% (Zhang et al., 2020). The different nanozymes and their application in wastewater treatment are illustrated in Table 6.2.

TABLE 6.2 Nanozymes and their application in wastewater treatment.

NANOZYMES	*MIMICS WITH NATURAL ENZYMES*	*TARGET POLLUTANTS*	*REFERENCES*
Graphene oxide quantum dots (GOQD)	Peroxide nanozymes	Catalytic degradation of rhodamine B dye	(Ibarbia et al., 2020)
		Xylenol orange dye	(Jain et al., 2019)
		Methylene blue	(Huang et al., 2020)
(CeO_2) nanoparticles	Peroxidase	Paraoxon and organophosphate pesticide	(Trenque et al. 2019)
Fe_3O_4@Cu/C nanozymes	Peroxidase	Methylene blue	(Huang et al., 2020
Fe_3O_4@Cu/GMP nanozymes	Laccase	Phenolics and naphthol	(Zhang et al., 2020
Ag/Fe_3O_4@h-BN nanozymes	Peroxidase/oxidase	Catalytic degradation of rhodamine B dye, methylene blue, methyl orange	(Xin et al., 2023)
CeO_2@ZIF-8 nanozyme	Peroxidase	Methyl orange	(Wang et al., 2023)
Co-MIONzyme	Oxidase/peroxidase	Methylene blue	(Mansur et al., 2022)
Magnetic Fe_3O_4-TiO_2/graphene nanocomposite	Peroxidase	Pesticide	(Boruah& Das, 2020)
COF@Cu-nanozyme	Laccase	Phenolic	(Yang et al., 2022)
Co_3O_4@Co-Fe oxide double-shelled nanocages	Peroxidase	Acid fuchsin	(Chen et al., 2020)

4.3. Adsorption

Adsorption is a physiochemical phenomenon linked with the accumulation of adsorbents on the surface of the adsorbent. In this process, various types of interaction such as ionic, covalent, hydrogen bond and Vander Waals forces of interactions are involved between the adsorbents and absorbate molecules (Jamshed et al., 2023; Prasad et al., 2023). This method is extensively used in wastewater treatment as it is rapid, inexpensive and competent in removal of a variety of pollutants. Presently, carbon-based materials such as graphene have been utilized due to its elegant properties such as high surface area, reaction rate and porosity (Varde, 2021).

4.3.1. Metal based nano-adsorbents

Metal-based nano-adsorbents are found to be highly efficient in removal of pollutants from wastewater. Metals and metal oxide including CdO, Fe_3O_4, MgO, MnO_2, TiO_2 and ZnO are already being used for the elimination of ions, dye and heavy metals from wastewater. They exhibit high surface area, specific affinity with pollutants, low solubility, less environmental toxicity and exclude secondary contaminations (Joudeh & Linke, 2022). The presence of an abundance of iron in the Earth's crust eases the synthesis of iron-based adsorbents of nano size, which are eco-friendly and economical. The nano-size Fe_3O_4 gets directly pumped to polluted areas without the risk of production of secondary contaminants. The Fe_3O_4 magnetic nano-adsorbents have the sorption efficiency of different heavy metal pollutants from wastewater including Cu^{+2}, Cd^{+2}, Cr^{+6} and Ni^{+2}. The efficiency of adsorption is greatly influenced by the process parameters such as temperature, concentration of the sorbents, pH and incubation time (Jun et al., 2018; Maksoud et al., 2020). It has also been observed that Fe_3O_4 nanoparticles are magnetic nano-absorbents that capture pollutants and metal ions via ion exchange processes, chelation and electrostatic interaction (Singh et al. 2011). Recently nano-absorbents have gained much attention in the field of wastewater remediation due to their unique properties, adaptability and easily turnable properties (Mondal et al., 2024). AgNPs (silver nanoparticles) and its various nanocomposites have been used as catalysis for wastewater treatment and disinfection for many years (Beyene et al., 2027). Another study revealed that the silver and titanium oxide nano-adsorbents which are doped in activated carbon, remove phenol from pharmaceutical effluents (Mustapha et al., 2021). Zinc oxide (ZnO) nanoparticles eliminate coliform from municipal wastewater treatments, NPs get aggregated in the cytoplasm of bacterial cells and hence inhibit their growth (Murcia et al., 2021). Fungal (*Acremonium potronii)* ZnO NPs have been shown to be economical as well as efficient in photocatalytic degradation of azo dyes (Ameen et al., 2021).

The interaction and mechanism of adsorption of heavy metal removal from wastewater is illustrated in Figure 6.3.

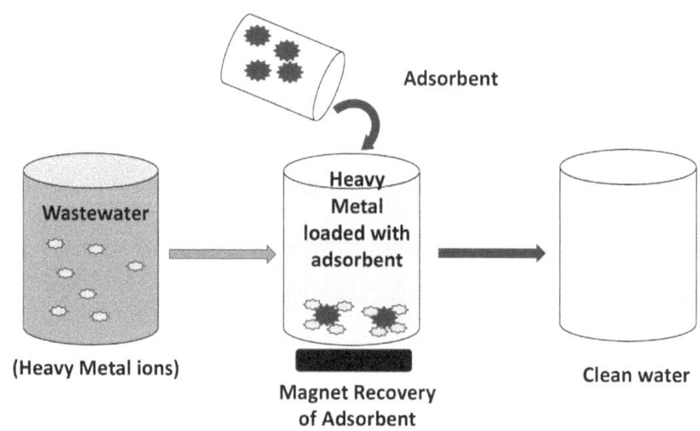

FIGURE 6.3 Mechanism of adsorption of heavy metal removal from wastewater.

4.3.2. Carbon-based nano-adsorbents

Carbon-based nano-adsorbents, including carbon black, carbon-nanofibers, carbon nano tubes, grapheme and fullerenes, can be used to treat wastewater. Fullerene (C60) is an important nanoparticle synthesized by sp2 hybridization and about 30–2000 carbon atoms that are held together in a globular assembly to produce a diameter of 8–9 mm of a single layer and around 538 nm in multi-layered fullerene (Sadegh et al., 2017). Nano-adsorbents like activated carbon (AC), graphene oxide (GO) and composite nano-absorbent AC-GO have been used for the removal of antibiotics, such as doxycycline and tetracycline, from wastewater. The adsorption mechanism obeys the Langmuir isotherm model and pseudo-second order kinetic reaction (Varde, 2021). Similarly, AC prepared from sawdust and coconut powder has efficiently removed Cr^{+6} from wastewater. AC grafted with sodium polyacrylate has been reported to remove heavy metal ions and dye from wastewater with higher efficiency (Wadhawan et al., 2020).

Carbon nanotubes (CNTs) are made up of enrolled graphite sheets with a special electronic, optoelectronic and semiconductor in addition to structural, physical, mechanical and chemical properties. CNTs have been reported to eliminate dye and heavy metal ions from wastewater. They have large sorption capacity for artificial dyes. Multi-walled carbon nanotubes (MWCNTs) have already been reported to remove safranin from wastewater (Wan et al., 2016). Multilayered polyethylene glycol-polyhydroxybutyrate carbon nanotubes (P-CNTs) were used for the industrial electroplating wastewater treatment and effectively eliminated toxic heavy metal ions (Tijani et al., 2024). Recently, another study shows that carbon nanotubes eliminated metallic pollutants from industrial effluents including As, Cd, Cr, Cu, Hg, and Zn (Sharma et al., 2009).

Currently, graphene and its derivatives have gained popularity in wastewater treatments due to their elegant quality such as high absorption rate, excellent molecular structure, unique stoichiometry and physiochemical properties. Graphene and their derivates form various adsorptive interaction with venomous pollutants, like electrostatic interaction, hydrogen bonding and π-π interactions. These adsorbents efficiently eliminated dyes, heavy metals, and polycyclic aromatic hydrocarbon (PAHs) from wastewater (Dhillon & Kumar, 2019). Graphene and graphene oxide (GO) comprise one or many layers of carbon atoms and hold two dimensional morphology with excellent thermal, electrical and mechanical properties. Graphene oxide nanosheets (GONSs), reduced graphene oxide nanosheets (rGONSs) and MWCNTs revealed superior adsorption capacity to remove synthetic organic pollutants including biphenyl, phenanthrene and soluble organic compounds (SOCs) (Ahmed et al., 2022).

4.3.3. Polymer-based nano-adsorbents

Polymer-based adsorbents have gained popularity in removal of heavy metal ions from wastewater due to many properties such as high mechanical strength, excellent pore size distribution, moderate surface area, easy to regenerate by mild treatment and flexible surface chemical framework (Jagaba et al., 2022). In a study, polymer-based nano-adsorbents namely polypyrrole-polyethyleneimine (PPy-PEI) nano-adsorbent was created and used for the successful elimination of Pb^{2+} ion from an aqueous solution (Birniwa et al., 2022). Chitosan, a biopolymer derived from chitin, possesses good chelation properties, biological compatibility, chemical stability, eco-friendly, superior pollutant affinity, high oxidation-reduction potential and high reactivity. The chitosan-based nano-adsorbents play an important role in the removal of metal ion from aqueous solution due to its natural ability and versatility (Haripriyan et al., 2022). Apart from the heavy metal, pharmaceuticals and dyes, the wastewater also consists of nutrient ions like phosphate and nitrate which influence wastewater quality and pose a serious threat to the environment if not removed before discharging to natural sources. The nano material-based adsorbents are specifically metal-based, carbon-based, silica-based, magnetic-based and polymer-based, reported to remove nitrate and phosphate from solutions (Suzaimi et al., 2021). In a similar study, it has been reported that multifunctional bi-metal doped micro- and nano-multi-functional polymeric adsorbents have been used for the elimination of arsenic and fluoride from wastewater (Kumar et al., 2011). Similarly, many adsorptive membranes such as polymer-ceramic membranes (PCMs), nano-enhanced membranes (NEMs), polymeric membranes (PMs) and electro-spinning nanofiber membranes (ENMs) have gained popularity for the elimination of heavy metal ions from wastewater (Vo et al., 2020).

4.4. Photocatalysis

Photocatalysis is a catalytic reaction or photo reaction in the presence of photocatalyst, and has been used for the elimination of arsenic and fluoride from wastewater. It is used specially for the wastes that are unable to be treated with conventional methods and possess toxic and resistant refractory organics. Another unique property of photocatalyst is that it repeatedly interacts with the pollutants, producing intermediates and regenerates itself after each cycle; also an advanced oxidation process having potential to degrade organic carbon present in wastewater (Chong et al., 2010, Ghaly et al., 2011). In the presence of light, TiO_2 acts as a photocatalyst and exhibits oxidative decomposition and super hydrophilic properties of O_2- and OH. These two forms putrefy the organic carbon of wastewater to intermediary forms and finally carbon dioxide and water.

$$\textbf{Organic matter} \xrightarrow[hv]{Tio_2+o_2} \textbf{Intermediate Compound} \xrightarrow[hv]{Tio_2+o_2} \mathbf{Co_2 + H_2O}$$

A semiconductor TiO photocatalyst has already been used for the purification of water and treatment of pharmaceutical wastewater (Kanakaraju et al., 2014). In a similar study, it has been reported that the TiO nanostructured photocatalyst is used for the degradation of wastewater pollutants; about 95% of the degradation is achieved in two hours of exposure to sunlight (Borges et al., 2016). The oxidative photocatalytic degradation of textile dye wastewater was achieved in the presence of $TiO_{2.}$ The UV-A induced photocatalytic oxidation over TiO_2 suspension was able to decolorize the effluent completely; reduction of COD varied between 40–90% depending on operating condition after four hours of treatment (Pekakis et al., 2006). Intimate coupling of photocatalysis and biodegradation (ICPB) is an improved and sustainable biological treatment for the breakdown of intractable pollutants by photocatalysis (Yu et al., 2020). Photocatalysis also inactivates microorganisms, and its efficiency depends on the concentration and physiological condition of the microorganisms (Rincon et al., 2004). TiO_2 nanoparticles potentially kill many bacteria including *Enterobacter cloacae*, *Escherichia coli*, *Pseudomonas aeruginosa* and *Salmonella typhimurium* (Ibanez et al., 2003); fungi including *Candida albicans* and *Fusarium solani*, protozoa (Acanthamoeba); and spores of *Bacillus subtilis* (Lonnen et al., 2005).

4.5. Disinfection and Pathological Control of Wastewater

Worldwide millions of deaths occur annually due to water borne illness, which are caused due to the consumption of contaminated water. The conventional available methods are not enough to remove the pathogenic microorganisms from drinking and wastewater. The biggest challenge in this regard is the development of novel and efficient techniques to control waterborne diseases (Rikta, 2019). Disinfection is very important in wastewater treatment, and it is done by many conventional methods such as chlorination, chlorine oxide, adsorption, precipitation, filtration bleaching powder and ozonization, microbial purification and advanced oxidation technologies (AOPs) to eliminate pathogenic microbes such as bacteria, fungi, parasites and viruses, and control many water borne diseases (Endo-Kimura & Kowalska, 2020, Impellizzeri et al., 2020). Water borne pathogens are responsible for 80% of illness in developing countries and are caused by bacteria, fungi and viruses (Wang et al., 2020). Recently, many nanomaterials and nanoparticles have been successfully employed to remove these pathogens from wastewater. Silver-nanoparticles are commonly in practice for disinfection due to their potential antimicrobial properties, broad spectrum and low toxicity. The mechanism behind silver nanoparticles is the generation of Ag^+ ion which has affinity with the thiol (-SH) groups of the enzymes/proteins and deactivate them. They also inhibit DNA replication and eliminate the microbes from wastewater (Feng et al., 2007). Similarly, CNTs were also reported to kill bacteria by putting oxidative stress on bacterial cell membrane. Graphite/graphene has also been reported to have antibacterial properties and used to disinfect wastewater. The toxicity of CNTs depends on its size and physical and chemical properties; the smaller the size, the greater the penetration and higher the toxicity (Tang & Lo, 2013; Bodzek et al., 2020). In a similar study, Ag-nanoparticles, CNTs and TiO_2 have anti-bacterial properties and are used as disinfectant for water treatment. This nanoparticles-based material has diverse modes of action, such as induction of photocatalytic production of reactive oxygen species that eliminate bacteria by disrupting bacterial cell envelope and other intracellular component. It also inactivates viruses and breaks down DNA of the microbes (Mahendra et al., 2009). AgNPs and other nanoparticles have been reported for their fungicidal, bactericidal and virucidal properties. AgNPs can be incorporated in various porous cellulosic materials for disinfection of water. Recently, AgNCs (silver-based porous nanocomposites) were efficiently used for the disinfection of drinking water and effectively removed 99–100% of bacteria and viruses. In various studies it has been reported that the silver-based nanoparticles can kill 99.99% of *E. coli* and MS2 bacteriophage (Bhardwaj et al., 2021). Ag-MNPs have been reported to 30.40% COD from wastewater and increase concentration, and contact time led to a decrease in microbial contamination and the disinfection process (Najafpoor et al., 2020). Another study reported that the 1% Ag-TiO_2 at 0.2g/l mediates photocatalysis for 3h removed the fecal and total coliform bacteria from effluent, however, it was unable to kill *Escherichia, Enterobacter, Citrobacter, Klebsiella* and *Salmonella* due to their capability of reactivation (Rodríguez-Méndez et al., 2017).

4.6. Sensing and Monitoring

The monitoring and detection of types of pollutants in wastewater and any water source is the major critical step towards its removal and treatment. The conventional methods are not very effective for tracing the quality and quantity of pollutants present in the wastewater. Recently, biosensors have attracted significant attention for sensing and monitoring of wastewater pollutants due to their selectivity, high sensitivity, simplicity, economic and reliability. These biosensors may detect heavy metals, organic chemicals and microbes in wastewater (Ejeian et al., 2018). Many nanomaterials, such as indium phosphide, iron oxide, zinc oxide, titanium dioxide, silica, platinum, nanowire, rhodium, manganese dioxide and graphene, have been used to combat environmental pollution due to their unique physical and chemical characteristics (Wassie et al., 2023). AuNPs and Au-AgNPs are successfully employed for the detection of pharmaceutical pollutants such as cocaine and methamphetamine (METH) by a simple rapid calorimetric biosensor method from pharmaceutical wastewater (Mao et al., 2019). Similarly in another study, gold-silver bimetallic nanoparticles have been reported to detect trivalent chromium (Cr^{3+}) and hexavalent (Cr^{6+}) by electrochemical biosensor in wastewater (Zhao et al., 2021). In another study, a magnetite nanoparticles-based potentiometric biosensor was used to monitor diclofenac in pharmaceutical wastewater (Elbalkiny et al., 2019). The nanoparticle-based sensors also detected pollutants in wastewater irrespective of their size and surface characteristics with high precision and accuracy (Rassaei et al., 2011). In another study, the hexa-valent chromium ion (Cr^{6+}) was detected in wastewater and human living cells by fluorescence sensor Gd (III) boehmite nanoparticles (Roy et al., 2019). A mesoporous silica nanospheres (MSN) was used to detect Cu^{2+} ions in wastewater samples. A derivative of fluorescein hydrazine was immobilized onto the mesoporous silica nanosphere (FHNS) and used to assist the monitoring of Cu^{2+} ions in wastewater and was detected by color change from yellow to brown in a few seconds (El-Sewify & Khalil, 2021).

4.7. Risk Associated with Nanotechnology

4.7.1. Production cost

Developed countries use the most advanced methods to deal with their wastewater for treatment and purification processes. Therefore, it must be ensured that the technologies used for wastewater treatment must be effective and should not have any adverse effect on biodiversity. If we look at the current scenario, developed nations use the latest and most sophisticated technologies for wastewater treatment, while developing nations sometimes only use conventional methods or even only basic treatment, i.e., disinfection.

Therefore, there is the need to develop novel wastewater technologies that are efficient but also cost effective so that each and every part of society can easily afford it. The development of nanomaterials-based wastewater treatment methods is high due to the requirement for high quality and pure nanomaterials for the fabrication of nanoparticles. Moreover, we have to focus on nanomaterials of lower purity for wastewater treatment to cut down the cost. For example, fullerene replaces the ultrapure C_{60} with almost the same effectiveness and reduces up to 90% of the cost (Pan et al., 2020; Qu et al., 2012). Additionally, we can also reduce the cost by reusing the nanomaterials over the long term. Many nanomaterials-based photocatalysts remain active for a long time and can be reused by regenerating and recharging them with suitable techniques (Kokkinos et al., 2020).

4.7.2. Eco-toxicity with nanomaterials

Conventional and advanced methods have been reported to cause undesirable effects on the ecosystem. Even the most common conventional method, chlorination, produces carcinogenic byproducts such as trihalomethane and N-nitroso-di-methylamine (Raza et al., 2016). Similarly, the use of nanomaterials may lead to toxicity of the environment. The resulting toxicity of the nanomaterials is also governed by their molecular structure, size and cellular uptake. Due to their minute size, NPs easily penetrate through endothelial barriers to reach blood and lymph nodes and then to various organs of the body, like kidneys, lungs, liver, heart, brain, spleen and bone marrow (Sukhanova et al., 2018). TiO_2-NPs have been reported to damage organs and DNA in rats. The engineered NPs also have serious health implications and may induce genotoxicity, carcinogenicity, pulmonary inflammation and cardiac problems (De Matteis & Rinaldi, 2018). AgNPs used for water treatment may induce oxidative stress and genotoxicity in *Labeo rohita,* a fresh fish (Khan et al., 2017). Similarly, MWCNTs had the toxic effect on two fish species: *Danio rerio* and *Astyanax altiparanae*. The CNTs crosslink with the DNA and interfere with DNA replication, induce sub-chronic neurotoxicity and induce oxidative stress in both species (Cimbaluk et al., 2018). Ceramic filters impregnated with AgNPs have shown detectable AgNPs in water, but their value was lower than the limit set by the World Health Organization (Albukhari et al., 2019). Many studies have revealed that metal-based nanoparticles (M-NPs) have the potential to accumulate in various organs and can cross blood–brain and pose serious health hazards to animals, human and environments. M-NPs interact with various components of the cell, such as proteins, DNA, mitochondria and Golgi apparatus, and trigger many abnormal cellular events including oxidative stress, cell membrane impairments, increased lysosomal hydrolases, mitochondrial dysfunction, apoptosis, fluctuation in ATP synthesis and cytoplasmic disruption (Wang et al., 2024).

5. CONCLUSION AND FUTURE OUTLOOK

Massive increases in population, rapid industrialization, urbanization and agricultural practices results in water pollution. Clean and pure water is the essential requirement

of every form of life for its sustainable growth and development. Across the globe there are many preliminary and advance practices that are used for the de-contamination of the polluted water so that it can be re-used. Currently new methods based on nanotechnology are employed to treat the wastewater and industrial effluents. Nanostructures including NPs, CNTs, graphene, nanofilters, nanopolymers, etc., are potentially applied for wastewater treatment due to their unique properties such as high adsorption capacity, reactivity, sensitivity and high surface-to-volume ratio. Owing these unique properties, this method is more superior that any traditional method. Due to their small size they easily enter to living tissue and accumulate and pose serious health hazards to mammals, plants and environments. Although nanotechnology is associated with the limitation of high production, it still has promising applications in wastewater treatment and management that could have immense prospects in the future. Nevertheless, its application in wastewater treatment demands government policies. At present, sincere and responsible efforts of scientists, compliance with government policies, systematic investigations of nano-materials regarding their biological, eco-toxicity, production cost and their remedial measures are desirable. As these concerns get resolved, nanomaterials and nanotechnological measures can be optimized, and elimination of micropollutants would be a great achievement in this regard.

REFERENCES

Abdel-Fatah, M. A. (2018). Nanofiltration systems and applications in wastewater treatment. *Ain Shams Engineering Journal*, *9*(4), 3077–3092. https://doi.org/10.1016/j.asej.2018.08.001

Agarwal, S., Darbar, S., & Saha, S. (2022). Challenges in management of domestic wastewater for sustainable development. *Current Directions in Water Scarcity Research, 6,* 531–552. https://doi.org/10.1016/B978-0-323-91838-1.00019-1.

Ahmad, N. N. R., Mohammad, A. W., Ang, W. L., & Teow, Y. H. (2023). Applying nanofiltration for sustainability: Case studies in various industries. In *Nanofiltration for Sustainability*, 15–31. CRC Press.

Ahmed, S. F., Mofijur, M., Rafa, N., Chowdhury, A. T., Chowdhury, S., Nahrin, M., ... & Ong, H. C. (2022). Green approaches in synthesising nanomaterials for environmental nanobioremediation: Technological advancements, applications, benefits and challenges. *Environmental Research*, *204*, 111967. https://doi.org/10.1016/j.envres.2021.111967

Albukhari, S. M., Ismail, M., Akhtar, K., & Danish, E. Y. (2019). Catalytic reduction of nitrophenols and dyes using silver nanoparticles@ cellulose polymer paper for the resolution of waste water treatment challenges. *Colloids and Surfaces A: Physicochemical and Engineering Aspects*, *577*, 548–561. https://doi.org/10.1016/j.colsurfa.2019.05.058

Aliste, M., Garrido, I., Hernández, V., Flores, P., Hellín, P., Navarro, S., & Fenoll, J. (2022). Assessment of reclaimed agro-wastewater polluted with insecticide residues for irrigation of growing lettuce (Lactuca sativa L) using solar photocatalytic technology. *Environmental Pollution*, *292*, 118367. https://doi.org/10.1016/j.envpol.2021.118367

Ameen, F., Dawoud, T., & AlNadhari, S. (2021). Ecofriendly and low-cost synthesis of ZnO nanoparticles from *Acremonium potronii* for the photocatalytic degradation of azo dyes. *Environmental Research*, *202*, 111700. https://doi.org/10.1016/j.envres.2021.111700

Babu, S., Rathore, S. S., Singh, R., Kumar, S., Singh, V. K., Yadav, S. K., … & Wani, O. A. (2022). Exploring agricultural waste biomass for energy, food and feed production and pollution mitigation: A review. *Bioresource Technology*, *360*, 127566. https://doi.org/10.1016/j.biortech.2022.127566

Beyene, H. D., Werkneh, A. A., Bezabh, H. K., & Ambaye, T. G. (2017). Synthesis paradigm and applications of silver nanoparticles (AgNPs), A review. *Sustainable Materials and Technologies*, *13*, 18–23. https://doi.org/10.1016/j.susmat.2017.08.001.

Bhardwaj, A. K., Sundaram, S., Yadav, K. K., & Srivastav, A. L. (2021). An overview of silver nano-particles as promising materials for water disinfection. *Environmental Technology & Innovation*, *23*, 101721. https://doi.org/10.1016/j.eti.2021.101721

Birniwa, A. H., Kehili, S., Ali, M., Musa, H., Ali, U., Kutty, S. R. M., … & Mahmud, H. N. M. E. (2022). Polymer-based nano-adsorbent for the removal of lead ions: kinetics studies and optimization by response surface methodology. *Separations*, *9*(11), 356. https://doi.org/10.3390/separations9110356

Bodzek, M., Konieczny, K., & Kwiecinska-Mydlak, A. (2020). Application of nanotechnology and nanomaterials in water and wastewater treatment: membranes, photocatalysis and disinfection. *Desalination and Water Treatment*, *186*, 88–106. https://doi.org/10.5004/dwt.2020.25231

Bodzek, M., Konieczny, K., & Kwiecińska-Mydlak, A. (2020). Nanotechnology in water and wastewater treatment. Graphene–the nanomaterial for next generation of semipermeable membranes. *Critical Reviews in Environmental Science and Technology*, *50*(15), 1515–1579. https://doi.org/10.1080/10643389.2019.1664258

Borges, M. E., Sierra, M., Cuevas, E., García, R. D., & Esparza, P. (2016). Photocatalysis with solar energy: Sunlight-responsive photocatalyst based on TiO2 loaded on a natural material for wastewater treatment. *Solar Energy*, *135*, 527–535. Https://Doi.Org/10.1016/J.Solener.2016.06.022

Boruah, P. K., & Das, M. R. (2020). Dual responsive magnetic Fe3O4-TiO2/graphene nanocomposite as an artificial nanozyme for the colorimetric detection and photodegradation of pesticide in an aqueous medium. *Journal of Hazardous Materials*, *385*, 121516. https://doi.org/10.1016/j.jhazmat.2019.121516

Chen, Q., Zhang, X., Li, S., Tan, J., Xu, C., & Huang, Y. (2020). MOF-derived Co3O4@ Co-Fe oxide double-shelled nanocages as multi-functional specific peroxidase-like nanozyme catalysts for chemo/biosensing and dye degradation. *Chemical Engineering Journal*, *395*, 125130. https://doi.org/10.1016/j.cej.2020.125130

Chong, M. N., Jin, B., Chow, C. W., & Saint, C. (2010). Recent developments in photocatalytic water treatment technology: A review. *Water Research*, *44*(10), 2997–3027. https://doi.org/10.1016/j.watres.2010.02.039

Cimbaluk, G. V., Ramsdorf, W. A., Perussolo, M. C., Santos, H. K. F., De Assis, H. C. D. S., Schnitzler, M. C., … & Cestari, M. M. (2018). Evaluation of multiwalled carbon nanotubes toxicity in two fish species. *Ecotoxicology and Environmental Safety*, *150*, 215–223. https://doi.org/10.1016/j.ecoenv.2017.12.034

De Matteis, V., & Rinaldi, R. (2018). Toxicity assessment in the nanoparticle era. *Cellular and Molecular Toxicology of Nanoparticles*, 1–19. doi:10.1007/978-3-319-72041-8_1

Dhillon, A., & Kumar, D. (2019). New generation nano-based adsorbents for water purification. *In Nanoscale Materials in Water Purification,* 783–798.

Ejeian, F., Etedali, P., Mansouri-Tehrani, H. A., Soozanipour, A., Low, Z. X., Asadnia, M., … & Razmjou, A. (2018). Biosensors for wastewater monitoring: A review. *Biosensors and Bioelectronics*, *118*, 66–79. https://doi.org/10.1016/j.bios.2018.07.019

Elbalkiny, H. T., Yehia, A. M., Safa'a, M. R., & Elsaharty, Y. S. (2019). Potentiometric diclofenac detection in wastewater using functionalized nanoparticles. *Microchemical Journal*, *145*, 90–95. https://doi.org/10.1016/j.microc.2018.10.017

El-Sewify, I. M., & Khalil, M. M., 2021. Mesoporous nanosensors for sensitive monitoring and removal of copper ions in wastewater samples. *New Journal of Chemistry*, *45*(5), 2573–2581.

Endo-Kimura, M., & Kowalska, E. (2020). Plasmonic photocatalysts for microbiological applications. *Catalysts*, *10*(8), 824.

Feng, Y., Yu, B., Xu, P., & Zou, M. (2007). The effective thermal conductivity of nanofluids based on the nanolayer and the aggregation of nanoparticles. *Journal of Physics D: Applied Physics*, *40*(10), 3164.

Frascaroli, G., Reid, D., Hunter, C., Roberts, J., Helwig, K., Spencer, J., & Escudero, A. (2021). Pharmaceuticals in wastewater treatment plants: A systematic review on the substances of greatest concern responsible for the development of antimicrobial resistance. *Applied sciences*, *11*(15), 6670. https://doi.org/10.3390/app11156670

Gao, Z., Xu, M., Hou, L., Chen, G., & Tang, D. (2013). Magnetic bead-based reverse colorimetric immunoassay strategy for sensing biomolecules. *Analytical chemistry*, *85*(14), 6945–6952. https://doi.org/10.1021/ac401433p

Ghaly, M. Y., Jamil, T. S., El-Seesy, I. E., Souaya, E. R., & Nasr, R. A. (2011). Treatment of highly polluted paper mill wastewater by solar photocatalytic oxidation with synthesized nano TiO2. *Chemical Engineering Journal*, *168*(1), 446–454. https://doi.org/10.1016/j.cej.2011.01.028

Haripriyan, U., Gopinath, K. P., & Arun, J. (2022). Chitosan based nano adsorbents and its types for heavy metal removal: A mini review. *Materials Letters*, *312*, 131670. –https://doi.org/10.1016/j.matlet.2022.131670

He, J. & Liang, M. (2020). Nanozymes for environmental monitoring and treatment. *Nanozymology: Connecting Biology and Nanotechnology*, 527–543. https://doi.org/10.1007/978-981-15-1490-6_16

Hu, M., Korschelt, K., Viel, M., Wiesmann, N., Kappl, M., Brieger, J., ... & Tremel, W. (2018). Nanozymes in nanofibrous mats with haloperoxidase-like activity to combat biofouling. *ACS Applied Materials & Interfaces*, *10*(51), 44722–44730. https://doi.org/10.1021/acsami.8b16307

Huang, Y.F., Zhang, L., Ma, L., Li, Y., & Zhong, C. (2020). Fe 3 O 4@ Cu/C and Fe 3 O 4@ CuO composites derived from magnetic metal–organic frameworks Fe3O4@ HKUST-1 with improved peroxidase-like catalytic activity. *Catalysis Letters*, *150*, 815-825.doi: 10.1007/s10562-019-02964-8

Ibáñez, J. A., Litter, M. I., & Pizarro, R. A. (2003). Photocatalytic bactericidal effect of TiO2 on Enterobacter cloacae: Comparative study with other Gram (−) bacteria. *Journal of Photochemistry and Photobiology A: Chemistry*, *157*(1), 81–85. https://doi.org/10.1016/S1010-6030(03)00074-1

Ibarbia, A., Sánchez-Abella, L., Lezama, L., Grande, H. J., & Ruiz, V. (2020). Graphene quantum dot-based hydrogels for photocatalytic degradation of organic dyes. *Applied Surface Science*, *527*, 146937. https://doi.org/10.1016/j.apsusc.2020.146937

Impellizzeri, F. M., Tenan, M. S., Kempton, T., Novak, A., & Coutts, A. J. (2020). Acute: Chronic workload ratio: Conceptual issues and fundamental pitfalls. *International Journal of Sports Physiology and Performance*, *15*(6), 907–913. https://doi.org/10.1123/ijspp.2019-0864

Jagaba, A. H., Kutty, S. R. M., Lawal, I. M., Aminu, N., Noor, A., Al-dhawi, B. N. S., ... & Yakubu, A. S. (2022). Diverse sustainable materials for the treatment of petroleum sludge and remediation of contaminated sites: A review. *Cleaner Waste Systems*, *2*, 100010. https://doi.org/10.1016/j.clwas.2022.100010

Jain, K., Patel, A. S., Pardhi, V. P., & Flora, S. J. S. (2021). Nanotechnology in wastewater management: a new paradigm towards wastewater treatment. *Molecules*, *26*(6), 1797. https://doi.org/10.3390/molecules26061797

Jain, S., Panigrahi, A., & Sarma, T.K. (2019). Counter anion-directed growth of iron oxide nanorods in a polyol medium with efficient peroxidase-mimicking activity for degradation of dyes in contaminated water. *ACS omega*, *4*(8), 13153–13164. https://doi.org/10.1021/acsomega.9b01201

Jamshed, A., Iqbal, A., & Ali, S. (2023). A quick review on the applications of nanomaterials as adsorbents. *MOJ Eco Environ Sci*, *8*(3), 86–89.

Joudeh, N. & Linke, D. (2022). Nanoparticle classification, physicochemical properties, characterization, and applications: a comprehensive review for biologists. *Journal of Nanobiotechnology*, *20(1)*, 262. doi:10.1186/s12951-022-01477-8

Jun, L. Y., Mubarak, N. M., Yee, M. J., Yon, L. S., Bing, C. H., Khalid, M., & Abdullah, E. C. (2018). An overview of functionalised carbon nanomaterial for organic pollutant removal. *Journal of Industrial and Engineering Chemistry*, *67*, 175–186.https://doi.org/10.1016/j.jiec.2018.06.028

Kanakaraju, D., Glass, B. D., & Oelgemöller, M. (2014). Titanium dioxide photocatalysis for pharmaceutical wastewater treatment. *Environmental Chemistry Letters*, *12*, 27–47. doi:10.1007/s10311-013-0428-0

Keškić, T., Petričević, M., Stamenić, T., Mandić, V., Delić, N., Pisinov, B., & Đurović, S. (2024). Agro-processing industry: Advanced technologies for meat industry wastewater treatment: A review. In 11th JEEP International Scientific Agribusiness Conference, Food for the Future–Vision of Serbia, Region and Southeast Europe, 02-04. February, Kopaonik, Serbia, *Proceedings*, *1*, 226–235.

Khalid, F., Tabish, M., & Bora, K. A. I. (2020). Novel poly (vinyl alcohol) nanofiltration membrane modified with dopamine coated anatase TiO2 core shell nanoparticles. *Journal of Water Process Engineering*, *37*, 101486. –https://doi.org/10.1016/j.jwpe.2020.101486

Khalil, M., Hanif, M. A., Rashid, U., Ahmad, J., Alsalme, A., & Tsubota, T. (2023). Low-cost novel nano-constructed granite composites for removal of hazardous Terasil dye from wastewater. *Environmental Science and Pollution Research*, *30*(34), 81333-81351. https://doi.org/10.1007/s11356-022-21367-8

Khan I., Saeed K., & Khan I. (2019). Nanoparticles: Properties, applications and toxicities. *Arabian Journal Chemistry, 12,* 908–931. doi: 10.1016/j.arabjc.2017.05.011.

Khan, M. S., Qureshi, N. A., & Jabeen, F. (2017). Assessment of toxicity in fresh water fish Labeo rohita treated with silver nanoparticles. *Applied Nanoscience*, *7*, 167–179. doi:10.1007/s13204-017-0559-x

Kokkinos, P., Mantzavinos, D., & Venieri, D. (2020). Current trends in the application of nanomaterials for the removal of emerging micropollutants and pathogens from water. *Molecules*, *25*(9), 2016. https://doi.org/10.3390/molecules25092016

Kumar, V., Talreja, N., Deva, D., Sankararamakrishnan, N., Sharma, A., & Verma, N. (2011). Development of bi-metal doped micro-and nano multi-functional polymeric adsorbents for the removal of fluoride and arsenic (V) from wastewater. *Desalination*, *282*, 27–38. https://doi.org/10.1016/j.desal.2011.05.013

Li, J., Xu, C., Long, J., Ding, Z., Yuan, R. & Li, Z., 2022. Lamellar MXene nanofiltration membranes for electrostatic modulation of molecular permeation: implications for fine separation. *ACS Applied Nano Materials*, *5*(5), 7373–7381. https://doi.org/10.1021/acsanm.2c01304

Lonnen, J., Kilvington, S., Kehoe, S. C., Al-Touati, F., & McGuigan, K. G. (2005). Solar and photocatalytic disinfection of protozoan, fungal and bacterial microbes in drinking water. *Water Research*, *39*(5), 877–883. https://doi.org/10.1016/j.watres.2004.11.023

Mahendra, S., Li, Q., Lyon, D. Y., Brunet, L., & Alvarez, P. J. (2009). Nanotechnology-enabled water disinfection and microbial control: merits and limitations. In *Nanotechnology Applications for Clean Water*, 157–166. William Andrew Publishing. https://doi.org/10.1016/B978-0-8155-1578-4.50021-4

Maksoud, M. A., Elgarahy, A. M., Farrell, C., Ala'a, H., Rooney, D. W., & Osman, A. I. (2020). Insight on water remediation application using magnetic nanomaterials and biosorbents. *Coordination Chemistry Reviews*, *403*, 213096. https://doi.org/10.1016/j.ccr.2019.213096

Mansur, A. A., Leonel, A. G., Krambrock, K., & Mansur, H. S. (2022). Bifunctional oxidase-peroxidase inorganic nanozyme catalytic cascade for wastewater remediation. *Catalysis Today*, *397*, 129–144. https://doi.org/10.1016/j.cattod.2021.11.018

Mao, K., Ma, J., Li, X., & Yang, Z. (2019). Rapid duplexed detection of illicit drugs in wastewater using gold nanoparticle conjugated aptamer sensors. *Science of the Total Environment*, *688*, 771–779. https://doi.org/10.1016/j.scitotenv.2019.06.325

Martínez-Zamudio, L. Y., González-González, R. B., Araújo, R. G., Hernández, J. A. R., Flores-Contreras, E. A., Melchor-Martínez, E. M., … & Iqbal, H.M. (2023). Emerging pollutants removal from leachates and water bodies by nanozyme-based approaches. *Current Opinion in Environmental Science & Health*, 100522. https://doi.org/10.1016/j.coesh.2023.100522

Meena, R. A. A., Sathishkumar, P., Ameen F., Yusoff, A. R. M, & Gu, F. L. (2018). Heavy metal pollution in immobile and mobile components of lentic ecosystems—A review. *Environmental Science Pollution Research*, 25, 4134–4148. https://doi.org/10.1007/s11356-017-0966-2

Mondal, B., Hasan, M. M., & Islam, M. D. (2024). Nano-adsorbent based solutions for wastewater treatment—an overview. *Development in Wastewater Treatment Research and Processes*, 401–424. https://doi.org/10.1016/B978-0-323-99278-7.00018-3

Moradi, G., Zinadini, S., & Rajabi, L. (2020). Development of high flux nanofiltration membrane using para-amino benzoate ferroxane nanoparticle for enhanced antifouling behavior and dye removal. *Process Safety and Environmental Protection*, *144*, 65–78. https://doi.org/10.1016/j.psep.2020.06.044

Mostafavi, S. T., Mehrnia, M. R., & Rashidi, A. M. (2009). Preparation of nanofilter from carbon nanotubes for application in virus removal from water. *Desalination*, *238*(1–3), 271–280.https://doi.org/10.1016/j.desal.2008.02.018

Murcia, J. J., Hernández Niño, J. S., Rojas, H., Brijaldo, M. H., Martín-Gómez, A. N., Sánchez-Cid, P., Navío, J.A., Hidalgo, M.C. & Jaramillo-Paez, C. (2021). ZnO/Ag3PO4 and ZnO–Malachite as Effective Photocatalysts for the Removal of Enteropathogenic Bacteria, Dyestuffs, and Heavy Metals from Municipal and Industrial Wastewater. *Water*, *13*(16), 2264. https://doi.org/10.3390/w13162264.

Mustapha, S. I., Aderibigbe, F. A., Adewoye, T. L., Mohammed, I. A., & Odey, T. O. (2021). Silver and titanium oxides for the removal of phenols from pharmaceutical wastewater. *Materials Today: Proceedings*, *38*, 816–822. https://doi.org/10.1016/j.matpr.2020.04.669.

Nagy, E. (2018). *Basic equations of mass transport through a membrane layer*. Elsevier.

Najafpoor, A., Norouzian-Ostad, R., Alidadi, H., Rohani-Bastami, T., Davoudi, M., Barjasteh-Askari, F., & Zanganeh, J. (2020). Effect of magnetic nanoparticles and silver-loaded magnetic nanoparticles on advanced wastewater treatment and disinfection. *Journal of Molecular Liquids*, *303*, 112640. https://doi.org/10.1016/j.molliq.2020.112640

Pan, Y., Liu, X., Zhang, W., Liu, Z., Zeng, G., Shao, B., … & Chen, M. (2020). Advances in photocatalysis based on fullerene C60 and its derivatives: Properties, mechanism, synthesis, and applications. *Applied Catalysis B: Environmental*, *265*. https://doi.org/10.1016/j.apcatb.2019.118579

Pandey, J., Dubey, R., Kate, A., Prasad, B., Sinha, A, & Mishra, M. S. (2022). Nanomedicines: A focus on nanomaterials as drug delivery system with current trends and future advancement. *Drug Research*, *72*(07), 355–366. https://doi.org/10.1055/a-1824-4619

Parham, H., Bates, S., Xia, Y., & Zhu, Y. (2013). A highly efficient and versatile carbon nanotube/ceramic composite filter. *Carbon*, *54*, 215-223. –https://doi.org/10.1016/j.carbon.2012.11.032

Pekakis, P. A., Xekoukoulotakis, N. P., & Mantzavinos, D. (2006). Treatment of textile dyehouse wastewater by TiO_2 photocatalysis. *Water Research*, *40*(6), 1276–1286. https://doi.org/10.1016/j.watres.2006.01.019

Pereira, L.S. Duarte, E., & Fragoso, R. (2014). Water use: Recycling and desalination for agriculture. 407–424. https://doi.org/10.1016/B978-0-444-52512-3.00084-X.

Polianciuc, S. I., Gurzău, A. E., Kiss, B., Ştefan, M. G., & Loghin, F. (2020). Antibiotics in the environment: causes and consequences. *Medicine and Pharmacy Reports*, *93*(3), 231. https://doi.org/10.15386/mpr-1742

Prasad, B., Gupta, S., & Kumar, A., 2023. Microbial remediation of lead: An overview. *Lead Toxicity: Challenges and Solution*, 175–195. https://doi.org/10.1007/978-3-031-37327-5_9

Qu, X., Brame, J., Li, Q., & Alvarez, P. J. (2013). Nanotechnology for a safe and sustainable water supply: Enabling integrated water treatment and reuse. *Accounts of Chemical Research*, *46*(3), 834–843. https://doi.org/10.1021/ar300029v

Rajasulochana P., & Preethy V. (2016). Comparison on efficiency of various techniques in treatment of waste and sewage water—A comprehensive review. *Resources Technology*, *2*, 175–184. doi: 10.1016/j.reffit.2016.09.004.

Ramalho, R. (2012). *Introduction to wastewater treatment processes*. Elsevier.

Rani, S. L. S., Satyannarayana, K. V. V., Arthanareeswaran, G., & Raja, V. K. (2024). Treatment of food processing industries wastewaters using a new clay-based inorganic membrane: Performance evaluation and fouling analysis. *Journal of the Taiwan Institute of Chemical Engineers*, 105439. https://doi.org/10.1016/j.jtice.2024.105439

Rassaei, L., Marken, F., Sillanpää, M., Amiri, M., Cirtiu, C. M., & Sillanpää, M. (2011). Nanoparticles in electrochemical sensors for environmental monitoring. *TrAC Trends in Analytical Chemistry*, *30*(11), 1704–1715. –https://doi.org/10.1016/j.trac.2011.05.009

Raza, M. A., Kanwal, Z., Rauf, A., Sabri, A. N., Riaz, S., & Naseem, S. (2016). Size-and shape-dependent antibacterial studies of silver nanoparticles synthesized by wet chemical routes. *Nanomaterials*, *6*(4), 74. https://doi.org/10.3390/nano6040074

Rikta, S. Y. (2019). Application of nanoparticles for disinfection and microbial control of water and wastewater. *Nanotechnology in Water and Wastewater Treatment*, 159–176. https://doi.org/10.1016/B978-0-12-813902-8.00009-5

Rincon, A. G., & Pulgarin, C. (2004). Bactericidal action of illuminated TiO2 on pure Escherichia coli and natural bacterial consortia: Post-irradiation events in the dark and assessment of the effective disinfection time. *Applied Catalysis B: Environmental*, *49*(2), 99–112. https://doi.org/10.1016/j.apcatb.2003.11.013

Rodríguez-Méndez, A., Guzmán, C., Elizalde-Peña, E. A., Escobar-Alarcón, L., Vega, M., Rivera, J. A., & Esquivel, K. (2017). Effluent disinfection of real wastewater by Ag–TiO2 nanoparticles photocatalysis. *Journal of Nanoscience and Nanotechnology*, *17*(1), 711–719.doi: https://doi.org/10.1166/jnn.2017.13066

Roy, S., Pal, K., Bardhan, S., Maity, S., Chanda, D. K., Ghosh, S., ... & Das, S. (2019). Gd (III)-doped boehmite nanoparticle: an emergent material for the fluorescent sensing of Cr (VI) in wastewater and live cells. *Inorganic Chemistry*, *58*(13), 8369–8378. https://doi.org/10.1021/acs.inorgchem.9b00425

Sadegh, H., Ali, G. A., Gupta, V. K., Makhlouf, A. S. H., Shahryari-Ghoshekandi, R., Nadagouda, M. N., ... & Megiel, E. (2017). The role of nanomaterials as effective adsorbents and their applications in wastewater treatment. *Journal of Nanostructure in Chemistry*, *7*, 1–14. doi:10.1007/s40097-017-0219-4

Samal, K., Mahapatra, S., & Ali, M. H. (2022). Pharmaceutical wastewater as emerging contaminants (EC): Treatment technologies, impact on environment and human health. *Energy Nexus*, *6*, 100076. https://doi.org/10.1016/j.nexus.2022.100076

Shalaby, M. S., Abdallah, H., Cenian, A., Sołowski, G., Sawczak, M., Shaban, A. M., & Ramadan, R. (2020). Laser synthesized gold-nanoparticles, blend NF membrane for phosphate separation from wastewater. *Separation and Purification Technology*, *247*, 116994. https://doi.org/10.1016/j.seppur.2020.116994

Sharma, Y. C., Srivastava, V., Singh, V. K., Kaul, S. N., & Weng, C. H. (2009). Nano-adsorbents for the removal of metallic pollutants from water and wastewater. *Environmental Technology*, *30*(6), 583–609. https://doi.org/10.1080/09593330902838080

Sheikh, M. S., Rahman, M. M., Rahman, M. S., Yildirim, K., & Maniruzzaman, M. (2023). Fabrication of nano composite membrane filter from graphene oxide (GO) and banana rachis cellulose nano crystal (CNC) for industrial effluent treatment. *Journal of Industrial and Engineering Chemistry*, *128*, 196–208. https://doi.org/10.1016/j.jiec.2023.07.048

Singh, S., Barick, K. C., & Bahadur, D. (2011). Novel and efficient three dimensional mesoporous ZnO nanoassemblies for envirnomental remediation. *International Journal of Nanoscience*, *10*(04n05), 1001–1005. https://doi.org/10.1142/S0219581X11008654

Sonune, A., & Ghate, R. (2004). Developments in wastewater treatment methods. *Desalination*, *167*, 55–63. –https://doi.org/10.1016/j.desal.2004.06.113

Sukhanova, A., Bozrova, S., Sokolov, P., Berestovoy, M., Karaulov, A., & Nabiev, I. (2018). Dependence of nanoparticle toxicity on their physical and chemical properties. *Nanoscale Research Letters*, *13*, 1–21. doi:10.1186/S11671-018-2457-X

Suzaimi, N. D., Goh, P. S., Nizam Nik Malek, N. A., Ng, B. C., & Ismail, A. F. (2021). Nano-adsorbents in wastewater treatment for phosphate and nitrate removal. *Environmental Nanotechnology, 5*, 339–370. doi:10.1007/978-3-030-73010-9_10

Tang, J., Chen, Q., Xu, L., Zhang, S., Feng, L., Cheng, L., ... & Peng, R. (2013). Graphene oxide–silver nanocomposite as a highly effective antibacterial agent with species-specific mechanisms. *ACS Applied Materials & Interfaces*, *5*(9), 3867–3874. https://doi.org/10.1021/am4005495

Tang, S. C., & Lo, I. M. (2013). Magnetic nanoparticles: essential factors for sustainable environmental applications. *Water Research*, *47*(8), 2613–2632.

Tang, Y., Jiang, S., Li, W., Shah, S. J., Zhao, Z., Pan, L., & Zhao, Z. (2022). Confined construction of COF@ Cu-nanozyme with high activity and stability as laccase biomimetic catalyst for the efficient degradation of phenolic pollutants. *Chemical Engineering Journal*, *448*, 137701. https://doi.org/10.1016/j.cej.2022.137701

Tijani, J. O., Abdulkareem, A. S., Mustapha, S., Ndamitso, M. M., Bada, S. O., & Sagadevan, S. (2024). Polyethyleneglycol-polyhydroxylbutyrate functionalized carbon nanotubes for industrial electroplating wastewater treatment. *Separation and Purification Technology*, *332*, 125736. https://doi.org/10.1016/j.seppur.2023.125736

Trenque, I., Magnano, G. C., Bolzinger, M. A., Roiban, L., Chaput, F., Pitault, I., … & Amans, D. (2019). Shape-selective synthesis of nanoceria for degradation of paraoxon as a chemical warfare simulant. *Physical Chemistry Chemical Physics*, *21*(10), 5455–5465. https://doi.org/10.1039/C9CP0017D

Varde, S. R. (2021). *Developing a framework for optimal performance of activated carbon/graphene oxide/activated carbon-graphene oxide nano adsorbent for the removal of antibiotics from wastewater* (Master's thesis).

Vecitis, C. D., Zodrow, K. R., Kang, S., & Elimelech, M. (2010). Electronic-structure-dependent bacterial cytotoxicity of single-walled carbon nanotubes. *ACS Nano*, *4*(9), 5471–5479. https://doi.org/10.1021/nn101558x

Vo, T. S., Hossain, M. M., Jeong, H. M., & Kim, K. (2020). Heavy metal removal applications using adsorptive membranes. *Nano Convergence*, *7*(1), 36. https://doi.org/10.1186/s40580-020-00245-4

Wadhawan, S., Jain, A., Nayyar, J., & Mehta, S. K. (2020). Role of nanomaterials as adsorbents in heavy metal ion removal from waste water: A review. *Journal of Water Process Engineering*, *33*, 101038. https://doi.org/10.1016/j.jwpe.2019.101038

Wan, S., Bi, H., & Sun L. (2016). Graphene and carbon–based nanomaterials as highly efficient adsorbents for oils and organic solvents. *Nanotechnology Reviews*, *5*(1), 3–22. https://doi.org/10.1515/ntrev-2015-0062

Wang, F., Zhang, Y., Liu, Z., Ren, J., & Qu, X. (2020). A mesoporous encapsulated nanozyme for decontaminating two kinds of wastewater and avoiding secondary pollution. *Nanoscale*, *12*(27), 14465-14471. https://doi.org/10.1039/D0NR3217D

Wang, J., Chu, L., Wojnárovits, L., & Takács, E. (2020). Occurrence and fate of antibiotics, antibiotic resistant genes (ARGs) and antibiotic resistant bacteria (ARB) in municipal wastewater treatment plant: An overview. *Science of the Total Environment*, *744*, 140997. https://doi.org/10.1016/j.scitotenv.2020.140997

Wang, L., Hou, J., Liu, S., Carrier, A.J., Guo, T., Liang, Q., ... & Zhang, X. (2019). CuO nanoparticles as haloperoxidase-mimics: Chloride-accelerated heterogeneous Cu-Fenton chemistry for H2O2 and glucose sensing. *Sensors and Actuators B: Chemical*, *287*, 180–184. https://doi.org/10.1016/j.snb.2019.02.030

Wang, Y. L., Lee, Y. H., Chou, C. L., Chang, Y. S., Liu, W. C., & Chiu, H.W.(2024). Oxidative stress and potential effects of metal nanoparticles: A review of biocompatibility and toxicity concerns. *Environmental Pollution*, 123617.

Wassie, A. T., Bachheti, R. K., & Bachheti, A. (2023). Recent advances in nanoparticles for environmental monitoring and sensing: An overview. *Nanomaterials for Environmental and Agricultural Sectors*, 107–122. doi:10.1007/978-981-99-2874-3_6

Xin, Z., Duan, K., Zhuo, Q., He, Q., Zhang, X., Zheng, C., ... & Zhao, X. (2023). Novel nanozyme Ag/Fe3O4@ h-BN with peroxidase-mimicking and oxidase-mimicking activities for dye degradation, As (V) removal and detection. *Chemical Engineering Journal*, *461*, 141589. https://doi.org/10.1016/j.cej.2023.141589

Yang, H., Lu, G., Yan, Z., Liu, J., Dong, H., Bao, X., ... & Sun, Y. (2020). Residues, bioaccumulation, and trophic transfer of pharmaceuticals and personal care products in highly urbanized rivers affected by water diversion. *Journal of Hazardous Materials*, *391*, 122245. https://doi.org/10.1016/j.jhazmat.2020.122245

Yu, M., Wang, J., Tang, L., Feng, C., Liu, H., Zhang, H., ... & Xie, Q. (2020). Intimate coupling of photocatalysis and biodegradation for wastewater treatment: mechanisms, recent advances and environmental applications. *Water Research*, *175*, 115673. https://doi.org/10.1016/j.watres.2020.115673

Zagklis, D. P., & Bampos, G. (2022). Tertiary wastewater treatment technologies: A review of technical, economic, and life cycle aspects. *Processes*, *10*(11), 2304. https://doi.org/10.3390/pr10112304

Zahoor, I., & Mushtaq, A. (2023). Water pollution from agricultural activities: A critical global review. *International Journal of Chemical and Biochemical Sci*ence, *23*(1), 164–176.

Zhang, S., Lin, F., Yuan, Q., Liu, J., Li, Y. and Liang, H. (2020). Robust magnetic laccase-mimicking nanozyme for oxidizing o-phenylenediamine and removing phenolic pollutants. *Journal of Environmental Sciences*, *88*, 103–111. https://doi.org/10.1016/j.jes.2019.07.008

Zhao, K., Ge, L., Wong, T. I., Zhou, X., & Lisak, G. (2021). Gold-silver nanoparticles modified electrochemical sensor array for simultaneous determination of chromium (III) and chromium (VI) in wastewater samples. *Chemosphere*, *281*, 130880. https://doi.org/10.1016/j.chemosphere.2021.130880

Zinicovscaia, I. (2016). Conventional methods of wastewater treatment. *Cyanobacteria for Bioremediation of Wastewaters*, 17–25. doi:10.1007/978-3-319-26751-7_3

7 Diversity, Biodegradation and Bioremediation of Industrial Based Polycyclic Aromatic Hydrocarbons (PAH) with Regard to Industrial Wastewater

Reyhan Ata and Günay Yıldız Töre

1. INTRODUCTION

Polycyclic aromatic hydrocarbons (PAH compounds) are dangerous organic pollutants found in petroleum and petroleum derivatives that are thrown into the environment through incomplete combustion of organic compounds such as coal, wood and fossil fuels used for heating and energy. As a group of persistent toxic substances, PAH pollutants are widely detected in air, soils, sediments, water bodies, as well as the

DOI:10.1201/9781003496014-7

tissues of animals and plants (Tang et al., 2018). PAH pollutants are harmful and recalcitrant organic pollutants due to their stability and hydrophobicity, while the high molecular-weight (HMW) PAH pollutants are even more detrimental to the environment and human health. Most of the PAH pollutants cause environmental pollution as a result of their long staying and accumulation properties in the environment and significantly affect the biological balance. The harmful effects of soot, tar and pitch have been known since the British Surgeon Percival Pott reported in 1775 that chimney swabbing wastes progressed testicular cancer. After more than 150 years, the carcinogenic compounds in the structure of pitch and tar were defined as PAH pollutants. PAH pollutants are known today as the largest class of chemical carcinogens (NIH, 2024). PAH pollutants are a group of 2 to 7 rings hydrocarbon compounds which are usually colourless or at white and light yellow-green colour physically. They may occur naturally through forest fires or volcanic eruptions. Most of the PAH pollutants are produced by industries anthropogenically, and are found in tar and crude oil, and some are used in the production of pharmaceuticals, paints, plastics and pesticides. In addition to industrial sources, motor vehicles and cigarettes are human sources of PAH pollutants. PAH pollutants are formed in the air, can pass into water and soil through rain and dust particles, thus dispersed almost everywhere They are divided into two groups: Low Molecular Weight (LMW) PAH pollutants and High Molecular Weight (HMW) PAH pollutants. LMW PAH pollutants, such as Naphthalene (NAP) Acenaphthylene (ACY) and Acenaphthene (ACP), have 2 to 3 rings, while HMW PAH pollutants, have 4 or more cyclic rings such as Pyrene (PYR), Benzo(e)pyrene (BeP), Dibenz(a, h)acridine (DhACR) and Dibenzo(a,l)pyrene (DlP). As the molecular weights of PAH pollutants increase, their solubility in water decreases and their carcinogenic and mutagenic effects increase. In parallel, the hydrophobic nature of PAH pollutants is related to the number of rings and carbons. Since PAH pollutants are in the form bound to particles in the atmosphere and water, they are resistant to degradation and are less likely to be oxidized in the atmosphere. These conditions increase their transportation through the atmosphere over long distances and make them ubiquitous in the environment. Since PAH molecules do not have different functional groups, they are in the class of high lipophilic compounds which means they are insoluble in water. They can dissolve in only oily contaminants present in water by binding to particles in an aqueous environment. Not only large amounts, even small amounts of PAH compounds, originated from different sources, are major pollutants that have detrimental, toxic, carcinogenic, strongly immunosuppressant, mutagenic and teratogenic effects on both the environment and living organisms. Therefore, there is an urgent need to control their amount in air, water and food. Studies have reported that the toxicity mechanism takes place through the pollutant's effects on cell membrane functions and membrane-dependent enzyme systems. There are also studies showing its dangerous effects on humoral immunity, immune system development and host resistance (Abdel-Shafy & Mansour, 2016; Mojiri et al., 2019).

The purpose of this chapter is to deeply understand the harmful effects of PAH pollutants on the environment and human beings, and to eliminate these effects by examining the important scientific studies on biodegradation and bioremediation in the last five to six years.

2. SOURCES AND DIVERSITY OF PAH POLLUTANTS

2.1. Natural and Anthropogenic Sources of PAH Pollutants

Sources of PAH pollutants can be considered in two categories: natural emission sources and anthropogenic emission sources (Mojiri et al., 2019). Volcanic eruptions, natural forest fires and steppe fires caused by lightning can be listed as natural emission sources (Abdel-Shafy & Mansour, 2016). Anthropogenic emission sources, which are the main determinants of PAH pollutants, are divided into four types: industrial, mobile, domestic and agricultural emission sources. Industrial activities are the main sources of PAH emissions such as incomplete combustion activities of iron and steel production, aluminum production, coal-tar production, cement production, asphalt industry, paint manufacturing, rubber tire manufacturing, fungicide and pesticide production, waste incineration, exhaust and electricity generation from refineries. (Abdel-Shafy & Mansour, 2016; Mojiri et al., 2019). Since the concentration of PAH pollution varies according to the seasons, PAH pollution is highest in the winter months and decreases further towards the spring, autumn and summer months, respectively. This is thought to be due to the incomplete combustion of fossil fuels used to heat dwellings due to atmospheric conditions through low temperature and light winds, poor diffusion and low photodegradation. On the other hand, agricultural emission sources arise as a result of open biomass combustion and agricultural waste incineration under non-optimal combustion conditions.

While domestic and agricultural activities increase PAH pollution in rural areas, industrial, domestic and mobile-based activities cause increasing PAH pollution in urban areas. Natural and anthropogenic sources of PAH pollutants are collectively demonstrated in Figure 7.1 (Abdel-Shafy & Mansour, 2016, Marris et al., 2020).

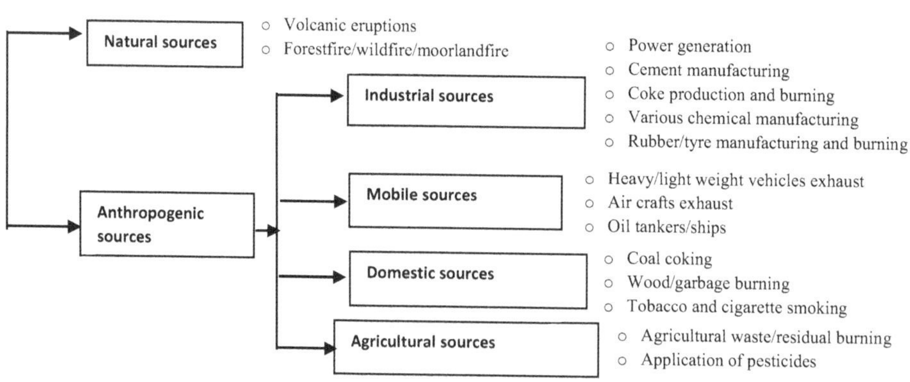

FIGURE 7.1 Sources of PAH pollutants.

Figure created by the authors.

Pyrogenic PAH pollutants usually have a chemical structure of 4 to 6 rings, and are formed as a result of incomplete combustion of fossil fuels at high temperatures. These are more concentrated in urban areas (Abdel-Shafy & Mansour, 2016; Marris et al., 2020; Moijri et al., 2019). Biogenic PAH pollutants may originate from plants and animals, or may have formed during early formation of sediments. Unlike petrogenic and pyrogenic PAH pollutants, biogenic PAH are usually found individually or in simple mixtures, and have the ability to be synthesized by biological entities such as microorganisms, phytoplankton, algae and plants during the slow biological transformation of organic materials (Moijiri et al., 2019). Figure 7.1 shows the different PAH emission sources.

2.2. Physical and Chemical Properties of PAH pollutants

Petroleum compounds are divided into two main classes: hydrocarbons and heteroatom compounds. Hydrocarbon molecules, which can be measured as total petroleum hydrocarbons (TPHs), contain only carbon (C) and hydrogen (H), while heteroatom compounds contain not only carbon and hydrogen, but also heteroatoms such as sulfur, nitrogen, and oxygen. PAH molecules are considered in the class of highly lipophilic compounds because they do not have distinct functional groups. Therefore, PAH compounds have low water solubility and their hydrophobic nature is directly related to the ring and carbon number. PAH compounds, which are bound to particles in the atmosphere and water, are less likely to oxidize, thus making them resistant to degradation and increasing their transport properties in the atmosphere to long distances, thereby causing them to be ubiquitous in the environment. PAH compounds, whose water solubility decreases as their molecular weight increases due to their hydrophobic properties, can accumulate in high concentrations in sediments, soils and aqueous media by being adsorbed on humic substances. Usually there are more than 100 PAH compounds in nature, not as a single compound but as a complex blend, and the health effects of each are not the same (Table 7.1).

TABLE 7.1 PAH compounds identified by the US Environmental Protection Agency.

COMPOUND	*CAS NUMBER*	*MOLECULAR WEIGHT (MW)*
Naphthalene-NAP	91-20-3	128 g/mol
Acenaphthylene-ACY	208-96-8	152 g/mol
Acenaphthene-ACP	83-32-9	154 g/mol
Fluorene-FLR	86-73-7	166 g/mol
9-Fluorenone-9-FL	486-25-9	180 g/mol
Dibenzothiophene-DBT	132-65-0	184 g/mol
Phenanthrene-PHE	85-01-8	178 g/mol
Anthracene-ANT	120-12-7	178 g/mol

(Continued)

TABLE 7.1 (Continued)

COMPOUND	*CAS NUMBER*	*MOLECULAR WEIGHT (MW)*
Fluoranthene-FLT	206-44-0	202 g/mol
Pyrene-PYR	129-00-0	202 g/mol
Retene-RET	483-65-8	234 g/mol
Benzo[c]phenanthrene-BcP	195-19-7	228 g/mol
Cyclopenta[c,d]pyrene-CPP	27208-37-3	226 g/mol
Benz[a]anthracene-BaA	56-55-3	228 g/mol
Chrysene-CHR	218-01-9	228 g/mol
Benzo[b]fluoranthene-BbF	205-99-2	252 g/mol
Benzo[j]fluoranthene-BjF	205-82-3	252 g/mol
Benzo[k]fluoranthene-BkF	207-08-9	252 g/mol
7,12-Dimethylbenz[a]anthracene-DMBA	57-97-6	256 g/mol
Benzo[e]pyrene-BeP	192-97-2	252 g/mol
Benzo[a]pyrene-BaP	50-32-8	252 g/mol
Perylene-PER	198-55-0	252 g/mol
3-Methylcholanthrene-3MC	56-49-5	268 g/mol
Dibenz[a,h]acridine-DhACR	226-36-8	279 g/mol
Dibenz[a,j]acridine-DjACR	224-42-0	279 g/mol
Indeno[1,2,3-cd]pyrene-IcP	193-39-5	276 g/mol
Dibenz[a,h]anthracene-DhANT	53-70-3	278 g/mol
7H-Dibenzo[c,g]carbazole-DBC	194-59-2	267 g/mol
Benzo[g,h,i]perylene-BgP	191-24-2	276 g/mol
Dibenzo[a,l]pyrene-DlP	191-30-0	302 g/mol
Coronene-COR	191-07-1	300 g/mol
Dibenzo[a,e]pyrene-DeP	192-65-4	302 g/mol

(USEPA, 2020).

2.3. Cycle of PAH Compounds in Air, Water and Soil

PAH pollutants in the soils can be transported to different distances through aquifers by infiltrating groundwater. The suspension times and transport distances of PAH compounds attached depend on the size of the particles, dry and oil deposition events, while their sweep depends on meteorological conditions (Abdel-Shafy & Mansour, 2016). After being released into the atmosphere, PAH compounds exist in two separate phases: as a vapor where it is absorbed into particulate matter, and as a solid. Since there is a significant correlation between the amount of dust in the air, the particle phase and the PAH concentrations, gas phase PAH concentrations increase in summer or tropical regions, while particle phase PAH pollutants predominate in winter or in

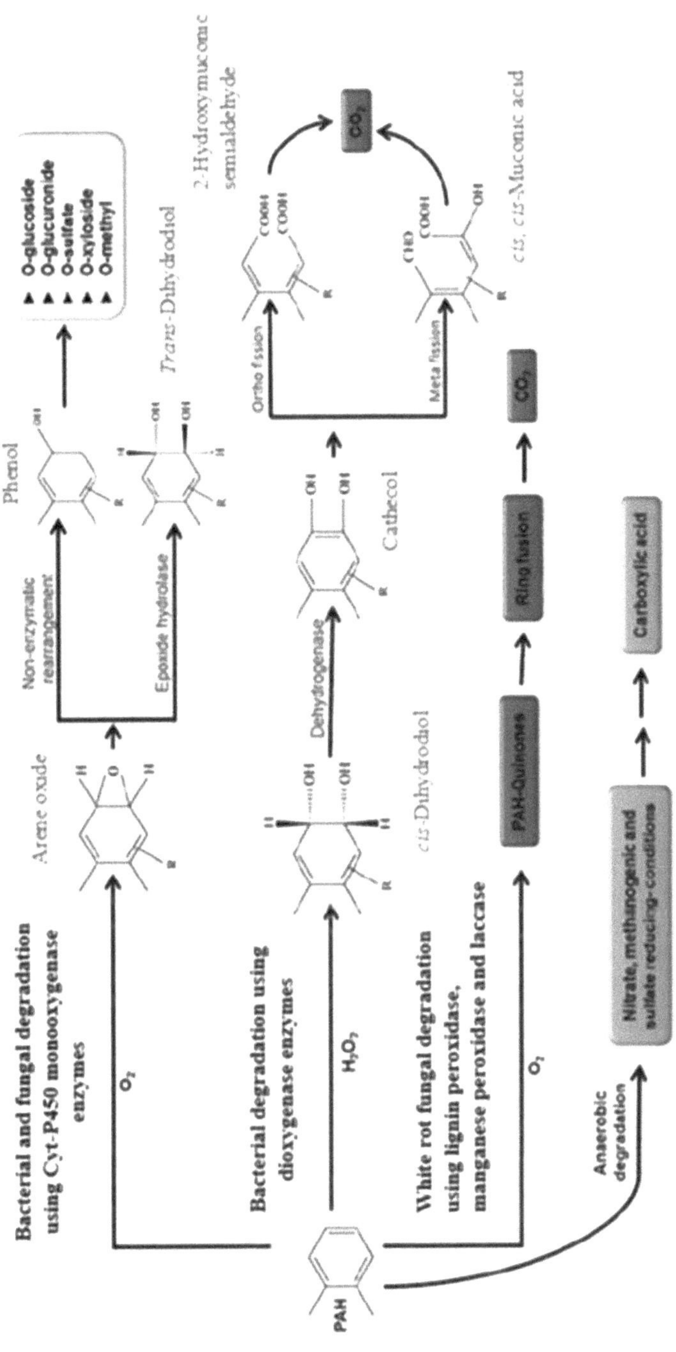

FIGURE 7.2 Degradation pathways of PAH compounds by bacteria and fungi (Haritash & Kaushik, 2009).

arctic regions in general. The decomposition and transformation of PAH pollutants in the gas phase in the atmosphere are carried out by reactions with NO_x, N_2O_5, OH^-, O_3, SO_2, and peroxyacetyl nitrate molecules and photolysis. As a result of these reactions, oxy-, hydroxy-, nitro and hydroxynitro-PAH compounds are formed. Dions are toxic for living organisms formed as a result of the reactions of O_3 and peroxyacetyl-nitrates with PAH pollutants, and dinitro-PAH compounds are formed as a result of their reactions with nitrogen oxide. Besides sulfonic acid is formed as a result of its reaction with sulfur dioxide. In light catalysis, they are transformed into oxidation products such as quinones, ketones and acids as a result of photolysis (Zodiatis et al., 2021). The degradation pathways of PAH compounds by bacteria and fungi are given at Figure 7.2.

3. HARMFUL EFFECTS AND ACCUMULATION

3.1. Adverse Effects on Environment and Accumulation Mechanisms

PAH pollutants cause harmful, toxic and carcinogenic effects to the environment and living organisms as well as humans. Considering their chemical properties, PAH pollutants have a long half-life, and they pollute the environment by releasing a wide variety of toxic derivatives during decomposition. Pyrogenic PAH pollutants, sourced from fires, forest fires, volcanic eruptions and incineration, tend to accumulate in high concentrations in soil, water and atmospheric environments. In winter, harmful effects, environmental pollution and transport of pollution to other environments increase by the atmospheric influences on smog clouds, from air to soil, and air to water reserves. Due to their hydrophobic properties, PAH pollutants, whose water solubility decreases as their molecular weight increases, can accumulate in high concentrations in precipitates, estuaries and sediments. PAH pollutants accumulating in soil and sediments can be transported to groundwater, plants and food. PAH compounds with three or more aromatic rings (HMW PAH pollutants) can very strongly adsorb to soil particles and be transported to plant roots and other parts of the plant due to their low vapor pressures and high hydrophobicity. The rate of accumulation of pyrogenic PAH concentrations in urban areas is naturally higher as they are formed by burning fossil fuels at high temperatures (AbdelShafy & Mansour, 2016; Mojiri et al., 2019).

3.2. Harmful Effects on Health

PAH compounds are widely used in industry. As industrialization increased, it became nearly impossible to avoid exposure to PAH pollutants in industry, manufacturing and personal life situations. In the industry, PVC and rubber-made products are given the desired elasticity and softness by using tar oils, extender oils and industrial soot with long chain and multi-ring PAH compounds. According to the German Environment Agency report, bath slippers, bicycle steering handles and saddles, many sports materials and children's toys are made of polymeric materials. PVC and rubber are the materials we use frequently in our daily lives, and studies have revealed that PVC bathroom slippers cause exposure to 546 mg/kg 16 PAH (highest 170 and 120 mg/kg fluorine and phenanthrene, respectively). Moreover, before the virtual environment became widespread, benzo (a) pyrene, pyrene, phenanthrene and naphthalene up to 52, 553, 778 and 283 mg/kg, respectively, were detected in printing inks used in newspapers. Acenaphthene and pyrene are used for the manufacture of pigments, dyes, plastics, pesticides and pharmaceuticals. On the other hand, anthracene is a kind of diluent chemical for wood preservatives and manufacture of dyes and pigments. While phenanthrene is used for the manufacture of resins and pesticides, fluoranthene is used for the manufacture of agrochemicals, dyes and pharmaceuticals. Fluorene is used for the manufacture of pharmaceuticals, pigments, dyes, pesticides and thermoset plastic (Abdel-Shafy & Mansour, 2016). Exposure to PAH pollutants generally occurs through smoking or inhalation of smoke from open fires and fossil fuels, food and agricultural plant leaves containing PAH pollutants, and skin contact (Abdel-Shafy & Mansour, 2016).

PAH pollutants have both acute and chronic health effects. Acute health effects include eye irritation, nausea, vomiting, diarrhea, confusion, skin irritation and inflammation. Chronic health effects include eye cataracts, kidney and liver damage, respiratory problems, decreased immune function, lung disorders and asthma-like symptoms (Abdel-Shafy & Mansour, 2016). Anthracene, benzo(a)pyrene and naphthalene, which are the important PAH pollutants, directly irritate the skin and cause allergic effects on the skin of animals and humans. Exposure of people to PAH pollutants is not only in the workplace, but also can occur in their normal lives by consuming contaminated water containing PAH, consuming grilled and smoked foods, smoking, etc. Smoking or being a passive smoker are the main ways to be exposed to PAH pollutants. One cigarette causes the human body to absorb 20–40 ng of benzo(a)pyrene. On the other hand, non-smokers may also be exposed to PAH pollutants through diet. Even the mere consumption of water can cause exposure to PAH pollutants if contaminated with PAH compounds. In addition to all, tea and coffee consumed, as much as water, are also polluted by PAH compounds due to the heating activities during the drying/roasting stage. A study also revealed the presence of PAH_4 [chrycen, benzo(a)pyrene, benzo(a)anthracene, benzo(b) fluorine] in 18 brands of tea and 13 coffee brands. The highest PAH_4 was detected in black tea (25–115 mg/kg) and instant coffee (2.2–5.1 mg/kg) (Brandt & Einhenkel-Arle, 2016). Most of PAH pollutants have been proven to have teratogenicity, carcinogenicity and mutagenicity that threaten human health. According to the International Agency for Research on Cancer (IARC), PAH pollutants are classified as group 1-carcinogenic to humans, group 2A-possibly carcinogenic to humans, group 2B-possibly carcinogenic to humans and group 3 carcinogens (Sun et al., 2019). While these harmful effects may vary depending on factors such as pre-effect status and age, long-term exposure to PAH pollutants can cause lung, skin, esophageal, colon, pancreatic, bladder and breast cancers. Since studies have shown that benzo(a) is the most carcinogenic PAH, it is used as an exposure indicator in risk assessments. The acceptable threshold limit for PAH pollutants is 10-6, and there are scientific studies reporting that the risk of cancer is 45% if this value is exceeded (Tong et al., 2018). To reveal teratogenic and mutagenic effects of PAH pollutants, studies have revealed that benzo(a)pyrene cause infertility in the male reproductive system. In 2017, Bolden et al. drew attention to the negative effects of PAH pollutants exposure on egg viability, ovaries and sperm quality, DNA damage in oocytes (causing mutagenic effect), non-cancerous (related to the reproductive system) diseases in both women and men. Similar studies, supporting that have also revealed embryo toxicity in experimental animals due to exposure to Naphthalene, benzo(a)anthracene and benzo(a)pyrene (Abdel-Shafy & Mansour, 2016; Bolden et al., 2017). Children, who tend to crawl on soil surfaces contaminated with PAH pollutants and tend to grab the dirty soil and bring it into their mouths, have the same or even more sensitivity to PAH pollutants than adults. It has been reported in studies that high PAH pollutants exposure causes an increase in childhood asthma and low IQ. According to Child Environmental Health Center reports, exposure to PAH pollutants during pregnancy causes adverse birth outcomes such as low weight, preterm birth and heart malformations. Although some exposure limit values have been determined on the basis of legal regulations, PAH pollutants, which are present in almost all products we use daily, including foodstuffs, have toxic effects; their exposure in our lives is almost inevitable. According to the World Health Organization (WHO), the highest exposure limit to benzo(a)pyrene is 0.7 $mg.L^{-1}$. The measurements made generally reveal that the concentration of PAH in drinking water varies between 1 $ng.L^{-1}$ and 11 $ng.L^{-1}$ (Seker & Bağcevan, 2020).

3.3. Toxicity of PAH Pollutants

In addition to being carcinogenic, teratogenic and mutagenic, PAH pollutants are also photo-toxic and immuno-toxic to microorganisms, animals and humans depending on the exposure mode, exposure time and exposure dose parameters (Bolden et al., 2017). The mechanism of toxicity is thought to be through affecting cell membrane functions and membrane-dependent enzyme systems. Studies have shown that they cause dangerous effects on immune system development, humoral immunity and host resistance. In addition to all these, the ecotoxic effects of PAHs on aquatic life and birds have also been demonstrated (Abdel-Shafy & Mansour, 2016). Since it is more explanatory to examine the toxicity of PAH pollutants as photo-toxic, geno-toxic and immuno-toxic effects, it was determined that PAH pollutants can be detoxified in the liver by forming water-soluble epoxide glutathione conjugates through the catalytic reactions of cytochrome P_{450} and many oxidase enzymes. However, some PAH pollutants are not disposable and produce non-polar reactive intermediates (e.g., diol epoxides, quinones, hydroxyalkyl derivatives); they can cause genotoxic effects by forming covalent bonds with nucleic acid (Abdel-Shafy & Mansour, 2016). The accumulation and bioavailability of some PAH pollutants in adipose tissue-rich viscera after exposure is due to their high lipophilicity. Most PAH pollutants are not geno-toxic on their own, and in order to have a geno-toxic effect, they must be metabolized into diol epoxides that react with DNA and cause geno-toxic damage (Abdel-Shafy & Mansour, 2016). PAH pollutants gain the ability to form reactive intermediates by electron transfer and the ability to react more easily with oxygen and other molecules. While the formation of these reactive intermediates occurs in cells, cellular components such as the cell membrane, nucleic acid or proteins are damaged. Moreover, PAH pollutants also have detrimental effects on the immune system, suppressing the immune system, and are regarded as suspected endocrine disrupting chemicals (EDCs). These immuno-toxic effects can be listed as pre-B, pre-T, inhibition of myeloid cell growth, B and T cell suppression, apoptosis of lymphoid tissues, disruption of myelopoiesis and altered cytokine production by macrophages and monocytes.

4. DETECTION OF PAH COMPOUNDS

4.1. Sampling and Determination of PAH Compounds

In the active PAH sampling method from air environment, air particles are adsorbed on filters or absorbent tubes as air sampler by using a pump. In another sampling method, which is the passive sampling method from indoor and outdoor air, polyurethane foam (PUF) is used as the collection material. PAH compounds deposited on filters or absorbents are sent to the laboratory for analysis in both methods. In order to take samples of PAH pollutions from soil or sediments, ISO 10381-2 standards are followed at the sampling stage and TS ISO 10381-5 rules are followed at the analysis stage. Accordingly, until the laboratory where the analysis will be carried out, 1 kg of contaminated soil samples taken from 5- and 30-cm depths were kept in the deep freezer at -18°C until they were extracted, and then kept in the oven and at 105°C for four days. After four days, 100 grams of these samples are taken and separated for extraction. Furthermore, for PAH analysis, water and wastewater sampling is carried out on the upper, middle and bottom level of water or wastewater, by using a sampling device enabling standard sampling. It would be appropriate to take water and wastewater samples into glass sample bottles and keep them at +2 to +4°C until analyzed (Tore et al., 2019). After a properly sampling, process soil and sludge samples are usually extracted into suitable conventional organic solvents such as water or any other suitable solvent. Usually, water is often used as an alternative solvent and reaction medium to many conventional organic solvents. Alternative extraction agents can be surfactants, bio surfactants, microemulsions, natural surfactants, cyclodextrins, vegetable oil and solution containing solid phase particles. Main extraction methods are supercritical fluids extraction (SCFE), superheated water extraction (SHWE), subcritical water extraction (SCWE), solid-phase extraction (SPE), suspended droplet micro extraction (SDME), sorptive extraction (SE), magnetic solid-phase extraction (MSPE), ultrasound-assisted liquid–liquid extraction (UALE), Tenax extraction (TE), ultrasound-assisted emulsification microextraction-UAEME), ultrasonic centrifuge solid-phase extraction (UCSPE), dispersive liquid-liquid microextraction (DLLME) and in-tube solid-phase microextraction (SPME) (Mukhopadhyay et al. 2020). For the subsequent purification process, extracted samples are usually separated from the interfering substances by adsorption columns chromatography method. The most commonly used sorbents are aluminum and silica gel. Dichloromethane, cyclohexane or a different solvent can be used as eluting solvent in varying amounts. It is used to separate PAH pollutants from nonaromatic, nonpolar compounds. Purification processes, in commercially prepared chemical cartridges (e.g., SPE cartridge), are time-and solvent-consuming and reproducible (Ata & Töre, 2019).

Finally, dissolved and extracted air, water and soil samples can become injectable into detector devices. Detection is carried out by gas chromatograph (GC), liquid chromatograph (LC), high or ultra-pressure liquid chromatograph HPLC or U-HPLC, depending on the situation. For mass determinations, gas chromatography and mass spectrometer (GC-MS, GC-MS/MS), liquid chromatography and mass spectrometry (LC-MS, LC-MS/MS), high pressure liquid chromatograph HPLC-MS, HPLC-MS/MS or UHPLC-MS, UHPLC-MS/MS, flame ionization detector (GC – FID, HPLC-FID) or fluorescent HPLC-FLD detector devices are equipped with mass spectrometer.

5. REMEDIATION TECHNOLOGIES FOR PAH POLLUTANTS IN SOILS, SEDIMENTS AND WATER/WASTEWATERS

5.1. Remediation Approaches to PAH Pollutants

Until today, there are applications in which both chemical or physical methods (incineration, base-catalysed de-chlorination, UV oxidation, fixation, solvent extraction, etc.) and biological methods are used for the removal of PAH pollutants. Many technologies are available to deal with PAH contaminated soils, sediments and water/wastewaters such as incineration and in situ thermal desorption (ISTD), solvent extraction, soil washing, chemical oxidation (ISCO), bioremediation, electrokinetic enhancement, multi-process phytoremediation, vermiremediation and integrated approaches. Remediation techniques are generally classified as either in situ or ex situ techniques. In situ remediation techniques are the realization of the treatment process on site without taking it to another area. These remediation techniques get more public consent due to less cost, minimum site interruption and high possibility of permanent waste elimination (Shah, 2014). In addition, since water or soil does not need to be transported or pumped, large contaminated areas can be treated. In situ approaches are based on the principle of stimulating and activating microorganisms in water or soil. Meanwhile, ex situ remediation techniques are the operations where the contaminated materials are physically removed to another area in the field (soil by digging, water by pumping) where remediation is performed (Shah, 2014). In situ bioremediation may not always be the best solution. This method is more effective in less compact soils such as sand, which allow microorganisms and fertilizing materials to spread rapidly. Solid clayey and dense stony soils are typical soil types not suitable for in situ bioremediation, and chemical contamination may take years to be treated by this method. In some soil treatment areas, ex situ bioremediation can be much faster and more effective than in situ approaches. Ex situ bioremediation may involve many different techniques depending on the type of soil to be treated, the amount and the type of chemical. The most common ex situ technique is slurry phase bioremediation. In this technique, contaminated soil is moved to another location and mixed with water and fertilizer (usually oxygen is added) in large bioreactors where

TABLE 7.2 Advantages and disadvantages of PAH removal technologies.

TYPES OF REMEDIATION	*REMOVAL TECHNIQUES*	*ADVANTAGES*	*DISADVANTAGES*	*REFERENCES*
Physical Remediation	Adsorption	• Green and sustainable chemistry • Low cost adsorbent • High porosity and efficiency	• Difficulty in biomass recovery after sorption • Complicated regeneration • Low mechanical strength and density • High cost when using activated carbon	(Adeola & Forbes, 2020; Gautam et. Al., 2014)
	Forward Osmosis Filtration	• Low external pressure demand • Less affected by fouling	• Requires an additional chemical to prepare draw solution • Fever membrane options	(Sabir et al., 2015; Guo et al., 2015; Jafari et al., 2015; Korenak et al., 2017)
Chemical Remediation	Fenton Reaction	• Performed at room temperature and ordinary temperature • Processes high performance and non-toxicity	• High cost of reagents • The large volume of iron sludge	(Huang et al., 2016)
	Zero Valent Iron (ZVI)	• Simple process • Low operating cost • Wide range of applications	• Removal effectiveness depends on the size and reactivity of ZVI particles • Not fully complete removal of PAHs	(Li & Yang, 2018)

(*Continued*)

TABLE 7.2 (Continued)

TYPES OF REMEDIATION	*REMOVAL TECHNIQUES*	*ADVANTAGES*	*DISADVANTAGES*	*REFERENCES*
Biological Remediation	Phytoremediation	• It is a green technology and environmentally friendly • Less expensive • It doesn't require expensive equipment or highly specialized personnel	• Bioavailability of plant and plant slow growth rate • Longer remediation time and cannot treat deep contaminants • Toxicity	(Malik et al., 2017; Fasani et al., 2018; Khodaver-diloo et al., 2020; Rao & Babu, 2014)
	Fungi	• The resistance to different environmental conditions • Able to grow on a variety of media • Able to produce enzymes for the degrading of organic pollutants	• Only can utilize the PAHs as a soul source of carbon and energy	(Ghosal et al., 2016)
	Bacteria	• They are already adapted to contaminated sites	• The slow rate of PAHs removal • Abiotic factors influential such as pH and temperature • Inefficient for the high molecular weight PAHs removal	(Kumari et al., 2018)
	Bacteria and Fungi	• Reduce the toxicity of media • High removal	• It needs to find the best combination	(Deveau et al., 2018)

Table created by authors.

biodegradation by soil microorganisms can be carefully monitored and controlled. The most popular remediation techniques for PAHs are summarized in Table 7.2.

5.2. Chemical Treatment of PAH Pollutants

Studies have demonstrated that chemical oxidation is a useful method for the rapid recovery of soil and ground-water contaminated with various solvents and organic pollutants, including petroleum-derived compounds referred to as total petroleum hydrocarbons (TPH). Chemical oxidants are widely used in chemical removal of PAH contaminants; dispersed down to a substrate to oxidize organic contaminants during the chemical oxidation treatment process, typical SCO treatment process can be listed as Fenton's reagent (ferrous iron-catalysed hydrogen peroxide at pH 2–4), modified Fenton's reagent (ferrous iron catalysed hydrogen peroxide at neutral pH with a chelator), permanganate, persulfate (alone or activated with metals, heat or alkaline conditions) or ozone injection (Siegrist et al., 2011).

5.3. Biological Treatment-Bioremediation of PAH Pollutants

Septic pits are an example of bioremediation that relies on natural microbes in the soil to decompose human waste. Contaminated environmental media such as soil, sediment, air and water can be treated physically or chemically, but these processes can be very expensive and chemical applications themselves can create additional pollution. The main advantage of bioremediation is being an approach to convert harmful pollutants into relatively harmless materials such as carbon dioxide, chlorine, water and simple organic molecules. Because living organisms are used in the cleaning process, bioremediation is generally cleaner than other types of cleaning approaches. Another advantage of bioremediation is that many cleaning approaches can be applied in situ. In situ biological systems (ISBS) require natural microbial processes to treat organic pollutants and are essential for the treatment of organic pollutants. In these systems, degradation is accomplished by indigenous microbial populations under natural conditions as part of their natural attenuation processes, or by technologies such as replacing it with inspiration that increase the rate of biodegradation, adjusting redox conditions, or bioaugmenting (Sakshi & Haritash, 2020). In bioremediation technologies, bio-regimen (providing bacteria that degrade dominant PAH pollutants) and additional methods such as bio-remediation, bio-augmentation (often an inoculum of microorganisms with known pollutant transformation abilities) and bio-stimulation (the addition of a bulking agent such as wood chips and/or nutrients such as N/P/K) are also widely used to solve the problem of low natural biodegradation efficiency of PAH pollutants by native microorganisms. In bioremediation processes, microbial remediation in PAH pollutants plays a dominant role in contaminated areas with bacterial, fungal and algae populations. The main and effective bacterial family used in bioremediation include *Pseudomonas, Alcanivorax, Microbulbifer, Sphingomonas, Micrococcus, Cellulomonas, Dietzia, Gordonia, Marinobacter, Mycobacterium,*

Haemophilus, Rhodococcus, Paenibacillus, Bacillus, Aeromonas, Burkcoholdum, Bacillus, Aeromonas and *Burkcoholdum.* In general, LMW-PAH compounds with 2 or 3 aromatic rings can be easily degraded by the native bacterial community, since PAH compounds with 4 or more aromatic rings (HMW-PAH compounds) are more stable and hydrophobic, so it is natural for them to be resistant to bacterial degradation (Sakshi & Haritash, 2020). A bioremediation mechanism of microorganisms (microbes) can convert many chemicals into harmless compounds, either through aerobic metabolism (reactions requiring oxygen (O_2) or anaerobic metabolism (reactions that do not require oxygen (O_2). Both reaction types include oxidation (oxidation) and reduction (reduction) reactions. In environments such as surface water and soil where oxygen is readily available, aerobic bacteria oxidize chemical compounds and decompose pollutants. In aerobic biodegradation reactions, O_2 can oxidize a variety of chemicals including organic molecules (molecules containing carbon atoms) such as petroleum products. In that process, O_2 is reduced to produce water. Microbes further decompose oxidized organic compounds into relatively simple and harmless molecules such as carbon dioxide (CO_2) and methane gas. Some aerobes also oxidize inorganic compounds (carbon-free mole-ash) such as metals and ammonium. Oxygen concentration can be very low in heavily contaminated areas and deeper environments such as aquifers. In the sub-surface soils, oxygen diffuses weakly into the ground, and the oxygen in this area is quickly consumed by the aerobes. Although it is possible to inject oxygen into application areas to promote aerobic biodegradation in low-oxidized environments, biodegradation can also occur naturally through anaerobic metabolism; however, anaerobic bacteria (anaerobes) need molecules other than oxygen as electron acceptors. Iron (Fe^{+3}), sulfate (SO_4^{-2}) and nitrate (NO_3^-) are common electron acceptors for redox reactions in anaerobes. Also, some microbes perform both aerobic and anaerobic metabolism. When the oxygen in the environment decreases, they turn to anaerobic metabolism to continue biodegradation. Aerobic bacteria (aerobes) use oxygen (O_2) as the electron acceptor molecule to oxidize PAH pollutants by oxidation. During this process, oxygen is reduced to form water (H_2O) and carbon dioxide (CO_2) is released from the oxidation of PAH pollutants. The energy obtained from the decomposition of the PAH pollutant is used in bacterial growth (biomass). Similar reactions are seen in anaerobic decomposition, but this time anaerobic bacteria (anaerobes) use iron (Fe^{3+}), sulphate (SO_4^{2-}), nitrate (NO_3^-) and other molecules as electron acceptors and oxidize the pollutant. In one study, to improve aerobic PAH bioremediation performance, enzymes such as lignin peroxidase, manganese peroxidase, laccases and phenol oxidases, and oxygenase-facilitated metabolism consisting of monooxygenase and dioxygenase were used (Gupta & Pathak, 2020; Sakshi & Haritash, 2020). According to this study, during bacterial aerobic biological PAH degradation, oxygen acts as both the final electron acceptor and also as an auxiliary substrate for hydroxylation and oxygen-mediated degradation of the aromatic ring. This aromatic ring is first hydroxylated by dioxygenase enzymes, and diol intermediates are released with the aid of dehydrogenase enzymes. These diol intermediates are degraded by the action of dioxygenase enzymes consisting of reductase, ferredoxin and terminal oxygenase subunits and transform into intermediates with the acid cycle (TCA) (Wu et al., 2020). On the other hand, anaerobic PAH degradation by cytochrome P_{450} gives more effective results in the presence of anoxic conditions in various environmental niches such as phreatic zone, deep water sedimentation and waterlogged soil under nitrate and sulphate reduction conditions (Dhar et al., 2020; Yang & Chen., 2021). In addition, there are studies with a consortium of mixed bacterial cultures combining aerobic and anaerobic catabolic activities (Haleyur et al., 2019; Patel et al., 2019; Yang & Chen, 2021). In addition to mesophilic bacteria, there are also PAH degradation studies with acidophilic and thermophilic bacteria populations. Besides bacterial populations, fungi (for example, *Aspergillus sp.*, *Trichocladium canadense, Fusarium oxysporum*) have attracted attention in recent years due to the ability of degradation character in the soil and releasing extracellular enzymes, biodegradation capability of high molecular weight (HMW) xenobiotics, having enzymatic superiority and multiplex pathways (Abo-State & El-Kelani, 2021; Mishraa et al., 2021). Fungal species are very competent PAH degraders, as they are tolerant of high concentrations of recalcitrant compounds, thrives in extreme conditions such as high temperature and low pH conditions, and produces lots of extracellular enzymes. Moreover, branching myceliums allow them to grow and spread through a solid matrix to degrade PAH pollutants in contaminated areas (in situ) by secreting extracellular enzymes. The mechanism of PAH bioremediation by fungi is different from bacteria, and all fungi metabolize PAH pollutants by collaborating to produce CO_2 and a series of oxidized products, rather than using PAH pollutants as a sole carbon source (Gupta & Pathak, 2020). Many soil fungi, ligninolytic (white rot fungi) (Ojha et al., 2019) or non-ligninolytic (Birolli et al., 2018) have been reported to remove both LMW and HMW PAH pollutants. PAH removal mechanism of ligninolytic fungi is based on the conversion of lignin to diphenol intermediates which are oxidized to quinones, at the same time oxidizing PAH pollutants and producing enzymes such as peroxidase, manganese peroxidase and laccases, additionally producing cytochrome P_{450} monooxygenase-like enzymes and producing arene oxide and water by oxidizing PAH pollutants (Gupta & Pathak, 2020). Algae cultures have attracted the attention of scientists for PAH bioremediation, although there are not many studies, hence the pathways and mechanisms of degradation have not been extensively experienced as bacteria populations. In a study conducted by Khemili-Talbi (2015), biosurfactant-producing halophilic archaeon *Natrialba sp. C21* was isolated to degrade phenol, naphthalene and pyrene at very high salinity conditions and to investigate degradation mechanisms. According to this study, degradation was detected by enzyme assays such as catechol 1,2-dioxygenase, catechol 2,3-dioxygenase, protocatechol 3,4-dioxygenase and protocatechol 4,5-dioxygenase, indicating maximum activity of catechol 1,2- dioxygenase in the ortho-cleavage pathway (Khemili-Talbi et al., 2015). Algae as well as bacteria and fungi have been used in PAH removal, cellular biodegradation and/or bioaccumulation in aquatic ecosystems. PAH biodegradation is mediated by production of hydroxylated and dihydroxylated intermediates via both monooxygenase and dioxygenase enzymes, respectively, depending on the algae type. Single-celled green microalgae cultures in freshwater have been preferred in scientific studies due to their ability to degrade even HMW PAH pollutants easily and also due to their widespread availability. To improve bioremediation performance, co-cultures, such as bacterial-fungal co-cultures, fungal-algae synergy and bacterial-algae synergy, have also proven to be the most effective bioremedial approaches in the laboratory as well as large-scale applications. In this consortium, algae cultures contain a variety of extra polymeric and lightweight

TABLE 7.3 PAH removal performance of PAH-degrading bacteria.

BACTERIA PAHS	*INITIAL CONCENTRATE*	*REMOVAL (%)*	*REFERENCES*
Rhodococcus opacus naphthalene	50	91.6	(Goswami et al., 2018)
phenanthrene		82.3	
fluoranthene		80.7	
Ochrobactrum halosaudis, phenanthrene	200	98	(Jamal & Pugazhendi, 2018)
Stenotrophomonas maltophilia, fluorene	50	100	
Achromobacter xylosoxidans, pyrene	50	69	
Micrococcus sp., phenanthrene	100	100	(Jamal, 2018)
Pseudomonas sp., pyrene	100	93	
Bacillus sp., benzo(k) fluoranthene *Bacillus subtilis, Bacillus megaterium*	10	77	
Microbacterium steraromaticum naphthalene	10	81.4	(Kumari et al., 2018)
Pseudomonas aeruginosa phenanthrene	3.5	67.1	
benzo(b)fluoranthene	6.5	61.0	
Stenotrophomonas maltophilia fluorene	1.9	47.9	
Pseudomonas aeruginosa fluorene	250	98	(Fathi & Ebrahimipour, 2018)
Serratia marcescens phenanthrene	200	97	
anthracene	250	94	
pyrene	150	45	
Ochrobactrum halosaudia AJH1, naphthalene	1000	97	(Pugazhendi et al., 2017)
Ochrobactrum halosaudia AJH2 phenanthrene	50	100	
Pseudomonas aeruginosa anthracene	100	90	
fluorene	100	98	
pyrene	50	90	
benzo(e)pyrene	10	85	
benzo(k)fluoranthene	10	81	

TABLE 7.3 (Continued)

BACTERIA PAHS	*INITIAL CONCENTRATE*	*REMOVAL (%)*	*REFERENCES*
Microbacterium maritypicum CB7 benzo(a)pyrene	10	69.0	(Mansouri et al., 2017)
Raoultella ornithinolytica acenaphthene	50	98.60	(Alegbeleye et al., 2017)
fluorene		99.90	
Serratia marcescens acenaphthene		95.70	
fluorene		97.90	
Bacillus megaterium acenaphthene		90.20	
fluorene		98.40	

Table created by authors.

compounds (such as lipids, proteins, nucleic acids, fermentation products) that promote bacterial activities, which increase fungal growth and PAH degradation (Subashchandrabose et al., 2019). The efficiencies of PAH-degrading bacteria to degrade specific PAH are described in Table 7.3.

5.3.1. Usage of biochar and activated carbon for the PAH bioremediation

Many studies of PAH bioremediation have reported that biochar, able to stimulate microbial activity by means of providing habitats and nutrients to microorganisms, improve the biodegradation of PAHs by bio-stimulation strategy. Properties such as organic carbon composition, porosity and mineral content used in PAH removal have different microbial responses in the soil, including diversity, abundance and community structure (Yang et al., 2021). There are studies reporting that biochar increased the biodegradation performance of aromatic pollutants by 14.2% to 16.1% in oil contaminated soils. Biochar has emerged as a material, considered to be the only sorbent, that can provide simultaneous effects on three main fate processes (volatility, absorption and biodegradation). Although biochar plays an important role in the removal of PAH pollutants, which can be cheaper and environmentally friendly, studies showed that only the biochar adsorption process may not be sufficient alone to remove pollutants due to decreasing N at above 700°C, which is necessary for microbial growth and metabolism (Wang et al., 2020). As a result, it can be said that a combination of biochar and other modifications give better results to improve remediation performance as a bio-stimulation strategy. As an alternative to biochar, Bonaglia et al. (2020) conducted incubation experiments to study the biodegradation potential of naphthalene in a variety of common microbial communities in the presence/absence of activated carbon (AC) under aerobic/anaerobic conditions. Radio-respirometry experiments with endogenous microorganisms showed that the degradation of naphthalene was strongly (96%) induced by the addition of AC under anaerobic conditions, but had no effect of AC on naphthalene biodegradation under aerobic conditions. Bio-augmentation tests with microbial population cultures grown on naphthalene revealed that activated carbon (AC) further stimulated naphthalene degradation (92%) and the relative abundance of *Geobacter, Thiobacillus, Sulfuricurvum* and methanogenic archaea increased sharply after modification with AC under anaerobic conditions. Explaining that AC particles promote direct electron transfer between microorganisms involved in PAH degradation, this study suggested that AC enrichment strategies could be used to facilitate bioremediation of anoxic oil contaminated sediments and soils (Bonaglia et al., 2020).

6. BIOREMEDIATION OF PAH POLLUTED WATER AND WASTEWATER

6.1. Bioremediation of PAH Pollutants in Aquatic Environments and Aquifers

Water (water, groundwater, aquifers and potable water), an important component of the ecosystem, is highly susceptible to PAH contamination from crude oil exploration, spills, spills and inadequate waste management. Continuous pollution of the existing freshwater sites across the globe by petroleum and their by-products has disastrous consequences for the animals, humans, plants and microorganisms. Terrestrial subterranean aquatic environments, including aquifers, which are the largest microbial reservoirs in the biosphere, make up about 98% of the world's freshwater and are becoming an increasingly important resource worldwide for domestic, agricultural and industrial use (Gleeson et al. 2016). Zhang et al. (2019) have determined 16 PAHs in China's freshwater to be two to three times higher than that of global freshwater in source water bodies of China. These water and groundwater sources, including PAH contaminants, can be treated biologically using both in situ and ex situ bioremediation procedures. Bioremediation processes involve the use of microorganisms, plants and enzymatic reactions and are regarded as an efficient means of treatment methods in the detoxification and degradation of environmental contaminants in water and other environmental compartments. Fungi (such as *Peniophora gigantea, Phanerochaete chrysosporium, Pycnosporus coccineus, Trametes versicolor*) and bacterial species (such as *Aspergillus sp., Trichocladium sp., Fusarium sp., Pseudomonas sp.*) are capable of degrading PAH pollutants during bioremediation processes in the aquatic environment. In a study by Alegbeleye et al. (2017), about bioremediation of acenaphthene and fluorene in two river systems in the Western Cape, South Africa, 19 bacterial isolates were used in batch scale experiments at the optimum temperature for efficient degradation of both compounds in a shaking incubator after 14 days, at 25°C, 30°C, 35°C, 37°C, 38°C,

40°C and 45°C. The degradation of fluorene was more efficient and rapid compared to that of acenaphthene (Alegbeleye et al., 2017). Dodor et al. (2020) studied and recommended an immobilized laccase mediator for catalysed oxidation of aqueous mixtures of PAH contaminant in aquatic environments. Bioremediation of PAH contaminants in water is already a complex problem, and the addition of heavy metal pollution requires further research to develop new and improved bioremediation techniques. In this sense, the inoculation of the polluted environment with biodegradative microorganisms (i.e., bio-augmentation) is a promising technique to improve the biodegradation of contaminated areas. Ma et al. (2018) reported the degradation of PAH mixtures of naphthalene, fluorine, phenanthrene, anthracene, and fluorant in water contaminated with heavy metals by the microbial consortium, including *Acremoniumsp* and *Bacillus subtilis*; microbial consortium showed higher tolerance to the heavy metals tested (Fe^{2+}, Al^{3+}, Ni^{2+}, Cu^{2+}, Mn^{2+} and Zn^{2+}) separately, with increasing metal concentrations decreasing the tolerance to individual metals. In the absence of heavy metals, the consortium decomposes naphthalene, fluorine, phenanthrene, anthracene and fluoranthene one by one (50 mmol/L) within 10 days efficiently, while the addition of Ni^{2+} suppresses the removal of phenanthrene and anthracene, Cu^{2+}, Zn^{2+}, Fe^{2+} and Al^{3+} supplementation showed a significant inhibition against phenanthrene, anthracene and fluoranthene removal, respectively. Cu^{2+} had a more significant effect on the degradation of these PAH compounds compared to other metals. On the other hand, Mn^{2+} has been disclosed to significantly increase the removal of fluorene, phenanthrene and fluorescent, while inhibiting anthracene degradation. The most common groundwater pollutants are BTEX (benzene, toluene, ethylbenzene, xylenes) compounds that require remediation. Although aerobic biodegradation of BTEX progresses faster than anaerobic biodegradation, an additional process is needed for natural attenuation processes at BTEX contaminated sites due to the low aqueous solubility and rapid consumption of oxygen. During anaerobic BTEX degradation, microbes transfer electrons from target compounds to different terminal electron acceptors such as nitrate, ferric iron, sulfate or carbonate. A plume gets constituted of different redox zones depending on the availability of electron acceptors. Among the BTEX compounds, benzene is the most difficult to degrade under anoxic conditions with various electron acceptors such as carbonate, sulphate, ferric iron or nitrate, at laboratory conditions. Previous studies have shown that in situ benzene biodegradation in groundwater gets stimulated by injecting nitrate and sulfate or nitrate, triethyl phosphate and persulfate. The ways and extent of anaerobic benzene degradation in contaminated aquifers are mainly controlled by the prevailing biogeochemical conditions (such as inorganic sulphur compounds sulphide and sulphate) and the presence of specific benzene degrading microorganisms. The presence of free sulfide in groundwater and aquifer indicates the presence of sulfate reducing conditions and precipitation of any iron ions as insoluble iron-sulfur compounds. Fe-S compounds precipitate very rapidly, while being electron donor in the presence of suitable electron acceptors such as nitrate. Upon oxidization they get transformed into another electron acceptor preferred in anaerobic degradation in the aquifer. However, they may also have an inhibitory effect on benzene degradation, as they can be toxic to aerobic and anaerobic organisms at high concentrations such as > 50 mg L^{-1}. Studies, that have also been conducted on making an alternative electron acceptor for anaerobic benzene oxidation, revealed that Fe-S-compounds are quickly oxidized to sulphate by nitrate-reducing, sulphide oxidizing in the presence of a wider range of BTEX-degrading denitrifying microorganisms for BTEX degradation (Han et al., 2021; Müller et al., 2021). Due to the high cost associated with the extraction and shipment, ex situ treatment of contaminated groundwater is not generally performed. Besides that, in situ bioremediation processes require understanding and engineering a set of subsurface process. In order to enhance in situ bioremediation, Marić et al. (2015) investigated the addition of nutrients, and stimulation of oxidation processes by injection of H_2O_2, into the aquifer in Serbia. In this study, bioaugmentation/reinoculation was achieved by injection of a zymogenous consortium of hydrocarbon degrading microorganisms isolated from the contaminated groundwater. The efficiency of the remediation was measured by changes in the content of total petroleum hydrocarbon (TPH), total chemoorganoheterotroph (TC) and hydrocarbon degrading (HD) microorganisms. After biostimulation and bioaugmentation, the number of TC and HD microorganisms started to increase. The applied remediation approach was highly efficient and very effective in reducing TPH to acceptable levels. Altogether these facts provide compelling evidence of the potential for remediation of groundwater contaminated with petroleum hydrocarbons. Other common groundwater pollutants are non-aqueous phase fluids (NAPLs) which are commonly used in industrial activities (Li et al., 2017). Isolated NAPL remediation technologies based on in situ plume treatment are not well suited for remediation of areas contaminated with PAH pollutants, due to their tendency to absorb into soil materials and their lower water absorption properties. Diesel is a common NAPL pollutant found in contaminated soils/waters, and can be used in chemical, physical and biological methods for its remediation. NAPL can be bioremediated by bacteria or phytoremediated using plants, but this is only suitable for shallow pollutants (Bajagain et al., 2020; Gautam et al., 2020). The efficiency of bioremediation using bacteria can also be enhanced by injecting bacteria, such as *Acinetobacter baumannii, Cellulosimicrobium cellulans* and *Pseudomonas fluorescens,* into the polluted zones (bioaugmentation). Integrated treatment technologies have been developed to improve the efficiency of NAPL's transfer from soils to the aqueous phase (Liang et al., 2017). In this direction, biological surfactants, with amphilic properties such as phospholipids, lipoproteins and glycolipids, can increase the biodegradation of diesel by increasing the bioavailability of NAPLs by dissolving in water. Accordingly in a study, NAPL remediation in soil and groundwater is achieved with on-site treatment technologies such as pump and treat system, on-site bioremediation and in situ chemical oxidation (ISCO) without excavation (Lamichhane et al., 2017). In recent years, ISCO has been integrated or completely replaced with environmentally friendly biological treatments, such as phytoremediation, surfactant enhanced bioremediation (SEBR) in terms of high operating costs and environmental damage (Saleh et al., 2019). To reduce cost, time and usage of chemicals, bacteria, fungi and algae have been applied to decompose pollutants dissolved in surfactant enhanced bioremediation (SEBR) processes integrated with in situ enzymatic oxidation by means of using oxygenase, dehydrogenase and ligninolytic enzymes. However, in these processes, a few days to several weeks are required for biotransformation of the pollutants due to the activity of some enzymes. In this context, the feasibility of anthracene and benzo(a)pyrene oxidation by fungal laccase was investigated for the first time in the literature in the present work, in a model and a real (groundwater) solution. In this study, four groundwater samples were treated in the laboratory by batch experiments with laccase to determine the efficiency of PAH oxidation and the product of laccase-transformation of PAH pollutants, by using anthracene as a model aromatic compound.

This study provided an opportunity to understand the effects of enzyme stability on pH and the presence of mediators on oxidation efficiency in some groundwater. Normally, NAPLs can inhibit biodegradation by forming barriers that prevent both groundwater and pollutant flow. In a study to address this, an improvement approach was presented using *Pseudomonas aeruginosa* with injectable cationic traps (filters) combined (Marshall et al., 2021). Moreover, seas have great importance among aquatic environments in terms of healthy consumption of seafood. Since PAH contaminants have the property of accumulating in seafood, they should be monitored and precautions should be taken. In this context, Çolakoğlu et al. (2020) conducted a study of bioaccumulation monitoring for the chemical contaminants in mussels *Mytilus galloprovincialis* from the southern coast of the Marmara Sea in Turkey. Aquatic creatures living in aquatic environments get their share of PAH pollutants. Among these are aquatic creatures, mangroves living on the tropical coastline and in deltas and bays where rivers bring fresh water and sediment to the sea, include many and various PAH degrading microorganisms. Any damage to ecologically sensitive mangrove habitats often takes a long time due to limited populations of local degradation microorganisms. On the other hand, once damaged, it often takes a long time to perform bio-stimulation processes relying on supplied nutrients or substrates to promote the degradation of natural microorganisms to bioremediate mangrove beds in which nutrients are found in low concentrations. For this purpose, to reduce oil contamination, microbial activity was stimulated by adding soluble inorganic fertilizers to mangrove deposits. A study conducted by Lang et al. (2016) emphasized that the number of benzene rings of PAH contaminants are quite effective for the bioremediation of PAH pollutants from mangrove sediments, as well as environmental factors such as pH, temperature, oxygen and the ability of endogenous or exogenous microflora to metabolize hydrocarbons. In this study, *Rhodococcus* and *Erythropolis* bacteria were used as an exogenous bacterial strain to evaluate the biodegradation of PAH mixtures from artificially contaminated mangrove sediments by applying each natural attenuation, bio-stimulation, bio-growth processes individually followed by combined bio-stimulation and bio-augmentation processes. They reported that the bio-stimulation/bio-augmentation combination has proven to be the most effective PAH degradation treatment (Lang et al., 2016). In addition to all these, PAH pollution in sediment environments, which is one of the important matters of aquatic environments, such as streams and rivers, is also of a great importance. In order to treat sedimental environments, sediment microbial fuel cells (SMFC) are used which are a type of bio-electro-chemical systems collecting electrical energy from organically rich sediments, such as benthic sediments and paddy soil, consisting of an anode embedded in the anaerobic bottom of the aquatic environment and a cathode submerged in overlying oxygen-rich water (Wang et al., 2020). In recent years, SMFCs became environmental remediation projects that can provide less aggressive and flexible electron donors or acceptors, cause the least disturbance to the aquatic environment, and ensure proper monitoring by optimizing electrochemical parameters (Gustave et al., 2020). Whether SMFC is used for power generation or environmental improvement, microorganisms living in sediments play a very important role in this process. They oxidize organic compounds and transfer the produced electrons to anodes to start the pollutant removal or electricity generation process (Yang et al., 2021).

6.2. Bioremediation of PAH Pollutants in Wastewater

In recent years, bioremediation studies, on wastewater, have come to the fore in addition to groundwater and aquifers, as the diversity of industrial activities has increased the diversity of PAH, causing industrial wastewater to contain highly complex PAH contaminants. The effects of different substrates on microorganism growth and the rate of PAH biodegradation can play an important role in such a mixed wastewater system. The impact of the simultaneous presence of different PAH contaminants on removal from wastewater, which usually contains a mixture of pollutants, is not well understood. Goswami et al. (2018), studied biological treatment of wastewater containing a mixture of polycyclic aromatic hydrocarbons using the oleaginous bacterium *Rhodococcus opacus*. In all experimental studies, the individual effect of PAH pollutants on biodegradation has shown that the interaction between substrates is more important. Compared to adsorption, volatilization, flocculation, reverse osmosis, photolysis, advanced oxidation processes (AOP) and chemical degradation methods, bioremediation processes of using a PAH degrading organism is considered to be more effective and economical for the proper removal of PAH pollutants. The simultaneous biodegradation of these PAH contaminants along with lipid accumulation by *Rhodococcus opacus* in a ternary substrate system. In this study, the inhibitory effect of PAH pollutants on biodegradation was strongly influenced by the respective initial PAH concentrations. Subsequently, an increase in the PAH pollution complexity (increasing number of rings) was reported. The more potent the PAH contaminants inhibited each other in terms of biodegradation. A 23 full factorial design of experiments was employed with the three PAH pollutants at two different levels. A maximum removal of 91.6%, 82.3% and 80.7% was achieved for naphthalene, phenanthrene and fluoranthene, respectively. PAH biodegradation efficiency in the mixture has been mainly affected by initial concentration and aromatic complexity of the PAH pollutants. This study revealed the potential utility of *R. opacus* for bioremediation and industrial wastewater treatment at the same time (Goswami et al., 2018). Industrial activities, such as agro-food, petroleum, textile, and leather industries, generate saline wastewater with a high content of organic pollutants such as aromatic hydrocarbons, phenols, nitroaromatics and azo dyes. Spillage and leakage of that wastewater into freshwater without any primary treatment also affects aquatic environment, potable/drinking water sources and agricultural environment. Castillo-Carvajal et al. (2014) have found that the use of halophilic bacteria provided effective results in industrial wastewater bioremediation due to their ability to degrade these hazardous substances efficiently in the presence of high salt. Moreover, the leakage of crude oil from tankers, oil spills from industrial activities, the discharge of gasoline or diesel into the seas, and the discharge of waste oil into the seas are the main events causing serious permanent and dangerous marine pollution. Cost-effective and efficient degradation methods have been developed for the treatment of soil and water polluted in this way (Alessandrello et al., 2017). For example, studies with biofilm-forming microorganisms (such as UltraTech microbes) presented efficient results in the degradation of these contaminants (Liu et al., 2019; Zheng et al.2020). Recently, the use of biocarriers to immobilize bacteria through biofilm formation to treat or to remove the dangerous component in crude oil has become a promising technology (Esfandiar et al., 2021). In this sense, there are remediation studies using carriers, which are bound by biofilm-forming microorganisms to enhance the survival rate, metabolic activity and resistance of bacteria to toxic effects of persistent compounds. Carriers such as sodium alginate-biochar-polyvinyl alcohol-diatomite mixture; peanut shell, cocopeat and rice husk powder, bamboo charcoal, peanut shell powder, PUF have been incorporated. In another related study, the efficiency of these immobilization carriers on supporting purple phototrophic bacterial (PPB) strains in hydrocarbon biodegradation of crude oil contaminated seawater was compared. The surface of these biocarriers was supplemented with crude oil polluted seawater and immobilized by PPB strains, *Rhodopseudomonas sp.* DD4, DQ41 and FO2. The bacteria-driven carrier systems degraded over 84.2% supplemented single polycyclic aromatic hydrocarbons (PAH contaminants) (Nhi-Cong et al., 2021). For the last decade, microbial surfactants have attracted the attention of researchers in the degradation of PAH pollutants from wastewater due to their diversity, selectivity, environmentally friendly nature, operation under harsh environmental conditions, easy large-scale production and potential applications in environmental protection (Joy et al., 2017; Liang et al., 2017).

6.3. Emerging Technologies for PAH Bioremediation

In the last decade, phytoremediation techniques, among microbial remediation (bioremediation) processes, are the emerging technologies being environmentally friendly and cost-effective. Phytoremediation is a passive technology for cleaning PAH contaminated soils and aquatic environments, even with heavy metal pollutants together, based on using of hyperacumulative plants. Approximately 350 plant species are naturally capable of removing toxic materials; poplar, juniper trees, some herbs and alfalfa have been used successfully in phytoremediation. During phytoremediation, chemical pollutants are absorped from contaminated water present in soil. After toxic PAHs enter the plant, hyperaccumulator plant cells can use enzymes to degrade the PAHs. In other cases, PAHs concentrate in plant cells and thus act as a kind of "plant sponge" to wipe away all plant pollutants. Contaminated plants are disposed of in other ways. Because high concentrations of pollutants kill most plants, phytoremediation works best when concentration is low in surface soil or groundwater. Heavy metals absorbed by plants in areas contaminated with metals either accumulate in the soil and become stable or are removed from the soil. Some hyperacumulative plants such as *Thlapsi, Urtica, Taraxacumofficinale, Chenopodium, Polygonumsachalase* and *Allyssim* have the ability to accumulate heavy metals such as cadmium, copper, lead, nickel and zinc (Tasleem et al., 2015). Compared to different remedial methods, the advantages are low cost, aesthetic appeal, easy application and short application time. Two main constraints of phytoremediation are that it can only be applied to the surface soil (up to a depth of about 50 cm) and cleaning typically takes years. Besides phytoremediation, rhizoremediation processes are also great options for hydrocarbon remediation because of their relatively low cost, energy efficiency, possible use on a wide range of pollutants, and minimal or near zero secondary waste generation. Rhizoremediation, based on phytoremediation, is an effective and feasible treatment technology between plants and microorganisms, in which in situ reactors are developed in sufficient rhizosphere to decompose pollutants by utilizing nutrients secreted by plants used by microorganisms to reproduce. Rhizoremediation provides a further step in bioremediation processes through the addition of plants, improving soil texture, aeration and promoting the development of soil microorganisms, while improving their catabolic metabolism. In this method, in addition to the use and stimulation of the disintegrating microbiota, rhizobacteria that promote plant growth are also used to alleviate pollutant-induced plant stress that support the growth of vegetation. There are on-site field-scale trials using this technology; further, the need to develop this strategy in the field is advocated, especially in semi-arid conditions. In this study, it is understood that there is limited information about the interaction mechanism of microbial functions and metabolic diversity during biological degradation. In this regard, it is essential to understand the metagenomics of bioremediation processes and to reveal and monitor microbial communities by gaining access to functional gene composition (Kachieng et al., 2018). Metagenomic analysis enables the species determination of bioremediation microorganisms obtained after DNA isolation of samples taken directly from the environment without culturing, and understanding the genetic diversity, population structure and the ecological roles of these microorganisms affecting human life. Metagenomic studies based on a gene-centred approach are studies aimed at evaluating the biological function of the gene rather than taxonomic identity. The genomic nature of microbial diversity may also change in the presence of different hydrocarbons in PAH contaminated environments. In this sense, metagenomic studies can provide excellent opportunities to find new microbial strains and genes that play a role in PAH bioremediation. In a related study conducted by Kachieng et al. (2018), the metagenome of the microbial community, in two different petroleum contaminated regions and microbial adaptations, were examined in oiled aquifer fields in South Africa. It was revealed that protozoa (62.04%) found to be the most dominant group, followed by fungi (24.49%) and other sequences. They stated that the adaptation and survival abilities of microorganism groups in oiled water fields should be investigated in more detail (Kachieng et al., 2018). Moreover, genomic studies provide the opportunity to obtain information about the genetics of bacteria and plants for easier removal of recalcitrant PAH contaminants. With the development of recombinant DNA technologies, scientists have begun to be able to make genetically modified (GM) organisms, which has had great potential to improve the bioremediation process. Scientists have added special gene encoding enzymes that can metabolize pollutants to reporter genes such as lux genes from bioluminescent marine bacteria, using recombinant DNA technology. As PAH pollutants degrade, bacteria emit light, which is used to monitor the rate of biodegradation. Using similar techniques, biosensors have been developed from recombinant bacteria with lux genes. To always stay one step ahead of environmental pollutants, researchers must continue to develop new cleaning techniques. One of the most practical approaches, that can be done to learn new bioremediation strategies is to create microcosms that are artificial test environments that mimic real-life environmental phenomena. Microcosms can be as small as a test tube or a few grams of soil in a bottle, or they are usually on a large scale to better reflect large areas. Some microcosms-containing soil, water, pollutants, and microbes for bioremediation capabilities-consist of small bioreactors measuring five gallons. For example, large pools in indoor or outdoor areas or plots that may prevent the contaminant from leaking from the test facility can be used as microcosmosis. When specific or genetically modified organisms with bioremediation potential are identified, studies in bioreactors or small isolated soil or water areas are important in determining the ability of these organisms to effectively clean up in wider applications. Scientists can test the ability of organisms to degrade different pollutants under different conditions by carefully managing environmental conditions such as humidity, temperature, nutrient amount, oxygen, pH and different soil types in the microcosmos. Even with microcosm tests, the destruction rate of experimental microbes placed in contaminated groundwater or soil microcosm can be monitored and the cleaning time can be calculated. In addition, this allows scientists to test the bioremediation of mixtures of different pollutants at the same time. By testing bioremediation studies in microcosm, both time and money can be saved by deciding whether a cleaning strategy to successfully clean the contaminated environment in the area is needed. Scientists have found that anaerobic bacteria in sediments can be an energy source. Therefore dirty sediments with PAH pollutants in sewage sludge, the bottoms of oceans and lakes may be used as undiscovered energy resources in bioremediation as an emerging study. Sediments in lakes and oceans are rich in organic material due to degradation of rotting leaves and dead organisms. Anaerobes found in these "swamp mud" heaps convert them into energy using organic molecules found in sediments. The term "Electrigens" is the name given to microbes that can transfer electrons to electrodes by oxidizing organic materials to carbon dioxide and thus generate electricity. When the necessary conditions are met, the electricity can be brought together and linked together to form electricity-conducting nanowires. These types of microbial strains can even be used to generate electricity from waste. For example, *Desulfuromonas acetoxidans* is an anaerobic marine bacterium that can oxidize organic molecules in sediments by using iron as an electron acceptor.

6.4. Conditions Affecting Bioremediation of PAH Pollutants

The presence of microorganisms in bioremediation studies does not always lead to in situ biodegradation, as many limitations, such as insufficient biomass, use of a wide range of substrates, competitive inhibition or catabolite pressure, may all inhibit the process. Hence the analysis and characterization of microorganisms and their correlation with biological degradation processes, geological, hydrogeological and geochemical properties of the environment have to be determined. Conditions affecting the performance of bioremediation depend on the relevant microbial population, degree of habituation, availability of contaminants, chemical structure of the compound, cellular transport properties, chemical partitioning in the growth medium and a favorable environment for recovery. These environments can also be provided with parameters such as pH, temperature, salinity, biosurfactants in the bioreactor. In the regard of pH, the environment in which PAH degrading microorganisms have the best performance is neutral pH. Bioremediation also causes problems, as areas with PAH contaminants are generally not at optimum pH. Lime, nutrients or fertilizers can be added to the environment to eliminate these problems and optimize the pH. Temperature is an important parameter affecting both the chemistry and degradation of the PAH compounds and the physiology of the microbial flora in the environment; increase in temperature also increases PAH solubility. Temperature also affects the metabolic activities of aerobic microorganisms. The optimum temperature in the soil varies between 30–40°C, 20–30°C in some freshwater environments,- and 15–20°C in marine environments. On the other hand, since there are microbial species that can grow at very low or very high temperatures, biodegradation of PAH contaminants can occur over a versatile temperature range. Some studies tend to focus on mesophilic temperatures and have shown that microorganisms can metabolize PAH pollutants at extreme temperatures. High salinity levels can disrupt the protein structure of microbes, denature enzymes, and thus decrease microbial metabolic rates. To counter high salt conditions, dilution to lower salinity, reverse osmosis prior to biological treatment, and salt removal by ion exchange or electrodialysis can be suggested as commonly used techniques. Bio-surfactants are a structurally diverse group of surfactants synthesized by various microorganisms and are used as additives to counteract the low water solubility of PAH pollutants and increase the efficiency of bioremediation. These are environmentally friendly, biodegradable, less toxic and non-hazardous, highly reactive and active at extreme temperatures, pH and salinity. Biological surfactants facilitate the transport of hydrophobic contaminants into the aqueous phase through specific interactions that result in dissolution, thus increasing their bioavailability, potentially making them more susceptible to biodegradation. Bio-reactors are aerobic or anaerobic designed containers in which biochemical processes are carried out. They can operate either in batch or continuous mode, or can be combined depending on the contaminant to be cured, the contaminated environment and cost constraints. Bioreactors have proven effective in improving the soil and in some cases improving water contaminated with organics (Varjani et al., 2014).

7. CONCLUSION

This chapter revealed that PAH pollutants have major harmful, toxic, carcinogenic, teratogenic and mutagenic effects on both the environment and the health of living things. Their remediation from air, soil, water and sediments is needed for the benefit of both the environment and human health. Moreover, it has also been proven that LMW PAH pollutants are relatively easy to reduce, but that most carcinogenic HWM PAH pollutants with more compact rings are very difficult to degrade. It is obvious that laboratory and pilot/reactor-scale studies for degradation of PAH pollutants and removal of heavy metals from the ecosystem have provided significant contributions and perspectives to the literature. However, the implementation of lab-scale experiments to field-scale industrial applications has to be critically examined and evaluated. Bioremediation of PAH pollutants, bio-stimulation followed by microbial degradation, and immobilization of the microbial consortium in microcapsules have become more interesting issues in recent years. As the natural environment is an interactive and interdependent system for the niche and nutrient cycle by microbial communities, synergistic hydrocarbon degradation mechanisms must also be understood in order to properly apply and scale up in situ bio-stimulation and bio-augmentation approaches. Understanding the dynamics and metagenomics of PAH degradation in the soil and aquatic environment will support improving genetically modified microbes to use as bio-sensors for the degradation of PAH pollutants more efficiently. In addition to all studies, composting, energy generation, green nanoremediation and biocatalyst technologies need to be explored. Further studies are necessary to create microcosms, artificial test environments that mimic real-life environmental phenomena, in order to develop more viable bioremediation strategies in the field.

REFERENCES

Abdel-Shafy, H. I., & Mansour, M. S. (2016). A review on polycyclic aromatic hydrocarbons: source, environmental impact, effect on human health and remediation. *Egypt. J. Pet., 25*, 107–123. doi:10.1016/j.ejpe.2015.03.011.

Abo-State, M. A. M., & El-Kelani, T. A. (2021). Bacterial biodegradation pathways of low and high molecular weight polycyclic aromatic hydrocarbons (PAHs). *Archives of Petroleum & Environmental Biotechnology, 5*, 158. doi: 10.29011/2574-7614.100058

Adeola, A. O., & Forbes, P. B. C. (2020). Advances in water treatment technologies for removal of polycyclic aromatic hydrocarbons: Existing concepts, emerging trends, and future prospects. *Water Environmental Research.* doi:10.1002/wer.1420.

Alegbeleye, O. O., Opeolu, B. O., & Jackson, V. (2017). Bioremediation of polycyclic aromatic hydrocarbon (PAH) compounds: (acenaphthene and fluorene) in water using indigenous bacterial species isolated from the Diep and Plankenburg rivers, Western Cape, South Africa. *Braz. J. Microbiol., 48*, 314–325. doi:10.1016/j.bjm.2016.07.027.

Alessandrello, M. J., Parellada, E. A., Tomás, M. S. J., Neske, A., Vullo, D.L., & Ferrero, M. A. (2017). Polycyclic aromatic hydrocarbons removal by immobilized bacterial cells using annonaceous acetogenins for biofilm formation stimulation on polyurethane foam. *Journal of Environmental Chemical Engineering, 5*(1), 189–195. doi:10.1016/j.jece.2016.11.037

Ata, R., & Töre, Y.G. (2019). Characterization and removal of antibiotic residues by NFC-doped photocatalytic oxidation from domestic and industrial secondary treated wastewaters in Meric-Ergene Basin and reuse assessment for irrigation. *Journal of Environmental Management, 233*, 673–680.

Bajagain, R., Gautam, P., & Jeong, S. W. (2020). Degradation of petroleum hydrocarbons in unsaturated soil and effects on subsequent biodegradation by potassium permanganate. *Environ. Geochem. Health,4*, 1705–1714.

Balati, A., Shahbazi, A., Amini, M. M., & Hashemi, S. H. (2015). Adsorption of polycyclic aromatic hydrocarbons from wastewater by using silica-based organic-inorganic nanohybrid material. *Journal of Water Reuse and Desalination, 5*(1), 50–63. doi:10.2166/wrd.2014.013

Birolli, W. G., Santos, D. D. A., Alvarenga, N., Garcia, A. C., Romão, L. P., & Porto, A. L. (2018). Biodegradation of anthracene and several PAHs by the marine-derived fungus Cladosporium sp. CBMAI 1237. *Mar. Pollut. Bull., 129*, 525–533. doi: 10.1016/j.marpolbul.2017.10.023.

Bolden, A. L., Rochester, J. R., Schultz, K., & Kwiatkowski, C. F. (2017). Polycyclic aromatic hydrocarbons and female reproductive health: a scoping review. *Reproductive Toxicology*,b*73*, 61–74. doi: 10.1016/j.reprotox.2017.07.012.

Bonaglia, S., Broman, E., Brindefalk, B., Hedlund, E., Hjorth, T., Rolff, C., … & Gunnarsson, J. S. (2020). Activated carbon stimulates microbial diversity and PAH biodegradation under anaerobic conditions in oil-polluted sediments. *Chemosphere, 248*, 126023.

Brandt, M., & Einhenkel-Arle, D. (2016). *Polycyclic aromatic hydrocarbons harmful to the environment! toxic inevitable?* German Environment Agency.

Castillo-Carvajal, L. C., Sanz-Martín, J. L., & Barragán-Huerta, B. E. (2014). Biodegradation of organic pollutants in saline wastewater by halophilic microorganisms: A review. *Environ Sci Pollut Res., 21*, 9578–9588. doi: 10.1007/s11356-014-3036-z

Çolakoğlu, S., Künili, I. E., & Çolakoğlu, F. (2020). Bioaccumulation monitoring of chemical contaminants in mussels mytilus galloprovincialis from the southern coast of the Marmara Sea, Turkey. *Turkish Journal of Veterinary & Animal Sciences, 44* (2), 235–243, doi: 10.3906/vet-1905-7.

Deveau, A., Bonito, G., Uehling, J., Paoletti, M., Becker, M., Bindschedler, S., …& Wick, L. Y. (2018). Bacterial–fungal interactions: ecology, mechanisms and challenges. *FEMS Microbiology Reviews, 42*(3), 335–352. doi:10.1093/ femsre/ fuy008.

Dhar, K., Subashchandrabose, S. R., Venkateswarlu, K., Krishnan, K., & Megharaj, M. (2020). Anaerobic microbial degradation of polycyclic aromatic hydrocarbons: a comprehensive review. *Rev. Environ. Contam. T., 251*, 25–108. doi: 10.1007/398_2019_29.

Dodor, D. E., Miyittah, M., & Ahiabor, B. D. K. (2020). Immobilized laccase mediator-catalyzed oxidation of aqueous mixtures of polycyclic aromatic hydrocarbons. *J. of Polycyclic Aromatic Compounds,40*(3), 563–573.

Esfandiar, N., Suri, R., & McKenzie, E.R. (2021). Simultaneous removal of multiple polycyclic aromatic hydrocarbons (PAHs) from urban stormwater using lowcost agricultural/industrial byproducts as sorbents. *Chemosphere*, 129812. doi:10.1016/j.chemosphere.2021.129812.

Fathi, Z., & Ebrahimipour, G. (2018). Isolation and identification of polycyclic aromatic hydrocarbons (PAHs) degrading bacteria from Arak Petrochemical Wastewater. *Journal of Microbiology Biotechnology and Food Sciences, 7*(5), 499–504. doi:10. 15414/ jmbfs. 2018.7. 5. 499- 504.

Gautam, P., Bajagain, R., & Jeong, S.W. (2020). Combined effects of soil particle size with washing time and soil-to-water ratio on removal of total petroleum hydrocarbon from fuel contaminated soil. *Chemosphere, 250*, 126206.

Gautam, R. K., Mudhoo, A., Lofrano, G., & Chattopadhyaya, M. C. (2014). Biomass-derived biosorbents for metal ions sequestration: Adsorbent modification and activation methods and adsorbent regeneration. *Journal of Environmental Chemical Engineering, 2*, 239–259. doi:10. 1016/j. jece. 2013. 12. 019

Ghosal, D., Ghosh, S., Dutta, T. K., & Ahn, Y. (2016). Current state of knowledge in microbial degradation of polycyclic aromatic hydrocarbons (PAHs): A review. *Frontiers Microbiology*. doi:10. 3389/ fmicb. 2016. 01369.

Gleeson, T., Befus, K.M., Jasechko, S.,Luijendijk, E., & Cardenas, M.B. (2016). The global volume and distribution of modern groundwater. *Nat Geosci., 9*, 161–167.

Goswami, L., Manikandan, N.A., Dolman, B., Pakshirajan, K., & Pugazhenthi, G. (2018). Biological treatment of wastewater containing a mixture of polycyclic aromatic hydrocarbons using the oleaginous bacterium Rhodococcus opacus. *Journal of Cleaner Production, 196*, 1282e1291.

Guo, J., Liu, J. H., Wang, L. Y., & Liu, H. (2015). Modification of ultrafiltration membranes with carbon nanotube buckypaper for fouling alleviation. *Membrane Water Treatment, 6*, 1–13. doi:10. 12989/ mwt. 2015.6. 1. 001.

Gupta, S., & Pathak, B. (2020). Mycoremediation of polycyclic aromatic hydrocarbons. *Abatement of Environmental Pollutants*, 127–149.

Gustave,W., Yuan, Z. F., Li, X. J., Ren, Y.X., Feng, W. J., Shen, H.B., & Chen, Z. (2020). Mitigation effects of the microbial fuel cells on heavy metal accumulation in rice (*Oryza sativa L.*). *Environ. Pollut., 260*.

Haleyur, N., Shahsavari, E., Jain, S. S., Koshlaf, E., Ravindran, V. B., Morrison, P. D.,… & Ball, A. S. (2019). Influence of bioaugmentation and biostimulation on PAH degradation in aged contaminated soils: response and dynamics of the bacterial community. *J. Environ. Manage., 238*, 49–58. doi: 10.1016/j.jenvman.2019. 02.115.

Han, K., Hong, U., Park, S., Kwon, S., & Kim, Y. (2021). In situ field method for evaluating biodegradation potential of BTEX by indigenous heterotrophic denitrifying microorganisms in a BTEX-contaminated fractured-rock aquifer. *Environ. Technol., 42*, 1326–1335.

Haritash, A. K., & Kaushik, C. P. (2009). Biodegradation aspects of Polycyclic Aromatic Hydrocarbons (PAHs): A review. *Journal of Hazardous Materials*. doi:10. 1016/j. jhazm at. 2009. 03. 137

Huang, D., Hu, C., Zeng, G., Cheng, M., Xu, P., Gong, X., … & Xue, W. (2016). Combination of Fenton processes and biotreatment for wastewater treatment and soil remediation. *Science of the Total Environmental*. doi: 10. 1016/j. scito tenv. 2016. 08. 199.

Ismail, N. A., Hamzah, N., Kasmuri, N., Jaafar, J., Ali, M. F., Khalil, K. A., & Singhal, N. (2020). *Optimization and interaction analysis of bacterial and fungal growth in the presence of benzo(a)pyrene in wastewater*. IOP Conf. Series: Earth and Environmental Science. doi:10. 1088/ 1755- 315/ 616/1/ 012067.

Jafari, A., Mahvi, A. H., Nasseri, S., Rashidi, A., Nabizadeh, R., & Rezaee, R. (2015). Ultrafiltration of natural organic matter from water by vertically aligned carbon nanotube membrane. *Journal of Environmental Health Science Engineering, 13*, 51. doi:10. 1186/ s40201- 015- 0207-x

Jamal, M. T. (2018). Application of hydrocarbon-degrading halophilic bacterial strains as an indicator to locate crude oil reservoirs in Atlantis II deep in the Red Sea. *International Journal of Advanced Research in Biological Science, 5*(12), 30–36. doi:10. 22192/ ijarbs. 2018.05. 12. 005.

Jamal, M. T., & Pugazhendi, A. (2018). Degradation of petroleum hydrocarbons and treatment of refinery wastewater under saline condition by a halophilic bacterial consortium enriched from marine environment (Red Sea), Jeddah, Saudi Arabia. *3 Biotech*. doi:10. 1007/ s13205- 018- 1296-x.

Joy, S., Butalia, T., Sharma, S., & Rahman, P.K. (2017). Biosurfactant production and concomitant hydrocarbon degradation potentials of bacteria isolated from extreme and hydrocarbon contaminated environments. *Chemical Engineering Journal, 317*, 232–241. doi:10.1016/ j.cej.2017.02.054

Kachieng, L., Jitendra, K., & Momba, M. (2018). Metagenomic profiling for assessing microbial diversity and microbial adaptation to degradation of hydrocarbons in two South African petroleumcontaminated water aquifers. *Scientific Reports, 8*, 7564. doi:10.1038/ s41598-018-25961-0.

Khemili-Talbi, S., Kebbouche-Gana, S., Akmoussi-Toumi, S., Angar, Y., & Gana, M. L. (2015). Isolation of an extremely halophilic arhaeon *Natrialba sp.* C21 able to degrade aromatic compounds and to produce stable biosurfactant at high salinity. *Extremophiles(pp.* 1109–1120). Springer. doi: 10.1007/s00792-015-0783-9.

Khodaverdiloo, H., Han, F. X., Taghlidabad, R. H., Karimi, A., Moradi, N., & Kazery, J. A. (2020). Potentially toxic element contamination of arid and semi-arid soils and its phytoremediation. *Arid Land Research and Management*. doi:10. 1080/ 15324 982. 2020. 17467 07.

Korenak, J., Basu, S., Balakrishnan, M., Hélix-Nielsen, C., & Petrinic, I. (2017). Forward osmosis in wastewater treatment processes. *Acta Chimica Slovenica*. doi:10. 17344/ acsi. 2016. 2852.

Kumari, S., Kumar Regar, R., & Manickam, N. (2018). Improved polycyclic aromatic hydrocarbon degradation in a crude oil by individual and a consortium of bacteria. *Bioresource Technology*. doi:10. 1016/j. biortech. 2018. 01. 075.

Lamichhane S., Krishna K. C. B., & Sarukkalige R. (2017). Surfactant-enhanced remediation of polycyclic aromatic hydrocarbons: a review. *J. Environ. Manage., 199*, 46–61. doi:10.1016/j.jenvman.2017.05.037.

Lang, F. S., Destain, J., Delvigne, F., Druart, P., Ongena, M., & Thonart, P. (2016). Biodegradation of polycyclic aromatic hydrocarbons in mangrove sediments under different strategies: Natural attenuation, biostimulation, and bioaugmentation with Rhodococcus erythropolis T902.1. *Water Air Soil Pollut.*, 227–297. doi:10.1007/s11270-016-2999-4

Li, J., Alves de Toledo, R., & Shim, H. (2017). Multivariate optimization for the simultaneous bioremoval of BTEX and chlorinated aliphatic hydrocarbons by Pseudomonas plecoglossicida. *J. Hazard. Mater., 321*, 238–246.

Li, Z., & Yang, P. (2018). *Review on physicochemical, chemical, and biological processes for pharmaceutical wastewater.* IOP Conf. Ser.: Earth Environmental Science. 113 012185. doi:10. 1088/ 1755- 1315/ 113/1/012185.

Liang, X., Guo, C., Liao, C., Liu, S., Wick, L.Y., Peng, D., ... & Dang, Z. (2017). Drivers and applications of integrated clean-up technologies for surfactant-enhanced remediation of environments contaminated with polycyclic aromatic hydrocarbons (PAHs). *Environ. Pollut. 225*, 129–140.

Liu, X., Liu, M., Chen, X., Yang, Y., Hou, L., Wu, S., & Zhu, P. (2019). Indigenous PAH degraders along the gradient of the Yangtze Estuary of China: Relationships with pollutants and their bioremediation implications. *Marine Pollution Bulletin, 142*, 419–427.

Ma, X., Li, T., Fam, H., Peterson, E. C., Zhao, W., Guo, W., & Zhou, B. (2018). The influence of heavy metals on the bioremediation of polycyclic aromatic hydrocarbons in aquatic system by a bacterial–fungal consortium. *Environmental Technology, 39*(16) 2128–2137, DOI: 10.1080/09593330.2017.1351492

Malik, Z., Ravindran, K., & Sathiyaraj, G. (2017). Phytoremediation: A novel strategy and eco-friendly green technology for removal of toxic metals. *International Journal of Agricultural and Environmental Research.* 1–18.

Mansouri, A., Abbes, C., & Landoulsi, A. (2017). Combined intervention of static magnetic field and growth rate of Microbacterium maritypicum CB7 for Benzo(a)pyrene biodegradation. *Microbial Pathogenesis.* doi:10. 1016/j. micpa th. 2017. 10. 008.

Marić, N., Ilic, M., Miletić, S., Gojgic-Cvijovic, G., Beškoski, V., Vrvic, M., & Papić, P. (2015). Enhanced in situ bioremediation of groundwater contaminated by petroleum hydrocarbons at the location of the Nitex textiles, Serbia. *Environmental Earth Sciences. 74.* doi:10.1007/s12665-015-4531-3.

Marris, C., Kompella, S. N., Miller, M., Incardona, J. P., Brette, F., Hancox, J. C., ... & Shiels, H. A. (2020). Polyaromatic hydrocarbons in pollution: a heart-breaking matter. *J. Physiol., 598,* 227–247. doi: 10.1113/jp278885.

Marshall, T., Paschos, A., Marangoni, A. G., Yang, F., & Pensini, E. (2021). Injectable cationic traps and sticky bacterial emulsifiers: A safe alliance during diesel bioremediation. *Colloids and Surfaces A: Physicochemical and Engineering Aspects, 613*, 126051.

Mojiri, A., Zhou, J. L., Ohashi, A., Ozaki, N., & Kindaichi, T. (2019). Comprehensive review of polycyclic aromatic hydrocarbons in water sources, their effects and treatments. *Sci. Total Environ., 2019*, 133971. doi: 10.1016/j. scitotenv.2019.133971.

Mukhopadhyay, S., Dutta, R., & Das, P. (2020). A critical review on plant biomonitors for determination of polycyclic aromatic hydrocarbons (PAHs) in air through solvent extraction techniques. *Chemosphere, 251*, 126441.

Müller, C., Knoller, K., Lucas, R., Kleinsteuber, S., Trabitzsch, R., Weiß, H., ... & Vogt, C. (2021). Benzene degradation in contaminated aquifers: Enhancing natural attenuation by injecting nitrate. *Journal of Contaminant Hydrology, 238*, 103759.

National Institutes of Health. (NIH), (2024). National Research Council (US) Committee on Pyrene and Selected Analogues. *Polycyclic aromatic hydrocarbons: evaluation of sources and effects.* Washington (DC): National Academies Press (US); 1983. 2, Polycyclic aromatic hydrocarbons from natural and stationary anthropogenic sources and their atmospheric concentrations. https://www.ncbi.nlm.nih.gov/books/NBK217758/

Nhi-Cong, L. T., Lien, D. T., Mai, C. T. N., Linh, N. V., Lich, N. Q., Ha, H. P., ... & Show, P. L. (2021). Advanced materials for immobilization of purple phototrophic bacteria in bioremediation of oil-polluted wastewater. *Chemosphere, 278*, 130464.

Ojha, N., Mandal, S. K., & Das, N. (2019). Enhanced degradation of indeno(1,2,3-cd)pyrene using *Candida tropicalis* NN4 in presence of iron nanoparticles and produced biosurfactant: a statistical approach. *3 Biotech*, *9*(3), 86. doi:10.1007/s13205-019-1623-x.

Patel, A. B., Singh, S., Patel, A., Jain, K., Amin, S., & Madamwar, D. (2019). Synergistic biodegradation of phenanthrene and fluoranthene by mixed bacterial cultures. *Bioresour. Technol.*, *284*, 115–120. doi: 10.1016/j.biortech. 2019.03.097.

Pugazhendi, A., Qari, H., Basahi, J. M. A., Godon, J. J., & Dhavamani, J. (2017). Role of a halothermophilic bacterial consortium for the biodegradation of PAHs and the treatment of petroleum wastewater at extreme conditions. *International Biodeterioration and Biodegradation.* doi:10. 1016/j. ibiod. 2017. 03. 015

Rao, T. V., & Babu, R. J. (2014). Phytoremediation. National seminar on impact of toxic metals, minerals and solvents leading to environmental pollution. *Journal of Chemical and Pharmaceutical Sciences.*

Sabir, A., Shafiq, M., Islam, A., Sarwar, A., Dilshad, M. R., Shafeeq, A., ... & Jamil, T. (2015). Fabrication of tethered carbon nanotubes in cellulose acetate/polyethylene glycol-400 composite membranes for reverse osmosis. *Carbohydrates Polymers.*, *132*, 589–597. doi: 10. 1016/j. carbp ol. 2015. 06. 035

Sakshi, A. K., & Haritash, A. K. (2020). A comprehensive review of metabolic and genomic aspects of PAH-degradation. *Archives of Microbiology, 202*, 2033–2058.

Saleh, H. M., Aglan, R. F., & Mahmoud, H. H. (2019). Qualification of corroborated real phytoremediated radioactive wastes under leaching and other weathering parameters. *Progr. Nucl. Energy*, *119*, 103178. doi: 10.1016/j.pnucene.2019. 103178.

Seker, F. A., & Bagcevan, B. (2020). Determination of polycyclic aromatic hydrocarbons (PAHs) in drinking water of Samsun and its surrounding areas, Turkey. *J Environ Health Sci Eng.*, *17*(2), 1205–1212. doi: 10.1007/s40201-019-00436-0. eCollection.

Shah, M. P. (2014). Environmental bioremediation: a low cost nature's natural biotechnology for environmental clean-up. *J. Pet. Environ. Biotechnol.*, *5*(4). doi: 10.4172/2157-7463.1000191.

Shahsavari, E., Adetutu, E. M., & Ball, A. S. (2015). Phytoremediation and necrophytoremediation of petrogenic hydrocarbon contaminated soils. *Phytoremediation: Management of Environmental Contaminants, 2*, 321. doi:10. 1007/ 978-3- 319- 10969-5_ 26.

Siegrist, R. L., Crimi, M., Simpkin, T. J., & Brown, R. A. (2011). In situ chemical oxidation for groundwater remediation. *Springer*, 3. doi:10.1007/978-1-4419-7826-4.

Subashchandrabose, S. R., Venkateswarlu, K., Venkidusamy, K., Palanisami, T., Naidu, R., & Megharaj, M. (2019). Bioremediation of soil long-term contaminated with PAHs by algal-bacterial synergy of *Chlorella sp.* MM3 and *Rhodococcus wratislaviensis* strain 9 in slurry phase. *Sci Total Environ., 659*, 724–731. doi: 10.1016/j.scitotenv.2018.12.453. Epub 2018 Dec 31.

Sun, Y., Wu, S., & Gong, G. (2019). Trends ofresearch on polycyclicaromatichydrocarbons in food:A 20-year perspectivefrom 1997 to 2017. *Trendsin Food Science&Technology, 83*, 86–98.

Tasleem, A., Jali, J. A, Haq, Q. M. R. H. (2015). Phytoremediation: A Promising Strategy On The Crossroads Of Remediation. In K. Rehman, H. Muhammad, S. Münir, O. A. Ruhi & Mermut (Eds.), *Soil Remediation and Plants* (pp. 63–84). Academic Press.

Tong, R., Yang, X., Su, H., Pan, Y., Zhang, Q., Wang, J., & Long, M. (2018). Levels, sources and probabilistic health risks of polycyclic aromatic hydrocarbons in the agricultural soils from sites neighboring suburban industries in Shanghai. *Sci. Total Environ., 616*, 1365–1373. doi:10.1016/j.scitotenv.2017.10.179

Tore, G. Y., & Ata, R. (2019). Assessment of recovery & reuse activities for industrial wastewaters in miscellaneous sectors. *European Journal of Engineering and Applied Sciences*. http://dergipark.gov.tr/EJEAS.

USEPA. (2020). Characterizing community exposure to atmospheric polycyclic aromatic hydrocarbons (PAHs) in the Memphis tri-state area. https://www.epa.gov/sites/production/files/202001/documents/memphis_pahs_study_final_report_08.pdf.

Varjani, S. J., Thaker, M. B., & Upasani, V. N. (2014). Optimization of growth conditions of native hydrocarbon utilizing bacterial consortium "HUBC" obtained from petroleum pollutant contaminated sites. *Indian J. Appl. Res., 4*, 474–476.

Wang, W., Zhang, Y., Li, M., Wei, X., Wang, Y., Liu, L., … & Shen, S. (2020). Operation mechanism of constructed wetland-microbial fuel cells for wastewater treatment and electricity generation: A review, Bioresour. *Technol., 314*.

Wu, Y., Xu, Y., & Zhou, N. (2020). A newly defined dioxygenase system from *Mycobacterium vanbaalenii* PYR-1 endowed with an enhanced activity of dihydroxylation of high-molecular-weight polyaromatic hydrocarbons. *Front. Env. Sci. Eng., 14*, 14. doi: 10.1007/s11783-019-1193-5.

Yang, L., Wu, Y., Wang, Y., Ana, W., Jin, J., Sun, K., & Wang, X. (2021). Effects of biochar addition on the abundance, speciation, availability, and leaching loss of soil phosphorus, *Sci. of Total Env., 758*, 143657.

Yang X., & Chen, S. (2021). Microorganisms in sediment microbial fuel cells: Ecological niche, microbial response, and environmental function. *Science of the Total Environment, 756*, 144145.

Zhang, L., Wang, Y., Tan, F., Yang, Y., Wu, X., Wang, W., & Liu, D. (2019). Tidal variability of polycyclic aromatic hydrocarbons and organophosphate esters in the coastal seawater of Dalian, China. *Science of the Total Environment, 708*. doi:10.1016/j.scitotenv.2019a.134441.

Zheng, M., Wang, W., & Papadopoulos, K. (2020). Direct visualization of oil degradation and biofilm formation for the screening of crude oil-degrading bacteria. *Bioremediation Journal, 24*(1), 1–11.

8 Treatment and Remediation of Bio-Medical Wastewater at Hospitals and Diagnostic Centres

Jyoti Sarwan, Vijaya Sood,
Pramod Malik, Komal Mittal,
Nazim Uddin, Jagadeesh Chandra Bose K.
and Heenu Sharma

1. INTRODUCTION

The effluents released from hospitals (general hospitals, specialist hospitals, and other sorts of hospitals) into the environment or urban sewer is referred to as hospital wastewater (HWW) (Kumari et al., 2020). Hazardous wastewater is not only released from hospital establishments but also from the other settings such as clinics, diagnostic centres, pharmaceuticals, radiographic laboratories, which are all referred to as biomedical wastewater (Khan et al., 2021). The actual concentrations of the physical, chemical, and biological components found in wastewater are referred to as its composition. The water quality varies according to the nature, magnitude, and location of each hospital/biomedical setting. For the sake of simplicity, we refer it to as hospital wastewater.

DOI:10.1201/9781003496014-8

Numerous studies and research projects demonstrate that hospital wastewater quality is comparable to average home wastewater quality based on measurement of pH, BOD, COD, TSS, and total coliforms. Numerous potentially dangerous substances, such as disinfectants, radioactive isotopes, microbiological pathogens, medications, chemical compounds, and medicines, can be found in hospital wastewater. It is true that human health and the environment may be negatively impacted by hospital wastewater. Thus, it is crucial to choose an appropriate treatment method for handling of hospital wastewater (Kinuthia et al., 2020). This chapter focuses on the hazards associated with discharge and inadequate release of hospital wastewater. It would benefit concerned persons with regard to the handling and treatment of hospital wastewater.

2. CONCERNS IN HOSPITAL WASTEWATER MANAGEMENT

Hospital wastewater treated on-site results in a sludge with significant helminthic and other pathogen concentrations. If treated wastewater is to be used for irrigation, the applicable World Health Organization (WHO) recommendations state that it should contain no more than one helminth egg per litre and no more than 1000 faecal coli forms per 100 mL (Symonds et al., 2018). Hospitals that are either not linked to public sewer or connected to sewer without a terminal sewage treatment plant are subject to the effluent restrictions under the Environmental (Protection) Act 1986. According to EPA 1986, hospital wastewater discharge requirements must be met (source: GSR 7, dated Dec. 22, 1998). India's sewage effluent discharge tolerance limit is governed by BIS standard IS: 4764:1973 (Mussarat et al., 2021). It is imperative to motivate environmental engineers, technologists, and economists to create and evaluate various treatment systems with respect to their efficiency, design, operational, financial, and overall risks. For the purpose of treating hospital wastewater, numerous treatment systems have been developed thus far (Yanbo et al., 2021) Membrane Bio-Reactor (MBR), Sequential Batch Reactor (SBR), Fluidized Bed Reactor (FBR), Extended Aeration (E.A.), Submerged Aeration Fixed Film (SAFF) Rector, and Activated Sludge Process (ASP) are a few of these treatment systems (Jani et al., 2020). As more user-friendly alternatives have emerged, the activated sludge process (ASP) is becoming outdated as a method of treating sewage (McCarty et al., 1962). With the exception of a longer hydraulic retention period to provide longer aeration for E.A. treatment technology is exactly the same as ASP treatment technology (Zhou et al., 2015).

SBR is a comparable treatment technology to an E.A. system, however, it only uses one tank for biodegradation, solids settling, and sludge removal from the same tank. Another name for it is an activated sludge treatment system that uses draw-and-fill (Steinmetz et al., 2002). The most recent development in attached and suspended growth aerobic biological treatment technique is called FBR. Influent is processed at a fast enough speed in a reactor to induce fluidization via a bed of tightly packed granular material. In order to enhance the performance of sewage treatment plants, SAFF, a recent advancement in attached growth process, has been added as a fixed film medium to activated sludge reactors (Mohan et al., 2005). MBR is also used to treat wastewater. It operates on the filtration of activated sludge principle and uses submerged membrane modules in bioreactors that are either flat sheet type or hollow fibre type (Naghizadeh et al., 2011).

3. METHODS FOR TREATMENT OF HOSPITAL WASTEWATER

3.1. Sludge Activation Process (SAP)

Hospital wastewater is traditionally treated by this process. The wastewater is treated in an open tank known as an aeration tank and air is provided to the aerobic bacteria by an air blower or a fixed or floating surface aerator. Wastewater treatment requires a hydraulic retention duration of around 12 to 15 hours. Utilizing the oxygen in the air, the microbes create new bacterial cells and stabilized, low-energy molecules like nitrates, sulphates, and carbonates from the organic waste (Sustarsic et al., 2009). Active microorganisms are present in the overflow that was conveyed to the nearby clarifying system, and it must be recycled back to the aeration tank in order to maintain certain important performance rating parameters like mixed liquor suspended solids (MLSS) and F/M ratio (Orhon et al., 2009).

Plug flow and complete mixing are the two mixing regime types that are often of primary importance in the activated sludge process. In the first, there is no mixing or mixed liquor present along the flow path, and the mixed liquor flows through the aeration tank in an ordered manner. The aeration tank's contents are thoroughly mixed and evenly distributed throughout. As a result, in steady state, the aeration tank's effluent and its contents are identical. Microorganisms make up the biological component of the activated sludge system. These microorganisms consist of between 70% and 90% organic matter and 10% and 30% inorganic matter. Protozoa, metazoa, rotifers, fungi, and bacteria make up the biological mass of activated sludge. Higher organism development is inhibited by the sludge re-circulation and the aeration tanks' continual agitation.

The type of microbe that predominates in a system is determined by the secondary influent wastewater's properties, the plant's operating mode, process design, and environmental factors (Lotito et al., 2014). Bacteria are the microorganisms with the most numerical significance in activated sludge. While some bacteria are anaerobes, which are only active in the absence of oxygen, others are stringent aerobes, which can only exist in the presence of oxygen. One crucial aspect in the survival of activated sludge when dissolved oxygen concentrations are low or possibly approaching depletion is the preponderance of facultative bacteria, which can exist in both the presence and absence of oxygen (Wang et al., 2020).

Both autotrophic and heterotrophic microbes can be found in activated sludge, however, the former are more prevalent. Heterotrophic bacteria use carbonaceous organic elements in influent wastewater as a source of energy to divide into new cells. They release energy in the process as they convert organic materials into molecules like carbon dioxide and water. Numerous significant genera, including *Arthrobacter, Citromonas, Flavobacterium, Pseudomonas*, and *Zoogloea* are home to heterotrophic bacteria (Carstea et al., 2018; Xie et al., 2012).

ASP is a very old technology that was widely used globally for wastewater treatment because there was no other accessible technology. The development of more user-friendly alternatives has rendered the activated sludge technology obsolete for treating sewage (Kalbar et al., 2013).

In activated sludge, microorganisms are both autotrophic and heterotrophic, but the former are more common. Carbonaceous organic materials in influent wastewater provide heterotrophic bacteria with energy for the production of new cells. In the process, they transform organic matter into molecules like water and carbon dioxide, which releases energy. Due to non-availability of technologies, ASP had been applied extensively for hospital wastewater treatment for a long period. The activated sludge technique is no longer a viable method for treating sewage due to the advancement of more user-friendly alternatives (Aziz et al., 2019).

3.1.1. Benefits

- Removal effectiveness of BOD5 > 90%.
- Easy to use.
- Require fewer labour specialists for upkeep and operation (O&M).
- Chemicals are not used to achieve nitrification or oxidation.
- Up to 97% maximum elimination of suspended solids.
- The most popular method of treating wastewater because there isn't a substitute technology.
- A moderately sized land area.
- Capacity to dilute hazardous materials and manage peak load.

3.1.2. Negative aspects of sludge activation process

- A larger volume of sludge without fully settling.
- Ineffective at removing colour from wastewater; oxidations might result in the creation of highly coloured intermediates, which can intensify the colour.
- Bad effluent quality and an issue with odour.
- More susceptible to temperature changes and stress loading.
- Ineffective at removing nutrients from wastewater; additional polishing requires tertiary treatment.
- Higher aeration costs and volume.
- Limited adaptability in operations.
- Instabilities in biomass, such as thickening of sludge.
- High requirement for chlorine and TSS in wastewater.
- High energy expenditure.
- Considering that more user-friendly and advanced methods are now available, this technology is now regarded as outdated (Chang et al., 2001).

3.2. Extended Aeration Process (E.A.)

The only difference between this type of treatment technique and the activated sludge method is the longer hydraulic retention period, which allows for prolonged aeration and full organic matter digestion. In order to allow for full aerobic biodegradation, the aeration tank typically has a retention time of 18 to 24 hours. The primary goal of extending the hydraulic retention period is to lessen the smell caused by partially digested sludge and to decrease the amount of sludge that is recycled into the aeration tank. When there are no space restrictions, this technique is perfect for huge installations (El Morabet et al., 2020).

Rather of being a hydraulically defined mode, extended aeration is a reaction-defined mode that might be theoretically plug flow or full mix. Due to the huge biomass volume's buffering effect, this process is rather immune to shock loadings in concentration, but it can be vulnerable to rapid increases in flow caused by the high MLSS loading to the final clarifier (Abu-ghararah et al., 1995). The BOD removal efficiency in this procedure is high due to high detention in the aeration tank, which allows the mixed liquid solid to perform significant endogenous respiration and stabilize well at low organic loading, long aeration period, high MLSS concentration, and low F/M. It is not necessary to separate digest the surplus sludge; instead, it can be handled with a centrifuge or filter press or dried immediately on sand beds (Jagaba et al., 2021).

Reducing adverse impacts on the environment mostly requires the treatment of hospital wastewater. Hospitals are major sources of these compounds. In some countries (like Japan, China, and Greece), one-third of the wastewater from large hospitals is pre-treated or biologically treated on-site; in many other countries, like Switzerland, the wastewater is connected directly to a municipal sewer and treated at municipal wastewater treatment plants (WWTP) (Lopsik et al., 2013) This results in a wide variety of micro-contaminants from diagnostic, laboratory, and research activities on the one hand, and medication excretion by patients on the other. The benefits of treating wastewater at the source include preventing losses into the environment via sewer leaks and combined sewer overflows, as well as dilution from mixing with urban sewage. Concerns with hospital wastewater include preventing the entry of medications, diagnostic tools, and disinfectants as well as the spread of (multi-resistant or pathogenic) bacteria, viruses, and parasite eggs (Walia et al., 2016).

Another method for designing compact wastewater treatment plants is the membrane bioreactor (MBR) system. The bioreactor is intended to extract particulate, colloidal, and dissolved contaminants from the solutions by integrating biological treatment procedures and a membrane system into a single apparatus (Deowan et al., 2019) Fouling is widely acknowledged to lower membrane performance. A multitude of factors, including the deposition of sludge particles, the adherence of macromolecules like extracellular polymeric substances (EPS), and the plugging of pores by small molecules, are commonly implicated in fouling (Judd et al., 2010). It is possible for soluble EPS (soluble macromolecule and colloid) to enter membrane pores and accumulate on the pore wall (Iorhemen et al., 2019). This reduces the membrane pore's total section area, plugs the hole, and ultimately raises the membrane resistance. Operating circumstances, membrane properties, and biomass characteristics all have a significant impact on the form of fouling. Numerous parameters, including membrane fouling, EPS production, treated effluent quality, biomass characteristics, and microbial activity, can be used to monitor the membrane's performance (Campo et al., 2021). Process configuration, membrane material, biomass concentration, hydraulic retention time (HRT), sludge retention time (SRT), and organic loading rate (OLR) all affect these variables (Gautam et al., 2022).

Proteins and polysaccharides are the primary extracellular polymeric substances (EPS) found in bacterial aggregates, as per earlier research. They can be created by wastewater or supplied by microbes, and they serve many purposes in aggregates. Due to their hydrolytic action, some of them can help remove organic debris and give the nutrients needed for bacteria to flourish, since prior research has demonstrated that confocal laser (Sheng et al., 2010).

3.3. Steps Prior to the Hospital Wastewater Treatment

3.3.1. Wastewater collection

Because biomedical effluent from hospitals and diagnostic centres may contain hazardous chemicals, medications, and microorganisms, proper handling is essential. In order to stop environmental contamination and safeguard public health, proper treatment and remediation are crucial. This is a summary of the procedures involved in collecting, treating, and cleaning biomedical wastewater. Sewage from operating rooms, labs, patient rooms, and other healthcare institutions is considered biomedical-wastewater (Rathi et al., 2022). It is imperative that collection systems are engineered to avert cross-contamination and guarantee the secure conveyance of wastewater to treatment plants (Parida et al., 2022).

3.3.2. Accumulation

Segregated liquid waste and other biomedical effluent are collected apart from other wastewater streams. Distinctive gathering mechanisms, such storage tanks and pipelines, are employed to avert blending with non-biomedical effluent.

3.3.3. Segregation

Healthcare institutions are the first places where segregation takes place in order to distinguish biomedical waste from other waste streams, such as household or industrial. The dangerous nature of many biomedical waste kinds, such as infectious waste, sharp objects, chemical waste, pharmaceutical waste, and radioactive waste, should be taken into consideration when sorting them. Healthcare personnel can more easily perform correct segregation by using colour-coded containers and labels (Khan et al., 2019).

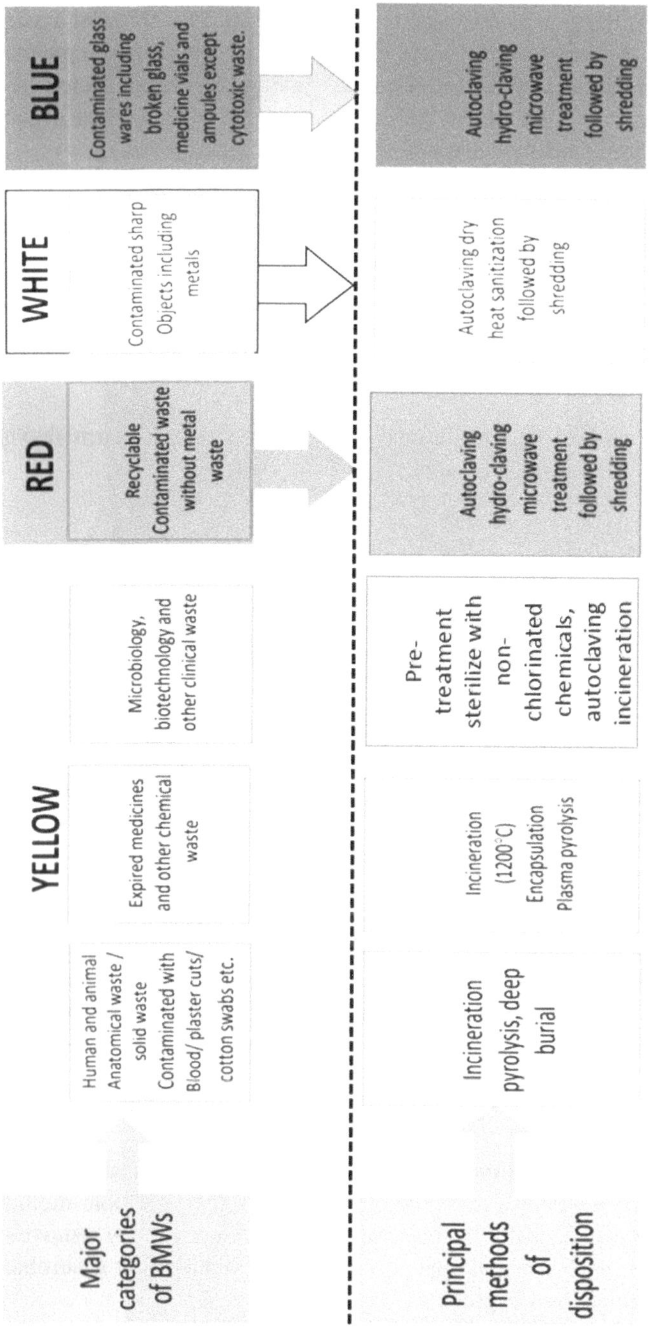

FIGURE 8.1 Methods for determination and treatment of hospital wastes.

Figure created by authors.

3.4. Pre-treatment of Hospital Waste Treatm

The following are primary steps for hospital waste treatment. Biomedic separated from other waste streams at the point of origin as part of th process. Before being processed further, liquid biomedical waste is colle and pre-treated to get rid of any major pollutants, and can also reduce t of the wastewater. Pretreatment techniques could consist of:

3.4.1. Screening

Coarse screens, sometimes called bar screens, are utilized to eliminat debris from wastewater, such as bandages, gauze, and syringes.

3.4.2. Grit removal

Grit chambers are utilized to remove heavy particles that could ha equipment, such as grit, gravel, and sand.

3.4.3. Grease traps

To avoid clogging and interfering with treatment operations, grease tı as grease interceptors, are used to collect fats, oils, and grease (FOG) food preparation facilities (Suarez et al., 2009).

3.4.4. Sedimentation

By allowing suspended materials to separate due to gravity, sediment the organic load and turbidity of wastewater.

3.4.5. Chemical procedure

Biomedical wastewater can be neutralized or disinfected using ch methods. To eradicate germs, chemical disinfectants like hydrogen or chlorine can be added. Through precipitation and settling, chemic flocculants can help remove organic debris and suspended particles.

3.4.6. Biological intervention

Microorganisms are used in biological treatment methods to bre contaminants found in wastewater. Through microbial activity, aer sequencing batch reactors (SBRs) and activated sludge can efficient debris and minerals. High-strength organic waste can be treated wit tion, which yields biogas as a byproduct (Figure 8.1).

3.5. Primary Treatment/ Principal Intervention

Wastewater is subjected to primary treatment following pretreatment in order to eliminate any remaining suspended organic matter and settle able solids. The following are examples of primary treatment processes.

3.5.1. Clarification

To produce clarified effluent, suspended solids are allowed to settle with the use of sedimentation tanks or clarifiers. Dissolved Air Flotation (DAF) systems use air bubbles to lift suspended solids out of wastewater and allow them to float to the top where they may be removed.

3.5.2. Filtration

Multimedia or sand filters can be used to get rid of organic debris and fine suspended solids from wastewater.

3.5.3. Biological degradation

Through biological degradation, secondary treatment procedures seek to lower the number of pathogens and organic contaminants present in wastewater.

3.6. Additional Measures to Be Taken for Treatment of Hospital Wastewater

3.6.1. Remedial care

Biological techniques like membrane bioreactors (MBRs), trickling filters, and activated sludge are frequently used in secondary treatment. Organic pollutants in wastewater are broken down by microorganisms, which release gases like carbon dioxide and methane as well as biomass. Pathogens and organic contaminants are further reduced at this stage.

3.6.2. Third-level care

To comply with strict discharge regulations, tertiary treatment attempts to eliminate any remaining contaminants, including microorganisms, medications, and other substances. Commonly used procedures include advanced oxidation, UV treatment, chlorination, and filtering (sand, carbon). For even finer purification, cutting-edge methods like membrane filtration and ozonation may be employed.

3.6.3. Removal of dangerous substances

To treat dangerous elements including heavy metals, medicines, and infections found in biomedical wastewater, specialized remediation methods could be required. Remedial techniques include ion exchange, chemical precipitation, adsorption onto activated carbon, and advanced oxidation processes (AOPs). Regulators should be followed when disposing of concentrated wastes or byproducts from these procedures.

3.6.4. Observation and adherence

Maintaining compliance with regulatory standards requires ongoing wastewater quality monitoring. Frequent analysis and sampling make it easier to spot abnormalities and adjust treatment procedures.

3.6.5. Recovering resources

Resources such as energy (biogas from anaerobic digestion), nutrients (phosphorus, nitrogen), and water for non-potable reuse should be recovered from treated wastewater.

3.6.6. Awareness

Campaigns for public awareness and educational initiatives can encourage patients, healthcare professionals, and the community to handle wastewater responsibly.

3.6.7. Improved medical care

To address persistent pollutants and toxins that conventional procedures are unable to properly remove, advanced treatment technologies could be required. Reverse osmosis (RO) and ultrafiltration (UF) are two membrane filtration techniques that can offer further purification to satisfy strict discharge requirements.

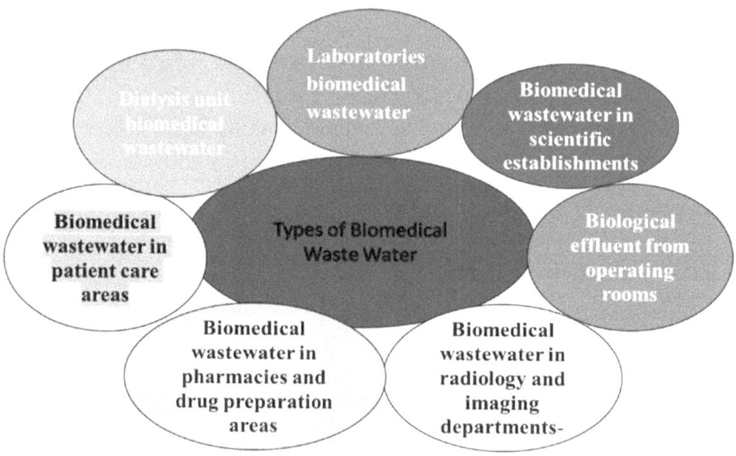

FIGURE 8.2 Describing the status of biomedical waste market and development.
Figure created by authors.

3.6.8. Inactivation of pathogens

To guarantee that all pathogens in biomedical wastewater are completely inactivated, specialized procedures such as pasteurization, autoclaving, or radiation may be used. For infectious waste streams, these treatments are especially crucial in stopping the spread of illness (Figure 8.2).

4. SURVEILLANCE AND QUALITY ASSURANCE OF HOSPITAL WASTEWATER

To guarantee adherence to regulatory requirements, treatment procedures and effluent quality must be continuously monitored. Regular monitoring is recommended for parameters such pH, turbidity, pathogen levels, chemical and biological oxygen demand (COD, BOD). It is important to implement quality control procedures in order to maximize treatment effectiveness and guarantee steady functioning (Servetas et al., 2022).

5. REUSING HOSPITAL WASTEWATER

If treated biomedical wastewater satisfies regulatory standards, it can be released into the environment without risk. As an alternative, treated wastewater might be appropriate for industrial processes, toilet flushing, irrigation, and other non-potable reuse applications. Regulations should be followed when managing and disposing of residual sludge or byproducts produced during treatment (Gupta et al., 2021).

6. TYPICAL TECHNIQUES FOR SECONDARY TREATMENT

Activated sludge is created when organic contaminants in wastewater are broken down by aerobic microorganisms, resulting in biomass and treated effluent. Further sequence for hospital waste treatment is as follows:

6.1. Trickling Filters

As wastewater percolates through a bed of media (such as rock, plastic, or textile), microbial biofilms break down organic materials.

Sequencing Batch Reactors (SBRs) systems provide flexibility and control over the treatment processes by treating wastewater in batches using a sequence of fill, react, settle, and decant cycles (Gautam et al., 2007).

6.2. Step-up Care

To ensure compliance with regulatory standards, tertiary treatment is used to further polish the effluent and eliminate any leftover pollutants. Options for tertiary treatment consist of filtration to get rid of pathogens, suspended particles, and trace pollutants, fine filtering techniques such membrane bioreactors (MBRs), ultra filtration, and microfiltration are utilized.

6.3. Disinfection

To inactivate pathogens and stop microbial re-growth, chemical disinfectants (such as chlorine, ozone, and UV radiation) are administered to the treated effluent.

6.4. Removal of Dangerous Substances

Hazardous substances found in biomedical wastewater, such as pharmaceutical residues, heavy metals, and persistent organic pollutants, may need to be remedied using specialized treatments. Remedial techniques include ion exchange, chemical precipitation, adsorption onto activated carbon, and AOPs.

6.5. Observation and Adherence

To guarantee adherence to regulatory requirements, it is crucial to continuously monitor the parameters pertaining to wastewater quality, such as pH, turbidity, pathogen levels, COD, and BOD. Frequent monitoring of effluent quality by sampling and analysis makes it easier to spot abnormalities and adjust treatment procedures accordingly ((El-Kanater et al., 2011).

7. DIFFERENT TYPES OF SET-UPS/ LABORATORIES PRODUCING BIOMEDICAL WASTEWATER

This is an overview of the cleanup and treatment procedure specifically designed for biomedical wastewater produced in laboratories. The vast variety of contaminants contained in biomedical wastewater created by hospitals and diagnostic centres, including pathogens, chemicals, medications, and biological materials, necessitates special considerations in its treatment and remediation. Waste should be separated into different categories at the point of generation. This includes solid waste such contaminated materials, glassware, and laboratory consumables, as well as liquid waste from sinks, drains, and equipment cleaning. Pretreatment of biomedical wastewater created in laboratories may entail the first steps of removing sediments, chemicals, and gross pollutants that could obstruct treatment procedures. To get rid of big particles, debris, and sediment, pretreatment techniques could include screening, sedimentation, and filtration. Wastewater from laboratories can be neutralized or disinfected using chemical treatment procedures, particularly if it includes harmful substances or microorganisms. To eradicate infections and cleanse wastewater, chemical disinfectants like hydrogen peroxide, ozone, and chlorine can be applied. Through precipitation and settling, chemical coagulants and flocculants can help remove organic debris and suspended particles. Degrading organic contaminants and lowering the BOD of laboratory wastewater can be accomplished with the use of biological treatment methods. Microorganisms are used in aerobic biological treatment techniques, like SBRs and activated sludge, to break down organic materials. High-strength organic waste can be treated with anaerobic digestion, which yields biogas as a byproduct.

7.1. Biological Effluent from Operating Rooms

To safeguard the environment and public health, biomedical wastewater produced by operating rooms in hospitals and diagnostic centres must be treated and remedied. Wastewater from operating rooms contains a variety of toxins, including blood, bodily fluids, surgical chemicals, and other potentially dangerous materials. Blood, saline solution, and surgical fluids are examples of liquid waste from operating rooms that should be collected separately from other wastewater sources. Screening, sedimentation, and filtering are examples of pretreatment techniques that can be used to get rid of big particles, trash, and surgical supplies. Through precipitation and settling, chemical coagulants and flocculants can help remove organic debris and suspended particles. High-strength organic waste can be treated with anaerobic digestion, which yields biogas as a byproduct. Microorganisms are used in aerobic biological treatment techniques, like SBRs and activated sludge, to break down organic materials (Gupta et al., 2021).

AOPs can break down complex chemical molecules and pharmaceutical leftovers by using UV light, hydrogen peroxide, or ozone (Hocaoglu et al., 2021). When treating wastewater that contains infectious agents and blood-borne pathogens from surgical procedures, these treatments are very crucial. If wastewater satisfies statutory standards, it can be released into the environment without risk (Verma et al., 2008).

7.2. Dialysis Unit Biomedical Wastewater

Because biomedical wastewater from dialysis units in hospitals and healthcare facilities contains impurities including urea, creatinine, electrolytes, and potentially hazardous compounds. These wastewaters need to be treated and remedied by specialists. To collect liquid waste from dialysis equipment, such as used dialysate, rinse water, and cleaning solutions, dedicated collection systems need be set up. Pretreatment techniques to get rid of big particles, blood clots, and dialyzer membranes can involve screening, sedimentation, and filtration. To neutralize or eliminate impurities like urea, creatinine, and other waste products found in dialysis effluent, chemical treatment procedures may be used. To disinfect wastewater and stop the spread of pathogenic agents, chemical disinfectants like per-acetic acid or chlorine can be utilized. Two membrane filtration techniques—reverse osmosis (RO) and nanofiltration (NF)—are frequently employed to eliminate ions, impurities, and dissolved pollutants from dialysis effluent. Urea, creatinine, and other solutes are efficiently removed by RO membranes, resulting in purified water that can be used again during dialysis (Symonds et al., 2018).

Dialysis effluent must be thoroughly disinfected in order to guarantee pathogen elimination and stop infection spread. To attain sufficient disinfection levels, ultraviolet (UV) radiation and chemical disinfection techniques like ozonation or chlorination may be used. Before wastewater is discharged, nutrients can be removed from it using biological treatment techniques like biological nutrient removal (BNR) or activated sludge. Pharmaceutical chemicals and organic contaminants can be efficiently degraded by AOPs that utilize UV light, hydrogen peroxide, or ozone. It is important to routinely check parameters like conductivity, pH, turbidity, total dissolved solids (TDS), and residual disinfectant levels (Verma et al., 2008).

7.3. Biomedical Wastewater in Patient Care Areas

To avoid environmental contamination and safeguard public health, biomedical wastewater produced in patient care areas of hospitals and diagnostic centres must be treated and cleaned up. Wastewater from patient care areas is contaminated by a range of substances, such as cleaning supplies, drugs, bodily fluids, and other potentially dangerous materials. To collect liquid waste from patient rooms, restrooms, sinks, and spaces where medical equipment is cleaned, separate collection systems need to be installed. The removal of big particles, hair, and other solids can be achieved using pretreatment techniques including as screening, sedimentation, and filtration. Through precipitation and settling, chemical coagulants and flocculants can help remove organic debris and suspended particles. High-strength organic waste can be treated with anaerobic digestion, which yields biogas as a byproduct (Verma et al., 2008).

Reverse osmosis (RO) and ultrafiltration (UF) are two membrane filtration techniques that can offer further purification to satisfy strict discharge requirements. Specialized treatments like pasteurization, autoclaving, or radiation may be used to ensure complete inactivation of germs in wastewater from patient care areas. These steps are essential when treating wastewater that contains bodily fluids and harmful bacteria (Rajan, 2021).

7.4. Biomedical Wastewater in Pharmacies and Drug Preparation Areas

To handle the disposal of pharmaceuticals, chemicals, and other potentially dangerous compounds in hospitals and diagnostic centres, biomedical wastewater generated in

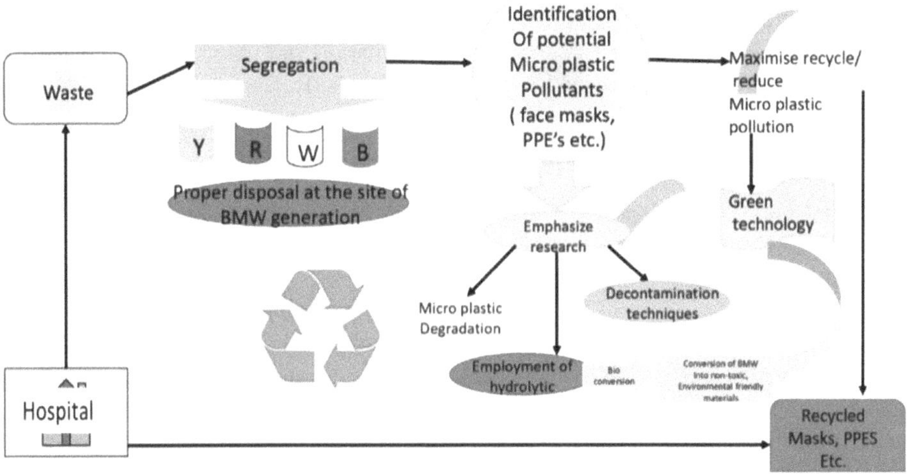

FIGURE 8.3 Different types of set-ups/laboratories producing biomedical wastewater. Figure created by authors.

pharmacies and drug manufacturing facilities must be treated and remedied. To get rid of big particles, packaging materials, and other solids, pretreatment techniques could involve screening, sedimentation, and filtration. Chemical treatment procedures can be utilized to eliminate or neutralize impurities including medications, chemicals, and cleaning supplies found in pharmacy effluent. Reverse osmosis (RO) and nanofiltration (NF) are two membrane filtration techniques that can be used to remove pollutants and dissolved contaminants from pharmaceutical wastewater. Wastewater from pharmacies must be thoroughly disinfected in order to guarantee pathogen elimination and stop the spread of infection. If treated pharmacy wastewater satisfies regulatory standards, it can be released into the environment without risk (Rajan, 2021) (Figure 8.3).

7.5. Biomedical Wastewater in Radiology and Imaging Departments

Because biomedical wastewater produced in radiology and imaging departments of hospitals and diagnostic centres contains contaminants such contrast agents, radioactive materials, chemicals, and organic substances, it needs to be treated and remedied with specific care. Radiology wastewater can be treated to eliminate organic pollutants and chemical residues by adsorption onto activated carbon or other adsorbent materials. Before disposing of excess chemicals and contrast agents, absorbent items like activated charcoal can be applied. Reverse osmosis (RO) and nanofiltration (NF) are two membrane filtration techniques that can be used to remove pollutants and dissolved chemicals from radiography wastewater. RO membranes efficiently eliminate organic compounds, metals, and dissolved ions to produce purified water that can be reused or discharged.

Radioactive materials and waste produced in radiology and imaging departments need to be handled by specialized treatment procedures. Before being disposed of, radioactive waste can be processed using ion exchange, precipitation, or evaporation techniques to eliminate radioactive isotopes and lower radiation levels (Rajan, 2021).

7.6. Biomedical Wastewater in Scientific Establishments

Biomedical wastewater produced in hospital and diagnostic centre research facilities frequently comprises a variety of contaminants, such as potentially infectious agents, dangerous chemicals, biological materials, and drugs. It is imperative that this effluent be properly treated and remedied to avoid contaminating the environment and to safeguard human health. Through precipitation and settling, chemical coagulants and flocculants can help remove organic debris and suspended particles. High-strength organic waste can be treated with anaerobic digestion, which yields biogas as a byproduct. To address persistent pollutants and toxins that conventional procedures are unable to properly remove, advanced treatment technologies could be required. Regular monitoring is necessary for parameters such BOD, COD, pH, turbidity, and pathogen levels (Verma et al., 2008).

8. CHALLENGES IN WASTEWATER TREATMENT

As the difficulties in eliminating pathogens grow more complicated and pressing, the sector of treating hospital wastewater is set to make major strides. There is a lot of promise in investigating cutting edge technologies like nanotechnology, membrane bioreactors, and sophisticated oxidation processes. The development of integrated systems that incorporate many different treatment procedures might greatly increase the effectiveness of pathogen eradication. Subsequent investigations have to concentrate on refining these technologies in order to maximize their cost-effectiveness and scalability, guaranteeing their broad adoption in various healthcare environments. Hospital wastewater treatment may significantly improve pathogen eradication, safeguard public health, and advance environmental sustainability by concentrating on these promising areas. To overcome the obstacles and realize the full potential of cutting-edge treatment technologies, cooperation and innovation must continue (Meshram, 2024).

9. FUTURE DIRECTIONS FOR BIOMEDICAL WASTEWATER TREATMENT

Hospitals are a major source of viruses, ARGs/ARBs, and pharmaceutical compounds (PhCs) that are released into surface waterways. Additionally, treating HWW at the site of production before it enters drainage systems may stop the spread of viruses, lower the concentration of ARGs and ARBs, and get rid of PhCs. Generally speaking, the issue cannot be resolved by a single HWW management technique. Research on HWW therapies is desperately needed because not all of them are effective in getting rid of all PhCs. In fact, a number of tactics are frequently used. At the pilot plant size, MBRs have shown promising results, removing 80–95% of some antibiotics, cardiovascular PhCs, and analgesics and anti-inflammatories. Nonetheless, the treated effluent still contains some resistant antibiotics, contrast agents, and psychiatric medications. The latter are highly significant because of how they impact ARBs and ARGs. Recalcitrant to biological and O_3 treatments, PhCs from many families, including the contrast media, can be effectively eliminated using advanced oxidation techniques. However, because of the high HWW flow rates, the cost of the chemicals and the energy used might be prohibitive. In light of this, it can be said that integrated technologies are the most effective way to remove PhCs, ARBs, ARGs, and even viruses like SARS-CoV-2 (COVID-19) from hospital effluents (Pariente, 2022).

10. CONCLUSION

Analytical investigation has led to the conclusion that hospital wastewater is frequently employed in field hospitals under harsh circumstances. Hospital effluent is considered high risk due to its vast range of harmful compounds, including disinfectants, radionuclides, and antibiotics, when compared to home sewage. The best field hospital wastewater treatment scenario combines MBR and technologies such as focused pulse, according to the currently used HWW treatment scenarios application, and will result in a reduction in the price, reactor size, and reagent requirement for HWW treatment.

REFERENCES

Abu-ghararah, Z. H. (1995). Oxygen uptake rate as an extended aeration process control parameter. *Journal of Environmental Science & Health Part A, 30*(5), 951–969.

Aziz, A., Basheer, F., Sengar, A., Khan, S. U., & Farooqi, I. H. (2019). Biological wastewater treatment (anaerobic-aerobic) technologies for safe discharge of treated slaughterhouse and meat processing wastewater. *Science of the Total Environment, 686*, 681–708.

Campo, R., Lubello, C., Lotti, T., & Di Bella, G. (2021). Aerobic granular sludge–membrane bioReactor (AGS–MBR) as a novel configuration for wastewater treatment and fouling mitigation: a mini-review. *Membranes, 11*(4), 261.

Carstea, E. M., Zakharova, Y. S., & Bridgeman, J. (2018). Online fluorescence monitoring of effluent organic matter in wastewater treatment plants. *Journal of Environmental Engineering, 144* (5). https://doi.org/10.1061/(ASCE)EE.1943-7870.0001360

Chang, G. R., Liu, J. C., & Lee, D. J. (2001). Co-conditioning and dewatering of chemical sludge and waste activated sludge. *Water Research, 35*(3), 786–794.

Deowan, S. A., Korejba, W., Hoinkis, J., Figoli, A., Drioli, E., Islam, R., & Jamal, L. (2019). Design and testing of a pilot-scale submerged membrane bioreactor (MBR) for textile wastewater treatment. *Applied Water Science, 9*, 1–7.

El-kanater, Q. (2011). Assessment of aquatic environmental for wastewater management quality in the hospitals: A case study. *Australian Journal of Basic and Applied Sciences, 5*(7), 474–782.

El Morabet, R., Khan, R. A., Mallick, J., Khan, N. A., Ahmed, S., Dhingra, A., ... & Bindajam, A. (2020). Comparative study of submerged membrane bioreactor and extended aeration process coupled with tubesettler for hospital wastewater treatment. *Alexandria Engineering Journal, 59*(6), 4633–4641.

Gautam, A. K., Kumar, S., & Sabumon, P. C. (2007). Preliminary study of physico-chemical treatment options for hospital wastewater. *Journal of Environmental Management, 83*(3), 298–306.

Gautam, R. K., Kamilya, T., Verma, S., Muthukumaran, S., Jegatheesan, V., & Navaratna, D. (2022). Evaluation of membrane cake fouling mechanism to estimate design parameters of a submerged AnMBR treating high strength industrial wastewater. *Journal of Environmental Management, 301*, 113867.

Geweid, G. G., Abdallah, M. A., Minakova, V., Pryakhin, A., Rathore, R., Milutinovic, M., ... & Elkhodragy, T. M. (2024). Optimizing dialysis water treatment based on medical planning requirements. *Expert Systems with Applications, 244*, 122974.

Gimeno, O., García-Araya, J. F., Beltrán, F. J., Rivas, F. J., & Espejo, A. (2016). Removal of emerging contaminants from a primary effluent of municipal wastewater by means of sequential biological degradation-solar photocatalytic oxidation processes. *Chemical Engineering Journal, 290*, 12–20.

Goto, Y., & Yoshida, N. (2017). Microbially reduced graphene oxide shows efficient electricity ecovery from artificial dialysis wastewater. *The Journal of General and Applied Microbiology, 63*(3), 165–171.

Hocaoglu, S. M., Celebi, M. D., Basturk, I., & Partal, R. (2021). Treatment-based hospital wastewater characterization and fractionation of pollutants. *Journal of Water Process Engineering, 43*, 102205.

Hospital wastewater as a source of environmental contamination: An overview of management practices, environmental risks, and treatment processes. *Journal of Water Process Engineering, 41*. https://doi.org/10.1016/j.jwpe.2021.101990.

Iorhemen, O. T., Hamza, R. A., Zaghloul, M. S., & Tay, J. H. (2019). Aerobic granular sludge membrane bioreactor (AGMBR): Extracellular polymeric substances (EPS) analysis. *Water Research, 156*, 305–314.

Jagaba, A. H., Kutty, S. R. M., Fauzi, M. A. H. M., Razali, M. A., Hafiz, M. F. U. M., & Noor, A. (2021, August). *Organic and nutrient removal from pulp and paper industry wastewater by extended aeration activated sludge system.* In IOP conference series: Earth and environmental science (Vol. 842, No. 1, p. 012021). IOP Publishing.

Jani, M., Arcos-Pareja, J.A., & Ni, M. (2020). Engineered zero-dimensional fullerene/carbon dots-polymer based nanocomposite membranes for wastewater treatment. *Molecules, 25*(21). doi: 10.3390/molecules25214934.

Judd, S. (2010). *The MBR book: Principles and applications of membrane bioreactors for water and wastewater treatment.* Elsevier.

Kalbar, P. P., Karmakar, S., & Asolekar, S. R. (2013). Assessment of wastewater treatment technologies: life cycle approach. *Water and Environment Journal, 27*(2), 261–268.

Khan, M. B., Lee, X. Y., Nisar, H., Ng, C. A., Yeap, K. H., & Malik, A. S. (2015). Digital image processing and analysis for activated sludge wastewater treatment. *Signal and Image Analysis for Biomedical and Life Sciences*, 227–248.

Khan, N. A., Ahmed, S., Vambol, S., Vambol, V., & Farooqi, I. H. (2019). Field hospital wastewater treatment scenario. *Ecological Questions, 30*(3), 57–69.

Khan, N. A., Vambol, V., Vambol, S., Bolibrukh, B., Sillanpaa, M., Changani, F., … & Yousefi M. (2021). Hospital effluent guidelines and legislation scenario around the globe: A critical review. *Journal of Environmental Chemical Engineering, 9*, 105874. doi: 10.1016/j.jece.2021.105874.

Kinuthia, G. K., Ngure, V., Beti, D., Lugalia, R., Wangila, A., & Kamau, L. (2020). Levels of heavy metals in wastewater and soil samples from open drainage channels in Nairobi, Kenya: Community health implication. *Sci Rep., 10* (1), 8434. doi: 10.1038/s41598-020-65359-5.

Kumari, A., Maurya, N. S., & Tiwari, B. (2020). Hospital wastewater treatment scenario around the globe. *Current Developments in Biotechnology and Bioengineering*. doi: 10.1016/B978-0-12-819722-6.00015-8.

Lopsik, K. (2013). Life cycle assessment of small-scale constructed wetland and extended aeration activated sludge wastewater treatment system. *International Journal of Environmental Science and Technology, 10*, 1295–1308.

Lotito, A. M., De Sanctis, M., Di Iaconi, C., & Bergna, G. (2014). Textile wastewater treatment: Aerobic granular sludge vs activated sludge systems. *Water Research, 54*, 337–346.

Majumder, A., Gupta, A. K., Ghosal, P. S., & Varma, M. (2021). A review on hospital wastewater treatment: A special emphasis on occurrence and removal of pharmaceutically active compounds, resistant microorganisms, and SARS-CoV-2. *Journal of Environmental Chemical Engineering*, *9*(2), 104812.

McCarty, P. L., & Brodersen, C. F. (1962). Theory of Extended Aeration Activated Sludge. *Journal Water Pollution Control Federation, 34*(11), 1095–1103. http://www.jstor.org/stable/25034723

Meshram, V. A., Singh A. R., & Verma, S. R. (2024). Water's vital role: Challenges and consequences of pollution: A review. *International Journal of Health Sciences and Research, 14*(2). doi:10.52403/ijhsr.20240243fatcat: w7dfurgbijh5hd3jf27heoea64

Mohan, S. V., Rao, N. C., Prasad, K. K., Madhavi, B. T. V., & Sharma, P. N. (2005). Treatment of complex chemical wastewater in a sequencing batch reactor (SBR) with an aerobic suspended growth configuration. *Process biochemistry, 40*(5), 1501–1508.

Mussarat, M., Jamal, W.A., Muhammad, D., Ahmad, M., Saleem, A., Khan, S., … & Malik, W. (2021). Risk of heavy metals accumulation in soil and wheat grains with waste water irrigation under different NPK levels in alkaline calcareous soil. *PLoS One, 16*(11). doi: 10.1371/journal.pone.0258724

Naghizadeh, A., Mahvi, A. H., Mesdaghinia, A. R., & Alimohammadi, M. (2011). Application of MBR technology in municipal wastewater treatment. *Arabian Journal for Science and Engineering, 36*, 3–10.

Orhon, D., Babuna, F. G., & Karahan, O. (2009). *Industrial wastewater treatment by activated sludge*. IWA Publishing.

Parida, V. K., Sikarwar, D., Majumder, A., & Gupta, A. K. (2022). An assessment of hospital wastewater and biomedical waste generation, existing legislations, risk assessment, treatment processes, and scenario during COVID-19. *Journal of Environmental Management, 308*, 114609.

Pariente, M. I., Segura, Y., Álvarez-Torrellas, S., Casas, J. A., de Pedro, Z. M., Diaz, E., … & Martínez, F. (2022). Critical review of technologies for the on-site treatment of hospital wastewater: from conventional to combined advanced processes.. *Journal of Environmental Management, 320*, 115769. https://doi.org/10.1016/j.jenvman.2022.115769.

Rajan, R., Robin, D. T., & Vandanarani, M. (2019). Biomedical waste management in Ayurveda hospitals–current practices and future prospectives. *Journal of Ayurveda and Integrative Medicine*, *10*(3), 214–221.

Rathi, R. (2022). Hospital wastewater treatment scenario in India. In N. A. Khan, V. Vambol, S. Vambol, N. Mozaffari, & N. Mozaffari (Eds.). IWA Publishing. https://doi.org/10.2166/9781789062625

Servetas, S. L., Parratt, K. H., Brinkman, N. E., Shanks, O. C., Smith, T., Mattson, P. J., & Lin, N. J. (2022). Standards to support an enduring capability in wastewater surveillance for public health: Where are we? *Case Studies in Chemical and Environmental Engineering, 6*, 100247.

Sheng, G. P., Yu, H. Q., & Li, X. Y. (2010). Extracellular polymeric substances (EPS) of microbial aggregates in biological wastewater treatment systems: A review. *Biotechnology Advances, 28*(6), 882–894.

Smital, T., Terzic, S., Zaja, R., Senta, I., Pivcevic, B., Popovic, M., … & Ahel, M. (2011). Assessment of toxicological profiles of the municipal wastewater effluents using chemical analyses and bioassays. *Ecotoxicology and Environmental Safety, 74*(4), 844–851.

Steinmetz, H., Wiese, J., & Schmitt, T. G. (2002). Efficiency of SBR technology in municipal wastewater treatment plants. *Water Science and Technology, 46*(4–5), 293–299.

Suarez, S., Lema, J. M., & Omil, F. (2009). Pre-treatment of hospital wastewater by coagulation–flocculation and flotation. *Bioresource Technology, 100*(7), 2138–2146.

Sustarsic, M. (2009, November). Wastewater treatment: understanding the activated sludge process. In Safety in Ammonia Plants and Related Facilities Symposium (pp. 26–29).

Symonds, E. M., Nguyen, K. H., Harwood, V. J., & Breitbart, M. (2018).Pepper mild mottle virus: A plant pathogen with a greater purpose in (waste) water treatment development and public health management. *Water Res., 144,* 1–12. doi: 10.1016/j.watres.2018.06.066.

Verma, L. K., Mani, S., Sinha, N., & Rana, S. (2008). Biomedical waste management in nursing homes and smaller hospitals in Delhi. *Waste Management, 28*(12), 2723–2734.

Walia, S., Murleedharn, C., Band, J., Kanwar, M., & Kumar, A. (2016). Quantitation of antibiotic resistance genes pollution in hospital waste water effluent and Urban Clinton River Water, Michigan, USA. *Current Medicine Research and Practice, 6*(4), 149–151.

Wang, S., Cui, Y., Li, A., Zhang, W., Wang, D., Chen, Z., & Liang, J. (2020). Deciphering of organic matter and nutrient removal and bacterial community in three sludge treatment wetlands under different operating conditions. *Journal of Environmental Management, 260*, 110159.

Xie, W. M., Ni, B. J., Seviour, T., Sheng, G. P., & Yu, H. Q. (2012). Characterization of autotrophic and heterotrophic soluble microbial product (SMP) fractions from activated sludge. *Water Research, 46*(19), 6210–6217.

Yanbo, J., Jianyi, J., Xiandong, W., Wei, L., & Lincheng, J. (2021). Bioaugmentation technology for treatment of toxic and refractory organic waste water based on artificial intelligence. *Front Bioeng Biotechnol, 9*. doi: 10.3389/fbioe.2021.696166.

Zhou, J., Ma, L., & Song, H. (2015 March). *Research advance of the model of activated sludge process to wastewater treatment*. Proceedings of the 2015 Information Technology and Mechatronics Engineering Conference. https://www.atlantis-press.com/proceedings/itoec-15/16817

Innovative Approaches for Sustainable Effluent Control in the Fertilizer Industry

9

Manish Pant, S.V. Chinna Swami Naik,
Manisha Duseja and Jabrinder Singh

1. INTRODUCTION

Water, a vital element for life, is increasingly becoming an economic asset globally (Karimi-Maleh et al., 2020). However, the relentless rise in water demand, coupled with the challenges of climate change, has escalated water scarcity to the forefront of global concerns in the 21st century. With the United Nations projecting a population surge to between 9.4 to 10.2 billion by 2050 (WWAP, 2018), a corresponding 20% to 30% increase in water demand is anticipated (UN, 2015). This surge in demand is exacerbated by rapid industrialization driven by overpopulation, leading to heightened production of industrial waste. Industrial waste, a byproduct of rapid industrialization, poses severe environmental threats by polluting water bodies, air, and soil (Ahmed et al., 2021). Alarmingly, globally ~80% of the wastewater is going to the ecosystems devoid of proper treatment, while in developing nations, ~8% of industrial effluents receives adequate treatment in (UNESCO, 2017). Forecasts from the European Investment Bank

DOI:10.1201/9781003496014-9

present a bleak scenario, suggesting that approximately 380 billion cubic meters of wastewater is produced each year, with an expected rise of 24% by 2030 and 51% by 2050 (European Investment Bank, 2022). Disconcertingly, the treatment of wastewater varies significantly between developed and developing nations, with high-income countries treating roughly 70% of wastewater compared to a mere 28% in developing economies.

Industrial effluents from agrochemical industries, including those involved in fertilizer, pesticide, and other agrochemicals, contain elevated levels of persistent contaminants (heavy metals). These pollutants pose various risks to human beings and ecosystems, ranging from toxicity and carcinogenicity to flammability. Improper disposal of industrial wastewater into aquifers can lead to hazardous impacts, underscoring the importance of effective management strategies (Duque et al., 2021). While agrochemical industries play a crucial role in supporting agriculture, their production processes often rely on substantial extents of non-replenishable resources. For instance, the production of nitrogen fertilizers through *Haber-Bosch* process consumes a substantial portion of the world's energy resources. Ammonia production alone, a vital source of nitrogen-fertilizer, reached millions of metric tons globally in the past decade (Cross et al., 2021).

Moreover, the extraction and processing of phosphorus and potassium fertilizers depend heavily on mineral resources, particularly phosphate rock and potash. This intrinsic reliance on finite resources underscores the need for sustainable practices in the agrochemical industry. Despite their importance, agrochemical manufacturing processes are also contributing to environmental pollution substantially, through the emission of gaseous, liquid, and solid pollutants. Effluents from these industries contains diversity of pollutants, including toxic ions, inorganic and organic compounds, heavy metals, noxious gases, and hazardous compounds, further emphasizing the urgent need for stringent pollution control measures (Ab Rahman et al., 2022; Santos et al., 2021).

Numerous methodologies have emerged to combat the heavy metal toxicity from effluents, with bio-adsorption standing out as a promising biotechnological solution (ElGendy & El-Bondkly, 2016; Sumalatha et al., 2023). Advanced oxidation processes (AOPs) have also showcased remarkable efficacy in eliminating a diverse array of metal pollutants (Mukherjee et al., 2023). However, it's crucial to acknowledge that these techniques entail considerable energy requirements and associated expenses for their deployment and upkeep. Prior to initiating the remediation process, it is imperative to conduct qualitative assay and quantitative measurements of pollutants. Even in cases where contaminants are present in low concentrations, suitable techniques must be employed for effective treatment (Abdullah et al., 2019). Various approaches encompass a spectrum of techniques aimed at controlling water pollution, spanning physical, chemical, and biological methods. These encompass decomposition, filtration, ion exchange, membrane filtration, and various other additional methods. Typically, a blend of treatment technologies has been deployed to effectively remediate industrial effluents. For instance, in the treatment of effluents, the process often begins with the removal of solid wastes, followed by successive steps including adsorption, ion exchange, bioremediation, and sedimentation, among other techniques.

The prime emphasis of this chapter is to evaluate sustainable approaches for treating wastewater from various perspectives. Within its pages, thorough discussions delve into the merits of biological treatment methods, such as phytoremediation and bioremediation, alongside other physical and chemical techniques. The overarching aim is to provide insights into effective strategies for mitigating metal pollution in industrial effluents while considering the balance between efficacy, energy demands, and economic feasibility.

2. HARMFUL EFFECT OF EFFLUENTS FROM THE FERTILIZER INDUSTRY

Fertilizers, whether they're natural or synthetic, are essential for modern agriculture, providing crucial nutrients that support plant growth. The evolution of fertilizer technology has been remarkable, with modern synthetic fertilizers mainly comprising nitrogen, phosphorous, potassium, and additional secondary nutrients. The global demand for these fertilizers is immense, with estimates suggesting the need for billions of tons annually (Spanoghe et al., 2020). Despite their pivotal role in sustaining agricultural productivity and meeting food demands, chemical fertilizers pose significant environmental challenges when improperly disposed of. Effluents from fertilizer industries contain a complex mix of organic compounds, nitrates, ammonia, phosphates, heavy metals like lead, mercury, cadmium, and particulate solids (Bhandari et al., 2016). Moreover, hazardous by-products like hexafluoro-silicic acid and phospho-gypsum compounds; pose threats by releasing toxic substances and radioactive elements into the environment (Ahmad et al., 2023; Cheremisinof, 1995). The adverse environmental impact of fertilizer industry effluents extends beyond water pollution. Ammonia emissions contribute to the eutrophication and acidification of catchment water resources (Hussain et al., 2018). Additionally, the excessive use of fertilizers can disrupt ecosystems and aquatic life, leading to disturbances in fish metabolism and increased mortality rates (Kaur et al., 2010; Roopadevi & Somasekhar, 2012). In response to these environmental risks, regulatory bodies like MOEFCC–Govt. of India have categorized fertilizer-producing industries as "red list industries" (2016), underscoring the urgent need for robust pollution control measures and the adoption of sustainable practices within the fertilizer industry. Eutrophication, biomagnification, methemoglobinemia are some other consequences faced by the environment due to the fertilizer industry.

The discharge of industrial effluents into fresh water resources is a menace for public health, with millions of lives at risk daily, as per World Health Organization (Uddin et al., 2014). Waterborne diseases, such as diarrhoea and gastrointestinal disorders, account for a substantial portion of health-related deaths globally (Schwarzenbach et al., 2010). Serious ailments including cholera, typhoid, dysentery, tuberculosis, and various organ disorders further compound the risks to human health (Khan & Ghouri, 2011). Industrial effluents also pose threats to animal health, leading to skin disorders, FMD-foot and mouth diseases, mastitis, and growth impediments (Hossain et al., 2011). Furthermore, the persistence of toxic metalloids in these effluents exacerbates their impact on aquatic and terrestrial life (Koetlisi & Muchaonyerwa, 2019). Some impacts are highlighted in Figure 9.1.

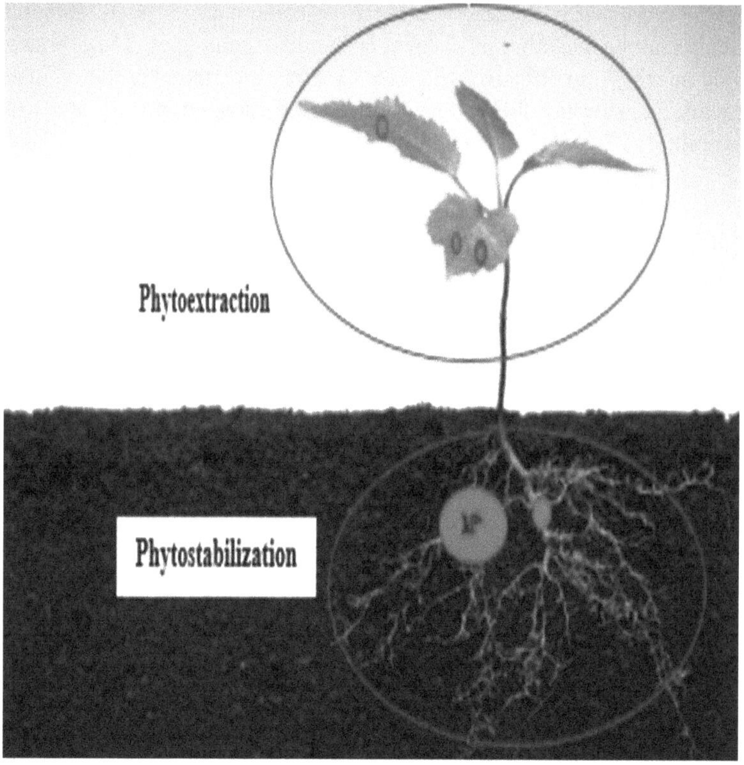

FIGURE 9.1 Effects of fertilizer effluents on the environment and human health.

The rapid pace of industrialization, driven by economic growth agendas, often outpaces efforts to effectively manage industrial effluents. Many developing countries, unable to afford expensive treatment technologies, face significant challenges in mitigating the adverse effects of industrial discharge. In this context, the current research underscores the importance of exploring natural and cost-effective effluent treatment technologies, especially in regions where financial resources are limited. A fertilizer industry effluent case study in Sultanpur, Uttar Pradesh, discovered frightening findings similar with observations from other sites. The effluent exhibited characteristics such as high pH, biochemical oxygen demand (BOD), EC-electrical conductivity, chemical oxygen demand (COD), total dissolved solids (TDS), nitrates, phosphates, sulfate content, low oxygen levels in water, and heavy metals the presence (Agrawal et al., 2020; Obire et al., 2008). These findings highlight the critical necessity for efficient and sustainable approaches to address industrial effluent pollution, particularly in areas where the impact on public health and the ecosystems is most severe.

After green revolution, more consumption of chemical fertilizers and its run-off into aquatic and terrestrial ecosystems without treatment brings hazardous impacts. Groundwater also becomes contaminated through the leaching of pesticides, insecticides, heavy metals, and fertilizer residues. Consumption of this polluted water can result in various waterborne diseases, including dysentery, typhoid, diarrhoea, liver and intestinal infections, hepatitis, and skin rashes (Divya et al., 2012). Wastewater treatment poses a significant challenge for complex fertilizer plants, contributing to environmental pollution (Bhandari et al., 2016). To address this issue, innovative approaches such as microbial-remediation, phytoremediation, and nanotechnology methods are being explored. These novel techniques aim to enhance the operation of effluent treatment plants (ETPs), improving their performance while reducing costs. By adopting these advanced methods, fertilizer plants can mitigate the environmental impact of their operations and safeguard public health

3. ADVANCED TECHNIQUES OF EFFLUENT TREATMENT

Bioremediation refers to the utilization of plant species and microbes to mitigate the toxicity or digest industrial wastes that pose significant hazards to humans and other organisms (Ullah et al., 2015). This bioprocess offers several advantages over traditional physio-chemical treatments, being enfvironmentally friendly, widely accepted by the public, cost-effective, and often conducted directly at the site of contamination. Bioremediation encompasses various techniques, such as bio-degradation, bio-mineralization, bio-augmentation, bioventing, biosorption, bioreactor employment, and bio-accumulation, collectively aimed at remedying toxic substances or converting them into non-toxic forms through microbial processes, a concept known as microbial remediation.

3.1. Phytoremediation

The method of phytoremediation, which involves harnessing microorganisms and plants for the purpose of nutrient recovery and environmental purification, including treating wastewater, is attracting considerable attention. Its allure stems from its eco-friendly approach, sustainable practices, as well as cost-effectiveness (Hu et al., 2020). Phytoremediation stands out as a dominant, effective, and extensively engaged technique for environmental restoration, relying on the degradation, stabilization, volatilization, and extraction of contaminants and microbes. It contributes to the breakdown of various harmful substances encompassing antibiotics, textile pigments and dyes, petroleum derivatives, explosives, heavy metals, and noxious gases (Senturk et al., 2023; Zhou et al., 2019).

Limited research has explored the capacity of timber plants to recover nitrogen from wastewaters. Timber species such as *Azadirachta indica*, *Acacia mangium*, and *Eucalyptus hybrid* sp. are recognized for their efficacy in removing nutrient pollutants from primary treated wastewaters (Thawale et al., 2006). Similarly, *Dalbergia sissoo L.* and *Khaya senegalensis* seedlings have demonstrated successful application in municipal effluent phytoremediation (Ali et al., 2013; Singh et al., 2010). Phytoremediation capabilities of teak, *Tectona grandis*, a prominent tropical timber species, in treating wastewater from urea manufacturing units, have been well established by research studies. The research focused on eight-month-old teak seedlings placed in bench-scale constructed wetlands, where they were exposed to four different nitrogen (N) concentration levels over an eight-week period, with replenishment every four days. The findings revealed the significant potential of teak seedlings in removing nitrogen, particularly when exposed to concentrations of up to 5 grams per pot volume (equivalent to 4 cubic decimetres) during the two-month experimental duration (Yavari et al., 2017).

In a recent study, researchers evaluated the efficacy of two water plant variants, *Salvinia sp.* and *Pistia sp.*, in removing dairy farming effluent phosphates. Additionally, they explored the perspectives of these dairy farms as organic fertilizer units. Both *Salvinia* and *Pistia sp.* have been extensively studied in phytoremediation research, stressing their established significance in eco-remediation efforts. The research study also evaluated amendments in water quality factors within an artificial wetland (AW) engineered for wastewater treatment (Kulangana et al., 2024). *Salvinia* exhibited a phosphate removal efficiency of 37.87% in AW I and 35.69% in AW II. Similarly, *Pistia* showed impressive removal rates of 84.32% in AW III and 47.51% in AW IV. These plants demonstrated significant abilities to absorb phosphate compounds and effectively reduce conductivity and total dissolved solids in cattle farm waste, emphasizing their crucial role in phytoremediation (Kulangana et al., 2024).

Phytoremediation encompasses various processes such as phytoextraction, phytovolatilization, rhizofiltration, rhizostimulation, phytodegradation, and phytostabilization (Figure 9.2). Among these applications, namely rhizofiltration, phytoextraction, and phytostabilization, are utilized for the purification of water bodies contaminated by

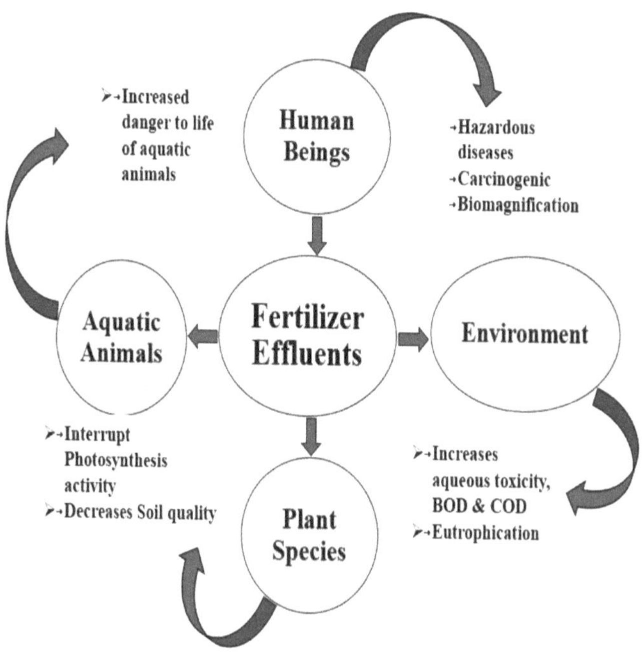

FIGURE 9.2 Phytoremediation – Phytostabilization and phytoextraction of effluent waste.

effluents discharged from nearby industries (Ali et al., 2020; OB & Muchie, 2010). Despite its effectiveness, phytoremediation encounters several challenges that require attention. For instance, pollutants negatively impact plant growth and development, slowing down the remediation process due to the limited availability of noxious waste in the rhizosphere and low biomass production. Moreover, the use of crops is often avoided to prevent pollutants from entering the food chain. Additionally, hyper accumulating plants typically accumulate only one specific pollutant, whereas industrial wastewater contains multiple pollutants. Another significant challenge is the prolonged duration required to clean contaminated sites, often exceeding 13 years for the remediation of a polluted site (Kanwar et al., 2020).

3.2. Microbial Remediation

Due to the limited effectiveness of phytoremediation, ecological ditches struggle to match the nitrogen (N) and phosphorus (P) removal capabilities of traditional municipal sewage treatment methods, particularly for irrigation and during monsoon season. Microbial treatment technologies, known for their efficacy, cost-effectiveness, and eco-friendliness, are increasingly favoured for managing diverse wastewater types, including domestic sewage, dyeing effluents, and animal waste (Fernandes et al., 2013; Wu et al., 2007). As a result, microbial technologies may offer a promising solution for advancing the next generation of ecological ditch systems.

Stimulated sludge methods employing A2/O (anaerobic-anoxic-aerobic) processes have demonstrated success in treating highly concentrated organic wastewater (Kassab et al., 2010). These micro-organisms possess the ability to remove nutrients (such as N and P) simultaneously and heavy metals (such as chromium and mercury) in an efficient and eco-friendly manner (Samaras et al., 2009; Ying et al., 2010). Wu et al. (2011) proposed an integrated method combining the A2/O technique with ecological ditches for treating diverse non-point source wastewater. This provision has the ability to treat highly concentrated N and P wastewater, and also revive the ecological waterways' micro-habitat. Having a wastewater load of 200 m^3 per day, the removal efficiencies of TN, TP, TDP, NO_3/N, and NH_4/N reached 82%, 81%, 74%, 79%, and 86%, respectively. As per Xia et al. (2020); microbial enhancement into traditional eco-ditches holds promise as a future research direction.

3.3. Nanotechnology in Effluent Control

Nanotechnology holds immense potential for wastewater treatment, offering the ability to manipulate substances at atomic and molecular levels to create innovative substances. Nano-structures, characterized by their small dimension, large surface area, and ease of utilization, present unparalleled opportunities for developing highly efficient catalysts and redox-active media for wastewater purification (Jain & He, 2018). Among the various nanotechnological approaches, nanostructured photocatalysts, nano adsorbents, nanoparticles, and nano-membranes shine as notable contenders for their effectiveness in wastewater treatment. Additionally, carbon nanotubes (CNTs) represent a noteworthy nanomaterial capable of effectively removing both inorganic and organic impurities from contaminated water, thereby rendering it suitable for consumption (Anjum et al., 2019).

Carbon nanotubes (CNTs) possess notable characteristics, including a substantial specific surface area and exceptional chemical, mechanical, and thermal stability. The distinctive qualities of carbon nanotubes make them a promising material for improving the adsorption capacity, as well as enhancing the mechanical and chemical stability of biomaterials in acidic environments during adsorption studies. Kinetic experiments focusing on fluoride removal from industrial wastewater were conducted at varying stirring rates, ranging from 100 to 300 rpm. Results indicated rapid adsorption, with equilibrium reached at 300 rpm within 20 minutes. The adsorption capacity was found to be approximately 975.4 mg/g, highlighting the hybrid material's potential for effectively removing fluoride from real-world matrices. The notable adsorption capacity was linked to the functional groups present in chitosan, along with the expansive interaction area facilitated by the sponge-like structure and carbon nanotubes. Investigation into the reuse and regeneration of CNTs demonstrated consistent adsorption capacity over five cycles (Affonso et al., 2020). However, concerns arise regarding the incorporation of nanomaterials in water and wastewater treatment, particularly regarding their toxicity. When utilizing these materials for water treatment, re-usability of nanomaterials and the durability nature in waste treatment are critical consideration.

Nanomaterials have garnered interest across diverse fields and scales, attracting attention from various stakeholders and organizations (Goh et al., 2016). However, financial constraints pose a significant obstacle to the advancement of nanotechnology, given its costly nature, potential hazards, and the need for specialized equipment, technical expertise, and infrastructure. Effective regulation, policy development, and program coordination are essential for the manufacturing and deployment of nanoparticles, yet this coordination is currently lacking. As an illustration, the widespread implementation of nanotechnology-enabled desalination may face a significant time lag before becoming a reality (Hofmann-Amtenbrink et al., 2015; Khan et al., 2021).

3.3.1. Membrane based methods

Membrane processes utilize a membrane obstacle to filter or separate particles from water, working at the cellular level. Each process features a specific membrane layer that functions as a filter, successfully eliminating impurities from water. Recent research has focused on nanotechnologies, such as graphene, aluminum oxide, and carbon nanotubes, to enhance the efficiency of wastewater treatment plants (WWTPs) (Thekkudan et al., 2017). The choice of materials for membrane fabrication depends on factors like pore size and cell size, with commonly used methods like forward osmosis (FO), microfiltration (MF), ultrafiltration (UF), nanofiltration (NF), and reverse osmosis (RO). Nano-membranes can be categorized into three main types: organic, inorganic, and organic-inorganic hybrids. Organic nano-membranes are composed of polymeric materials such as cellulose acetate, polyethersulfone, polyacrylonitrile, cellulose nitrates, polyvinyl alcohol, and biomacromolecules (Wang et al., 2018). Inorganic nanomembranes consist of materials like metals, metal oxides, carbon nanotubes (CNTs), and graphene (Zhang et al., 2019). The integration of inorganic materials into a polymeric system results in the formation of inorganic-organic hybrid nanomembranes (Goh et al., 2015). Ideally, the membranes employed should exhibit specific favourable characteristics, such as uniform velocity distributions without stagnant regions, simple maintenance and cleaning procedures, affordability, compactness, and minimal tangential fluid resistance (Saravanan et al., 2021; Shehata et al., 2023).

3.3.2. Photocatalysis

The term photocatalysis originates from the combination of two Greek words, "photo" and "catalysis," signifying the complex degradation of compounds in the light presence (Yaqoob et al., 2020). Essentially, photocatalysis refers to a process where materials are stimulated or activated by light, which can be visible light, sunlight, or ultraviolet (UV) light. Unlike thermos-catalysts, which are activated by heat, photocatalysts rely on light energy photons for activation (Gomes et al., 2019). The nanophotocatalysts are commonly utilized in the treatment of contaminated water because of their shape-dependent properties and increased surface area-to-volume ratio, which enhance catalytic activity (Chen et al., 2019).

Advanced technologies such as reduced graphene oxide and graphene oxide have been effectively employed in developing both homogeneous and heterogeneous photocatalysts in recent years. These catalysts aim to improve degradation rates and lower operational costs in large-scale practical applications (Das et al., 2021; Iyyappan et al., 2023). While photocatalysis is effective in oxidizing organonitrogen compounds, it can also facilitate the conversion of inorganic nitrogen compounds such as NH_4, NO_2, and NO_3 into usable fertilizers (Al Juboury et al., 2022). In batch experiments conducted under sunlight irradiation, the photocatalytic breakdown of organonitrogen compounds resulted in the production of inorganic nitrogen constituents like ammonia, nitrite or nitrate which can be utilized as fertilizers. For instance, following 12 hours of circulation, the inorganic nitrogen concentration reached NH_4 = 17.2 g/L, NO_2^- = 18.1 mg/L, and NO_3^- = 15.9 mg/L (Al Juboury et al., 2022). In a study by Joshi and Srivastava (2011), wastewater from various industries was analysed for chromium content. The removal of chromium using a photocatalyst under UV light and in the dark was investigated across different pH ranges. The results indicated that the maximum amount of removal of Cr(VI) occurred at pH 2. Among the photocatalysts tested, TiO_2 demonstrated the highest removal capacity for Cr (VI) compared to TiO_2 thin film. Subsequent research has focused on enhancing the photocatalytic abilities of metal oxides by incorporating modifications with other materials like metal ions, metal dye sensitizers, and carbonaceous materials. However, further modifications are needed to optimize the performance of these photocatalysts (Joshi & Srivastava, 2011).

3.3.3. Adsorption processes

Adsorption processes are promising for nutrient recovery from wastewater due to their simplicity, cost-effectiveness, and economic viability (Loganathan et al., 2014). Nevertheless, these methods are susceptible to changes in pH and temperature conditions (Fidel et al., 2018). The incorporation of biochar derived from waste materials not only cuts down on expenses linked to adsorption methods but also produces tailored, purpose-built sorbents (Panwar & Pawar, 2022). Researchers have developed various adsorbents with exceptional properties for adsorbing fertilizer effluents from water (Gizaw et al., 2021). Nano-adsorbents, in particular, have garnered significant interest in wastewater treatment due to higher surface area, desired surface charge, and specificity towards targeted pollutants (Suzaimi et al., 2019). Ammonium ions (NH_4–N) are prevalent in water, and studies have shown that biochars produced at different temperatures exhibit varying affinities for NH_4–N ions. For instance, PS-300 demonstrated the greatest affinity for NH_4–N ions, adsorbing them at a rate of 5.38 mg/g (Yang et al., 2018). Reduced adsorption values were found for PS-550 (3.37 mg/g) and WS-550 (2.08 mg/g) compared to PS300, indicating variations in biochar affinity for NH_4–N ions based on production temperature (Yang et al., 2018). The authors elucidated that NH_4–N adsorption is predominantly influenced by electrostatic and chemical bond interactions between the negatively charged biochar surface and the positively charged nitrogen compounds (NH_4–N). In contrast, NO_3^- anions present a challenge for biochar adsorption due to electrostatic repulsion from the negatively charged surface (Zhang et al., 2020). Modifications of biochar, like with chitosan and lanthanum, have demonstrated enhanced PO_4–P adsorption capacities, attributed to Lewis acid-base interactions, electrostatic interactions, and ligand exchange (Huang et al., 2020). Additionally, natural biodegradable waste, such as *Posidonia oceanica*, has been explored for ammonium recovery, suggesting its potential use as bio-compost in agriculture (Jellali et al., 2011). Similarly, sawdust has been utilized for ammonium ion recovery, with the resulting ammonium-loaded sawdust proposed for application as bio-compost and fertilizers (Wahab et al., 2010). The adsorbents efficiency for heavy metal removal is influenced by various factors, including temperature, pH, contact time, stirring speed, and initial concentration (Agarwal & Singh, 2017). While effective in removing pollutants, adsorption processes have drawbacks including the generation of secondary pollutants detrimental to human health and the environment, as well as the generation of substantial sludge due to excessive chemical usage, constraining its broad application (Velusamy et al., 2021).

3.3.4. Nanofiber

Electrospinning emerges as a cost-effective and highly efficient technique for producing ultra-fine fibres from diverse materials such as ceramics, polymers, and even metals. These resulting nanofibers boast exceptional properties including minute pore size, large surface area, and high porosity, rendering them ideal for various applications in wastewater treatment. For instance, electrospun polyacrylonitrile nanofiber mats have been successfully employed for heavy metal ions removal (Li & Xia, 2004). In industrial wastewater treatment, an advanced approach integrates nanofiber technology with biological processes to efficiently remove toxic xenobiotics. A key advantage of this technology lies in the formation of biofilms by bacteria, which develop both on the surface of the carrier material and within its core. This dual-layered biofilm structure provides protection to the bacteria against harmful environmental factors, enhancing their efficacy in pollutant removal (Joseph et al., 2023; Mahvi et al., 2011).

4. OTHER APPROACHES INCLUDING FLOCCULATION AND COAGULATION

Both, coagulation and flocculation play essential roles in solid-liquid separation within industrial wastewater treatment processes. During coagulation, coagulants are introduced to destabilize colloidal suspensions by neutralizing their charges, leading to the accumulation of smaller particulate matter (Sun et al., 2020). Subsequently, in flocculation, mechanical agitation is employed to gather these destabilized particles into larger clusters. In certain cases, the addition of flocculants helps expedite particle aggregation and improve settling efficiency. Various coagulants and flocculants exhibit diverse properties influenced by structural properties such as charge, functional group, molecular weight, and ionic strength (Kolya & Tripathy, 2013).

The efficiency of the process significantly depends on the choice of coagulants, which often work in tandem during coagulation and flocculation. For instance, flocculants like long-chain anionic or non-ionic polymers complement the role of coagulants such as cationic inorganic salts, primarily due to the prevalence of negatively charged suspended particles (Sillanpaa et al., 2018). Upon the addition of coagulants, metal salts rapidly hydrolyze at the iso-electric point, generating cations that are absorbed by negatively charged colloidal particles in wastewater. This leads to the concurrent decrease in surface charge and the generation of micro-flocs. Further application of flocculants assists in consolidating micro-flocs into more compact and sizable flocs, easing their removal through subsequent physical methods such as filtration or sedimentation. The principal mechanisms included in this process involve charge neutralization and bridging (Saravanan et al., 2021; Wei et al., 2018).

5. CHALLENGES REGARDING EFFLUENTS FROM THE FERTILIZER INDUSTRY

5.1. Deep Learning Techniques in Wastewater Management

Delve deeper into the potential of deep learning techniques in addressing specific challenges in urban wastewater management, such as predictive modelling for optimizing treatment processes and real-time monitoring for early detection of pollution events. Discuss recent advancements and case studies showcasing successful applications of deep learning in wastewater management.

5.2. Innovative Solutions for Sludge Treatment

Explore innovative approaches for addressing the challenges associated with sludge treatment in textile industry effluent plants. Consider technologies like anaerobic digestion, thermal hydrolysis, and sludge-to-energy transformation to improve sludge management practices and reduce environmental impact.

5.3. Sustainable Corporate Practices

Provide practical recommendations for improving corporate performance in the fertilizer industry by adopting sustainable practices. This could include implementing resource-efficient production processes, investing in renewable energy sources, and incorporating circular economy principles into business strategies.

5.4. Rural Wastewater Management

Propose tailored solutions for overcoming the challenges faced by rural communities in constructing and operating wastewater treatment plants. This could involve decentralized treatment systems, community-based management approaches, and capacity-building initiatives to empower local stakeholders.

5.5. Advanced Analytical Techniques for Contaminant Detection

Discuss emerging analytical techniques, such as mass spectrometry and molecular biomarkers, for the detection and identification of micro-plastics, nanoparticles, and other emerging contaminants in water. Highlighting the importance of investing in research and development to enhance analytical capabilities and address knowledge gaps in contaminant detection.

5.6. Global Collaboration for Wastewater Management

Advocate for increased international cooperation and knowledge-sharing platforms to facilitate global coordination in wastewater management. Emphasize the importance of harmonizing standards and best practices to address transboundary water pollution and promote sustainable development goals.

5.6.1. Incentivizing Sustainable Practices

Suggest policy measures and incentives to encourage industries to prioritize sustainability in their wastewater management practices. This could include tax incentives for investments in water-saving technologies, subsidies for green infrastructure projects, and regulatory frameworks that promote resource recovery and reuse.

6. LIMITATIONS REGARDING EFFLUENTS FROM THE FERTILIZER INDUSTRY

The environmental impact of industrial activities often overlooks the crucial aspect of water footprints and their calculation. Failure to include water footprint analysis results in a lack of comprehensive understanding regarding the sustainability challenges faced by industries (Singh et al., 2023).

7. CURRENT STATUS AND FUTURE PERSPECTIVES

Currently, approximately 80% of the world's water resources are at risk, endangering human health, the environment, and climate stability. It's crucial to restore water quality, nutrients, and energy within these resources to combat these urgent challenges. Untreated wastewater discharge not only contaminates food sources but also poses health hazards and greatly contributes to greenhouse gas emissions, notably nitrous oxide and methane. It produces three times more pollution than conventional agricultural practices, underscoring the imperative for action amid a rapidly expanding global population. Cities worldwide are taking proactive measures to transform resources, supported by both governmental and private sectors. In the Gulf of Aqaba, Aqaba, the largest city, has implemented a "zero-emission" policy for industrial wastewater, aiming to safeguard marine environmental quality. Currently, 90% of wastewater is collected and treated, demonstrating the city's commitment to environmental conservation. Similarly, Bangkok has implemented efficient systems for collecting, treating, and converting sludge into fertilizer, while exploring wastewater resource utilization for power generation to mitigate land pollution.

Amid rapid population and industrial expansion, Beijing is addressing environmental challenges by investing in large-scale infrastructure projects and embracing wastewater recycling technologies. These initiatives have notably boosted treatment efficiency. Chennai, grappling with water shortages, has achieved 100% sewage collection through robust water reuse strategies, meeting 15% of its water demand through recycling initiatives. Experts in the field are continuously improving regulations and safety standards to protect human health. Traditional water treatment facilities face challenges from increasing pollution levels and limited space availability, driving the adoption of innovative treatments such as membrane bioreactors, ballasted reaction processors, and biological aerated filters, which require less land area while ensuring efficient treatment. Achieving high-quality water purification involves complex processes requiring sophisticated technologies and precise control mechanisms. Automation through analytical tools, intelligent programmatic controls, and human-machine interfaces enhances system performance and reduces administrative burdens. In conclusion, ongoing efforts in industrial wastewater management are essential to safeguard environmental and public health. By implementing innovative technologies and stringent regulations, sustainable water management practices can be ensured, mitigating the adverse impacts of wastewater discharge on ecosystems and human well-being (Saravanan et al., 2021; Younas et al., 2021).

- Revolutionary wastewater collection framework designs provide real-time environmental assessment data to inform resource management decision-making.
- Cutting-edge technologies or advancements in existing technology aim to remove toxic pollutants with minimal costs and energy footprint.

Top of Form

- Further exploration is necessary to gauge the efficiency of the discussed adsorbents, especially those derived from carbon, when combined with alternative treatment approaches.
- Exploring microbial interactions within aquatic environments is imperative for better understanding.
- Developing cost-effective methods, strategies, cycles, and devices for energy management is essential.

8. CONCLUSION

In this chapter, we discuss the complex landscape of industrial wastewater management within the fertilizer industry. Through a thorough examination of hurdles and drivers, we emphasized the vital significance of environmental factors, while sustaining resource consumption. We shed light on pivotal regulatory, technological, and operational aspects, setting the cornerstone for the development of sustainable solutions. These solutions not only bolster economic viability but also fortify ecological resilience. Serving as a comprehensive manual, our chapter furnishes invaluable insights for stakeholders grappling with the intricacies of industrial wastewater management, presenting efficacious and conscientious practices tailored to the industrial realm. Moreover, we underscore the paramount importance of interdisciplinary collaboration among industries, researchers, policymakers, and society at large in tackling the multifaceted challenges inherent in industrial wastewater treatment. With the evolution of industries, there arises an imperative to embrace innovative strategies to curb pollution, embrace circular economy principles, and ensure strict adherence to laws and regulations. The delineated challenges—ranging from water scarcity to pollution abatement and sustainable resource management—underscore the pressing need for proactive interventions.

To address the harmful consequences of industrial wastewater emancipation on ecosystems and public health, we advocate for embracing emerging technologies, optimizing processes, and deploying effective treatment methods. Despite the challenges, identified facilitators provide promising avenues for progress. Through advocating for supportive policies, engaging stakeholders, advancing technology, and embracing sustainable practices, we can spearhead the transition to a more responsible and efficient approach to managing industrial wastewater. Industries must emphasize the incorporation of ecologically conscious practices not only to comply with regulations but also to promote societal and environmental sustainability. Our comprehensive review serves as a guide for informed decision-making in industrial wastewater management within the fertilizer manufacturing sector. By evaluating challenges and opportunities, our goal is to catalyze a renewed commitment to accountable and sustainable approaches, fostering synergy among the economy, industry, and the environment.

REFERENCES

Abdullah, N., Yusof, N., Lau, W. J., Jaafar, J., & Ismail, A. F. (2019). Recent trends of heavy metal removal from water/wastewater by membrane technologies. *Journal of Industrial and Engineering Chemistry*, *76*, 17–38. https://doi.org/10.1016/j.jiec.2019.03.029

Ab Rhaman, S. M. S., Naher, L., & Siddiquee, S. (2021). Mushroom quality related with various substrates' bioaccumulation and translocation of heavy metals. *Journal of Fungi*, *8*(1), 42. https://doi.org/10.3390/jof8010042

Affonso, L. N., Marques Jr, J. L., Lima, V. V., Gonçalves, J. O., Barbosa, S. C., Primel, E. G., … & Cadaval Jr, T. R. (2020). Removal of fluoride from fertilizer industry effluent using carbon nanotubes stabilized in chitosan sponge. *Journal of Hazardous Materials*, *388*, p. 122042. https://doi.org/10.1016/j.jhazmat.2020.122042

Agrawal, K., Bhatt, A., Bhardwaj, N., Kumar, B., & Verma, P. (2020). Integrated approach for the treatment of industrial effluent by physico-chemical and microbiological process for sustainable environment. *Combined Application of Physico-Chemical & Microbiological Processes for Industrial Effluent Treatment Plant*, 119–143. https://doi.org/10.1007/978-981-15-0497-6_7

Ahmad, N., Usman, M., Ahmad, H. R., Sabir, M., Farooqi, Z. U. R., & Shehzad, M. T. (2023). Environmental implications of phosphate-based fertilizer industrial waste and its management practices. *Environmental Monitoring and Assessment*, *195*(11), 1326. https://doi.org/10.1007/s10661-023-11958-4

Ahmed, J., Thakur, A., & Goyal, A. (2021). Industrial wastewater and its toxic efects. In M. P. Shah (Ed.). *Biological treatment of industrial wastewater* (pp. 1–14). The Royal Society of Chemistry, London. https://doi.org/10.1039/9781839165 399-00001

Ali, H. M., Siddiqui, M. H., Khamis, M. H., Hassan, F. A., Salem, M. Z., & El-Mahrouk, E. S. M. (2013). Performance of forest tree Khaya senegalensis (Desr.) A. Juss. under sewage effluent irrigation. *Ecological Engineering*, *61*, 117–126. https://doi.org/10.1016/j.ecoleng.2013.09.051

Ali, S., Abbas, Z., Rizwan, M., Zaheer, I. E., Yavaş, İ., Ünay, A., ... & Kalderis, D. (2020). Application of floating aquatic plants in phytoremediation of heavy metals polluted water: A review. *Sustainability*, *12*(5), 1927. https://doi.org/10.3390/su12051927

Al Juboury, M. F., Abdulredha, M., & Nile, B. K. (2022). Photocatalysis and flocculation processes for recycling aquaculture effluent into nutrient-rich irrigation water. *Water Supply*, *22*(3), 3103–3113. https://doi.org/10.2166/ws.2021.417

Anjum, M., Miandad, R., Waqas, M., Gehany, F., & Barakat, M. A. (2019). Remediation of wastewater using various nano-materials. *Arabian Journal of Chemistry*, *12*(8), 4897–4919. https://doi.org/10.1016/j.arabjc.2016.10.004

Bhandari, V. M., Sorokhaibam, L. G., & Ranade, V. V. (2016). Industrial wastewater treatment for fertilizer industry—A case study. *Desalination and Water Treatment*. Advance online publication. https://doi.org/10.1080/19443994.2016.1186399

Chen, W., Liu, Q., Tian, S., & Zhao, X. (2019). Exposed facet dependent stability of ZnO micro/nano crystals as a photocatalyst. *Applied Surface Science*, *470*, 807–816. https://doi.org/10.1016/j.apsusc.2018.11.206

Cheremisinoff, N. P. (1995). *Handbook of emergency response to toxic chemical releases: A guide to compliance*. Elsevier.

Cross, L., & Gruere, A. (2022). Public Summary Medium-Term Fertilizer Outlook 2022–2026. IFA Market Intelligence Service A/22/82, p. 13.

Das, A., Adak, M. K., Mahata, N., & Biswas, B. (2021). Wastewater treatment with the advent of TiO2 endowed photocatalysts and their reaction kinetics with scavenger effect. *Journal of Molecular Liquids*, *338*, 116479. https://doi.org/10.1016/j.molliq.2021.116479

Divya, J., & Belagali, S. L. (2012). Impact of chemical fertilizers on water quality in selected agricultural areas of Mysore district, Karnataka, India. *International Journal of Environmental Sciences*, *2*(3), 1449–1458.

Duque, A.F., Campo, R., Val del Rio, A., & Amorim, C.L. (2021). Wastewater valorization: Practice around the world at pilot-and full-scale. *Int J Environ Res Public Health*, *18*(18), 9466. https://doi.org/10.3390/ijerph18189466

European Investment Bank. (2022, May). *Wastewater as a resource*. https://www.eib.org/en/publications/wastewater-as-a-resource.

El-Gendy, M. M. A. A., & El-Bondkly, A. M. A. (2016). Evaluation and enhancement of heavy metals bioremediation in aqueous solutions by Nocardiopsis sp. MORSY1948, and Nocardia sp. MORSY2014. *Brazilian Journal of Microbiology*, *47*(3), 571–586. https://doi.org/10.1016/j.bjm.2016.04.029

Fernandes, H., Jungles, M. K., Hoffmann, H., Antonio, R. V., & Costa, R. H. (2013). Full-scale sequencing batch reactor (SBR) for domestic wastewater: performance and diversity of microbial communities. *Bioresource Technology*, *132*, 262–268. https://doi.org/10.1016/j.biortech.2013.01.027

Fidel, R. B., Laird, D. A., & Spokas, K. A. (2018). Sorption of ammonium and nitrate to biochars is electrostatic and pH-dependent. *Scientific Reports*, *8*(1), 17627. https://doi.org/10.1038/s41598-018-35534-w

Gizaw, A., Zewge, F., Kumar, A., Mekonnen, A., & Tesfaye, M. (2021). A comprehensive review on nitrate and phosphate removal and recovery from aqueous solutions by adsorption. *AQUA—Water Infrastructure, Ecosystems and Society*, *70*(7), 921–947. https://doi.org/10.2166/aqua.2021.146

Goh, P. S., Matsuura, T., Ismail, A. F., & Hilal, N. (2016). Recent trends in membranes and membrane processes for desalination. *Desalination*, *391*, 43–60. https://doi.org/10.1016/j.desal.2015.12.016

Goh, P. S., Ng, B. C., Lau, W. J., & Ismail, A. F. (2015). Inorganic nanomaterials in polymeric ultrafiltration membranes for water treatment. *Separation & Purification Reviews*, *44*(3), 216–249. https://doi.org/10.1080/15422119.2014.926274

Gomes, J., Lincho, J., Domingues, E., Quinta-Ferreira, R. M., & Martins, R. C. (2019). N–TiO2 photocatalysts: A review of their characteristics and capacity for emerging contaminants removal. *Water*, *11*(2), 373. https://doi.org/10.3390/w11020373

Hofmann-Amtenbrink, M., Grainger, D. W., & Hofmann, H. (2015). Nanoparticles in medicine: Current challenges facing inorganic nanoparticle toxicity assessments and standardizations. *Nanomedicine: Nanotechnology, Biology and Medicine*, *11*(7), 1689–1694. https://doi.org/10.1016/j.nano.2015.05.005

Hu, H., Li, X., Wu, S., & Yang, C. (2020). Sustainable livestock wastewater treatment via phytoremediation: Current status and future perspectives. *Bioresource Technology*, *315*, 123809. https://doi.org/10.1016/j.biortech.2020.123809

Huang, Y., Lee, X., Grattieri, M., Yuan, M., Cai, R., Macazo, F. C., & Minteer, S. D. (2020). Modified biochar for phosphate adsorption in environmentally relevant conditions. *Chemical Engineering Journal*, *380*, 122375. https://doi.org/10.1016/j.cej.2019.122375

Hussain, F., Mustufa, G., Zia, R., Faiq, A., Matloob, M., Shah, H., & Irfan, J. A. (2018). Constructed wetlands and their role in remediation of industrial effluents via plant-microbe interaction–a mini review. *J Bioremediat Biodegrad*, *9*(4), 1–8. https://doi.org/10.4172/2155-6199.1000447

Iyyappan, J., Gaddala, B., Gnanasekaran, R., Gopinath, M., Yuvaraj, D., & Kumar, V. (2023). Critical review on wastewater treatment using photo catalytic advanced oxidation process: Role of photocatalytic materials, reactor design and kinetics. *Case Studies in Chemical and Environmental Engineering*, 100599. https://doi.org/10.1016/j.cscee.2023.100599

Jain, A., & He, Z. (2018). "NEW" resource recovery from wastewater using bioelectrochemical systems: moving forward with functions. *Frontiers of Environmental Science & Engineering*, *12*, 1–13. https://doi.org/10.1007/s11783-018-1052-9.

Jellali, S., Wahab, M. A., Anane, M., Riahi, K., & Jedidi, N. (2011). Biosorption characteristics of ammonium from aqueous solutions onto Posidonia oceanica (L.) fibers. *Desalination*, *270*(1–3), 40–49. https://doi.org/10.1016/j.desal.2010.11.018

Joseph, T. M., Al-Hazmi, H. E., Śniatała, B., Esmaeili, A., & Habibzadeh, S. (2023). Nanoparticles and nanofiltration for wastewater treatment: From polluted to fresh water. *Environmental Research*, 117114. https://doi.org/10.1016/j.envres.2023.117114

Joshi, K. M., & Shrivastava, V. S. (2011). Photocatalytic degradation of Chromium (VI) from wastewater using nanomaterials like TiO2, ZnO, and CdS. *Applied Nanoscience*, *1*(3), 147–155. https://doi.org/10.1007/s13204-011-0023-2

Kanwar, V. S., Sharma, A., Srivastav, A. L., & Ram, L. (2020). Phytoremediation of toxic metals present in soil and water environment: A critical review. *Environmental Science and Pollution Research*, *27*, 44835–44860. https://doi.org/10.1007/s11356-020-10713-3

Karimi-Maleh, H., Karimi, F., Malekmohammadi, S., Zakariae, N., Esmaeili, R., Rostamnia, S., & Gupta, V. K. (2020). An amplified voltammetric sensor based on platinum nanoparticle/polyoxometalate/two-dimensional hexagonal boron nitride nanosheets composite and ionic liquid for determination of N-hydroxysuccinimide in water samples. *Journal of Molecular Liquids*, *310*, 113185. https://doi.org/10.1016/j.molliq.2020.113185

Kaur, A., Vats, S., Rekhi, S., Bhardwaj, A., Goel, J., Tanwar, R. S., & Gaur, K. K. (2010). Physico-chemical analysis of the industrial effluents and their impact on the soil microflora. *Procedia Environmental Sciences*, *2*, 595–599. https://doi.org/10.1016/j.proenv.2010.10.065

Kassab, G., Halalsheh, M., Klapwijk, A., Fayyad, M., & Van Lier, J. B. (2010). Sequential anaerobic–aerobic treatment for domestic wastewater–a review. *Bioresource Technology*, *101*(10), 3299–3310. https://doi.org/10.1016/j.biortech.2009.12.039

Khan, M. A., & Ghouri, A. M. (2011). Environmental pollution: Its effects on life and its remedies. *Researcher World: Journal of Arts, Science & Commerce*, *2*(2), 276–285. https://ssrn.com/abstract=1981242

Khan, S., Naushad, M., Al-Gheethi, A., & Iqbal, J. (2021). Engineered nanoparticles for removal of pollutants from wastewater: Current status and future prospects of nanotechnology for remediation strategies. *Journal of Environmental Chemical Engineering*, *9*(5), 106160. https://doi.org/10.1016/j.jece.2021.106160

Koetlisi, K. A., & Muchaonyerwa, P. (2019). Sorption of selected heavy metals with different relative concentrations in industrial effluent on biochar from human faecal products and pine-bark. *Materials*, *12*(11), 1768. https://doi.org/10.3390/ma12111768

Kolya, H., & Tripathy, T. (2014). Biodegradable flocculants based on polyacrylamide and poly (N, N-dimethylacrylamide) grafted amylopectin. *International Journal of Biological Macromolecules*, *70*, 26–36. https://doi.org/10.1016/j.ijbiomac.2014.06.028

Kulangana, T., Ariyarathne, W. M. T. P., Priyantha, H. M. D. N., & Iqbal, M. C. M. (2024). Enhancing water quality and sustainable organic-phosphate fertilizer production through phytoremediation of cattle farm effluent using artificial wetlands. *Sri Lankan Journal of Biology*, *9*(1). https://doi.org/10.4038/sljb.v9i1.135

Li, D., & Xia, Y. (2004). Electrospinning of nanofibers: Reinventing the wheel? *Advanced Materials*, *16*(14), 1151–1170. https://doi.org/10.1002/adma.200400719

Loganathan, P., Vigneswaran, S., Kandasamy, J., & Bolan, N. S. (2014). Removal and recovery of phosphate from water using sorption. *Critical Reviews in Environmental Science and Technology*, *44*(8), 847–907. https://doi.org/10.1080/10643389.2012.741311

Mahvi, A. H., Sarmadi, M., Sanaei, D., & Abdolmaleki, H. (2020). Removal of lead ion from aqueous solutions by adsorption onto phosphate-functionalized treated waste papers (PF-TWPs). *Desalin. Water Treat.*, *200*, 205–216. doi: 10.5004/dwt.2020.26130

Ministry of Environment, Forest and Climate Change. (2016). Environment Ministry releases new categorisation of industries,Press Information Bureau, Government of India. http://pib.nic.in/newsite/PrintRelease.aspx?relid=137373

OB, A., & Muchie, M. (2010). Remediation of heavy metals in drinking water and wastewater treatment systems: processes and applications. *Int. J. Phys. Sci*, *5*, 1807–1817.

Obire, O., Ogan, A., & Okigbo, R. N. (2008). Impact of fertilizer plant effluent on water quality. *International Journal of Environmental Science & Technology*, *5*, 107–118. https://doi.org/10.1007/BF03326003

Pawar, A., & Panwar, N. L. (2022). Biomass derived activated biochar for wastewater treatment. In *Handbook of sustainable development through green engineering and technology* (pp. 213–230). CRC Press.

Renu, Agarwal, M., & Singh, K. (2017). Heavy metal removal from wastewater using various adsorbents: a review. *Journal of Water Reuse and Desalination*, *7*(4), 387–419. https://doi.org/10.2166/wrd.2016.104

Roopadevi, H., & Somashekar, R. K. (2012). Assessment of the toxicity of waste water from a textile industry to Cyprinus carpio. *Journal of Environmental Biology*, *33*(2), 167.

Samaras, P., Papadimitriou, C. A., Vavoulidou, D., Yiangou, M., & Sakellaropoulos, G. P. (2009). Effect of hexavalent chromium on the activated sludge process and on the sludge protozoan community. *Bioresource Technology*, *100*(1), 38–43. https://doi.org/10.1016/j.biortech.2008.05.036

Santos, A. F., Almeida, P. V., Alvarenga, P., Gando-Ferreira, L. M., & Quina, M. J. (2021). From wastewater to fertilizer products: Alternative paths to mitigate phosphorus demand in European countries. *Chemosphere*, *284*, 131258. https://doi.org/10.1016/j.chemosphere.2021.131258

Saravanan, A., Kumar, P.S., Jeevanantham, S., Karishma, S., Tajsabreen, B., Yaashikaa, P. R., & Reshma, B., 2021. Effective water/wastewater treatment methodologies for toxic pollutants removal: Processes and applications towards sustainable development. *Chemosphere*, *280*, 130595. https://doi.org/10.1016/j.chemosphere.2021.130595

Schwarzenbach, R. P., Egli, T., Hofstetter, T. B., Von Gunten, U., & Wehrli, B. (2010). Global water pollution and human health. *Annual Review of Environment and Resources*, *35*, 109–136. https://doi.org/10.1146/annurev-environ-100809-125342

Şentürk, İ., Eyceyurt Divarcı, N. S., & Öztürk, M. (2023). Phytoremediation of nickel and chromium-containing industrial wastewaters by water lettuce (Pistia stratiotes). *International Journal of Phytoremediation*, *25*(5), 550–561. https://doi.org/10.1080/15226514.2022.2092063

Shehata, N., Egirani, D., Olabi, A. G., Inayat, A., Abdelkareem, M. A., Chae, K. J., & Sayed, E. T. (2023). Membrane-based water and wastewater treatment technologies: Issues, current trends, challenges, and role in achieving sustainable development goals, and circular economy. *Chemosphere*, *320*, 137993. https://doi.org/10.1016/j.chemosphere.2023.137993

Sillanpaa, M., Ncibi, M. C., Matilainen, A., & Vepsalainen, M. (2018). Removal of natural organic matter in drinking water treatment by coagulation: A comprehensive review. *Chemosphere*, *190*, 54–71. http://doi.org/10.1016/j.chemosphere.2017.09.113

Singh, B. J., Chakraborty, A., & Sehgal, R. (2023). A systematic review of industrial wastewater management: Evaluating challenges and enablers. *Journal of Environmental Management*, *348*, 119230. https://doi.org/10.1016/j.jenvman.2023.119230.

Singh, G., Bhati, M., & Rathod, T. (2010). Use of tree seedlings for the phytoremediation of a municipal effluent used in dry areas of north-western India: Plant growth and nutrient uptake. *Ecological Engineering*, *36*(10), 1299-1306. https://doi.org/10.1016/j.ecoleng.2010.06.006

Spanoghe, J., Grunert, O., Wambacq, E., Sakarika, M., Papini, G., Alloul, A., & Vlaeminck, S. E. (2020). Storage, fertilization and cost properties highlight the potential of dried microbial biomass as organic fertilizer. *Microbial Biotechnology*, *13*(5), 1377-1389. https://doi.org/10.1111/1751-7915.13554

Sumalatha, V., Ayodhya, D., & Balchander, V. (2023). Facile synthesis of hexagonal-shaped CuO NPs from Cu (II)-Schiff base complex for enhanced visible-light-driven degradation of dyes and antimicrobial studies. *Inorganica Chimica Acta*, *548*, 121358. https://doi.org/10.1016/j.ica.2022.121358

Suzaimi, N. D., Goh, P. S., & Malek, N. A. N. N. (2019). Modified-nano-adsorbents for nitrate efficient removal: A review. *Journal of Applied Membrane Science & Technology*, *23*(2). https://doi.org/10.11113/amst.v23n2.161

Thawale, P. R., Juwarkar, A. A., & Singh, S. K. (2006). Resource conservation through land treatment of municipal wastewater. *Current Science*, 704-711.

Thekkudan, V. N., Vaidyanathan, V. K., Ponnusamy, S. K., Charles, C., Sundar, S., Vishnu, D., … & Subramanian, S. (2017). Review on nanoadsorbents: a solution for heavy metal removal from wastewater. *IET Nanobiotechnology*, *11*(3), 213-224. https://doi.org/10.1049/iet-nbt.2015.0114

Uddin, M. A., Begum, M. S., Ashraf, M., Azad, A. K., Adhikary, A. C., & Hossain, M. S. (2023). Water and chemical consumption in the textile processing industry of Bangladesh. *PLOS Sustainability and Transformation*, *2*(7), e0000072. https://doi.org/10.1371/journal.pstr.0000072

Ullah, A., Mushtaq, H., Ali, H., Munis, M. F. H., Javed, M. T., & Chaudhary, H. J. (2015). Diazotrophs-assisted phytoremediation of heavy metals: a novel approach. *Environmental Science and Pollution Research*, *22*, 2505-2514. https://doi.org/10.1007/s11356-014-3699-5

UNESCO. (2017).*The United Nations world water development report 2017: Wastewater: the untapped resource; facts and figures*. https://unesdoc.unesco.org/ark:/48223/pf0000247553

United Nations. (2015). *Transforming our world: The 2030 Agenda for Sustainable Development*. United Nations. https://daccess-ods.un.org/access.nsf/Get?OpenAgent&DS=A/RES/70/1&Lang=E

Velusamy, S., Roy, A., Sundaram, S., & Kumar Mallick, T. (2021). A review on heavy metal ions and containing dyes removal through graphene oxide-based adsorption strategies for textile wastewater treatment. *The Chemical Record*, *21*(7), 1570-1610. https://doi.org/10.1002/tcr.202000153

Wahab, M. A., Jellali, S., & Jedidi, N. (2010). Ammonium biosorption onto sawdust: FTIR analysis, kinetics and adsorption isotherms modeling. *Bioresource Technology*, *101*(14), 5070-5075. https://doi.org/10.1016/j.biortech.2010.01.121

Wang, Z., Wu, A., Colombi Ciacchi, L., & Wei, G. (2018). Recent advances in nanoporous membranes for water purification. *Nanomaterials*, *8*(2), 65. https://doi.org/10.3390/nano8020065

Wei, H., Gao, B., Ren, J., Li, A., & Yang, H. (2018). Coagulation/flocculation in dewatering of sludge: A review. *Water Research*, *143*, 608-631. https://doi.org/10.1016/j.watres.2018.07.029

WWAP. (2018). *The United Nations World Water Development Report 2018: Nature-based solutions for water*. UNESCO. https://unesdoc.unesco.org/ark:/48223/pf0000261424

Wu, H., Wang, S., Kong, H., Liu, T., & Xia, M. (2007). Performance of combined process of anoxic baffled reactor-biological contact oxidation treating printing and dyeing wastewater. *Bioresource Technology*, *98*(7), 1501-1504. https://doi.org/10.1016/j.biortech.2006.05.037

Wu, Y., Hu, Z., Yang, L., Graham, B., & Kerr, P. G. (2011). The removal of nutrients from non-point source wastewater by a hybrid bioreactor. *Bioresource Technology*, *102*(3), 2419-2426. https://doi.org/10.1016/j.biortech.2010.10.113

Xia, Y., Zhang, M., Tsang, D. C., Geng, N., Lu, D., Zhu, L., & Ok, Y. S. (2020). Recent advances in control technologies for non-point source pollution with nitrogen and phosphorous from agricultural runoff: current practices and future prospects. *Applied Biological Chemistry*, *63*, 1-13. https://doi.org/10.1186/s13765-020-0493-6

Yang, H. I., Lou, K., Rajapaksha, A. U., Ok, Y. S., Anyia, A. O., & Chang, S. X. (2018). Adsorption of ammonium in aqueous solutions by pine sawdust and wheat straw biochars. *Environmental Science and Pollution Research*, *25*, 25638-25647. https://doi.org/10.1007/s11356-017-8551-2

Yaqoob, A. A., Parveen, T., Umar, K., & Mohamad Ibrahim, M. N. (2020). Role of nanomaterials in the treatment of wastewater: A review. *Water*, *12*(2), 495. https://doi.org/10.3390/w12020495

Yavari, S., Malakahmad, A., & Sapari, N. B. (2017). Phytoremediation of urea fertilizer factory wastewater by teak (Tectona grandis). *Environment Protection Engineering*, *43*(4), 5-15. http://dx.doi.org/10.5277/epe170401

Ying, C., Umetsu, K., Ihara, I., Sakai, Y., & Yamashiro, T. (2010). Simultaneous removal of organic matter and nitrogen from milking parlor wastewater by a magnetic activated sludge (MAS) process. *Bioresource Technology*, *101*(12), 4349-4353. https://doi.org/10.1016/j.biortech.2010.01.087

Younas, F., Mustafa, A., Farooqi, Z.U.R., Wang, X., Younas, S., Mohy-Ud-Din, W., … & Hussain, M. M. (2021). Current and emerging adsorbent technologies for wastewater treatment: trends, limitations, and environmental implications. *Water*, *13*(2), 215. https://doi.org/10.3390/w13020215

Zhang, A., Li, X., Xing, J., & Xu, G. (2020). Adsorption of potentially toxic elements in water by modified biochar: A review. *Journal of Environmental Chemical Engineering*, *8*(4), 104196. https://doi.org/10.1016/j.jece.2020.104196

Zhang, B., Huang, G., Wang, L., Wang, T., Liu, L., Di, Z., & Mei, Y. (2019). Rolled-up monolayer graphene tubular micromotors: Enhanced performance and antibacterial property. *Chemistry–An Asian Journal*, *14*(14), 2479-2484. https://doi.org/10.1002/asia.201900301

Zhou, Q., Li, X., Lin, Y., Yang, C., Tang, W., Wu, S., & Lou, W. (2019). Effects of copper ions on removal of nutrients from swine wastewater and on release of dissolved organic matter in duckweed systems. *Water Research*, *158*, 171-181. https://doi.org/10.1016/j.watres.2019.04.036

10 Futuristic Development and Optimization Methods for Phytoremediation of Industrial Wastewater

Sadhana Y. More, Yogeshwar R. Baste and Rohit Rai

1. INTRODUCTION

The localized environment, where untreated water is discharged freely, is most affected by wastewater, making it difficult to regulate the difficulties of providing clean water to rural and urban areas. Wastewater effluents are the primary source of problems with water contamination since they induce eutrophication (Ahmed, 2021). On the other hand, these could promote algae growth, increase the expense of purification, pose health hazards to humans and animals, and result in excessive oxygen loss, all of which could alter the population dynamics of aquatic systems. During the Industrial Revolution, humans released a wide range of potentially harmful chemicals and non-chemical substances into the environment (Srivastava & Giri, 2021). Dyes, heavy

DOI:10.1201/9781003496014-10

metals, organic and inorganic chemicals, and hazardous inorganic compounds are a few of the pollutants that can seriously endanger human health. Diverse approaches and strategies can be employed to avert, eliminate, and rectify the adverse effects of contaminants discharged into the surroundings. One low-cost way to lessen the danger that contaminated areas pose to the ecosystem and public health is to lower the contamination level in rhizosphere soil. Water containing toxic substances damages the aquatic ecosystem by preventing light from penetrating, which prevents aquatic plants from photosynthesis. Numerous techniques, including membrane filtration, reverse osmosis, chemical precipitation, oxidation, adsorption, and flotation, might lessen the effects of hazardous metal ions (Adams et al., 2000).

However, because metal can be absorbed in tiny concentrations and has economically viable qualities, adsorption is highly precise and widespread. Toxic effects are avoided since the plants absorb most of the pollutants through their roots. Additionally, root systems have a wide surface area that collects and absorbs non-significant impurities as well as the nutrients and water needed for growth. This helps to clean up and remediate wastewater from toxins. Experimental investigations have demonstrated that *Salvinia molesta* and other plants are more effective at eliminating harmful dyes and contaminants; as a result, this promising method is currently being applied in a variety of sectors (Adams et al., 2000).

Common methods for eliminating contaminants include adsorption treatments and plant-based green technologies. During the last years, there has been a great deal of advancement in this discipline. Experiments with varying adsorption capacities, operating conditions, and application forms demonstrated the remarkable influence and recyclable nature of biomass adsorbents. Leaf, peel, and other plant parts are effective in eliminating pollutants. The purpose of this chapter is to talk about how plants help clean wastewater by removing chemicals, heavy metals, colors, inorganic elements, pesticides, hydrocarbons, and pharmaceutical containments. Moreover, it highlights the critical function that plants play in this context, a topic that has been covered in recent publications (Srivastava & Giri, 2021). This chapter addresses a novel discussion on the phytoremediation of dyes, heavy metals, hydrocarbons, pharmaceuticals, inorganic elements, and pesticides, which are the most abundant pollutants in wastewater.

Using plants, green vegetation, trees, aquatic plants, and grasses, phytoremediation is a significant process that eliminates, transfers, stabilizes, and eliminates harmful contaminants from wastewater, sediments, soils, surface water, groundwater, and the surrounding atmosphere (Asante-Badu et al., 2020). Enhanced rhizospheric biodegradation, phytoextraction (also known as phytoaccumulation), phytostabilization, and phytodegradation are the mechanisms of phytoremediation (Adams et al., 2000; Ali et al., 2013). Metals, explosives, solvents, hazardous organic pollutants (such as pesticides, polycyclic aromatic hydrocarbons (PAHs), and crude oil), and landfill leachates can all be taken up and remedied by phytoremediation. Certain plant species have the propensity to retain contaminants, including metals, in their roots, where they might collect them (Tangahu et al., 2011). To remove metals and other contaminants from tainted water or wastewater, these plant species can be planted elsewhere. The roots ought to be taken as soon as they are saturated with metals and/or pollutants. Hyper-accumulator plants can often eliminate, gather, and store a significant amount of metallic contaminants (Article et al., 2022). Currently, research is being done on trees to see if they may remove organic contaminants from groundwater. The organic

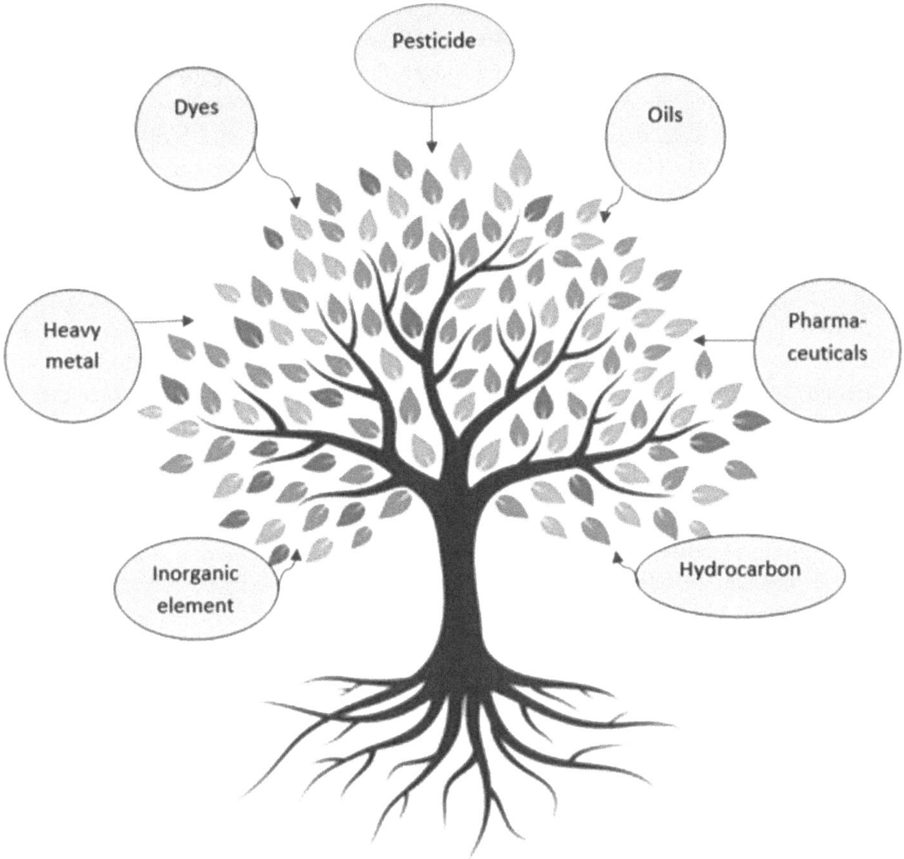

FIGURE. 10.1 Removal of different types of pollutants by plants.
Figure created by authors.

pollutants' transport, accumulation, transpiration, and potential metabolization to plant tissue or CO_2 are all included in the study. In the meantime, plants shield against rain, dust, and wind (Babu et al., 2021). Additionally, they stop pollutants from being carried by groundwater from the site to lower depths or the nearby surroundings (Chandanshive et al., 2018) (Figure 10.1).

2. METHOD AND MECHANISM OF PHYTOREMEDIATION

A new environmentally friendly method for identifying, dissolving, and eliminating different kinds of toxins from the environment is called phytoremediation technology.

Plant species are used in the removal of various toxins that have negative impacts on human health and other biological systems. These plant species absorb these toxins from the surroundings and eliminate their harmful effects. This method has an advantage over traditional procedures since it is environmentally friendly, whereas traditional techniques have negative impacts on the environment and biological system (Imron et al., 2019). The remediation of pollutants from water, particularly metal contaminants, involves several processes that transform them into harmless molecules and ultimately remove waste from the water. These processes include photodegradation, phytoaccumulation, phytostabilization, rhizodegradation, phytofiltration (also known as rhizofiltration), phytoextraction, and photodegradation (Adams et al., 2000). The mechanisms of accumulation or adsorption are applied in phytostabilization (Ali et al., 2013). To stop pollutants from moving throughout the ecosystem, this method uses soil or groundwater toxins that are adsorbed onto the root or build up in the rhizosphere. In contrast, rhizodegradation occurs when contaminants, particularly organic waste, and heavy metals, break down and degrade in the rhizosphere to produce less or no hazardous chemicals. The utilization of different kinds of microbes improves this process as well (Imron et al., 2019).

The phytoremediation method and the rhizofiltration technique both employed the same mechanism. However, the contaminants are absorbed by the plants' roots in this situation. Wastes and contaminants are broken down by plants via metabolic processes. Through the use of several enzymes, the plant absorbs metals and trash from the surrounding environment or wastewater and breaks them down into harmless chemicals; the process is also termed as phytotransformation (Imron et al., 2019). Under the phytovolatilization method, many wastewater kinds are absorbed by plants and transformed into harmless substances. Later, through transpiration, these harmless substances are discharged into the atmosphere through leaves. Similarly, the process known as phytoaccumulation occurs when wastes are deposited in different plant sections like roots, shoots, and leaves. Various plant species have been employed to eliminate various heavy metals, organic wastes, and other kinds of pollutants (Adams et al., 2000; Ali et al., 2013; Asante-Badu et al., 2020).

Certain plants can accumulate heavy metals (HMs) (Ag, Cd, Hg, and Pb), which are known to have no biological role, in addition to important elements like Cu, Fe, Mg, Mo, Mn, Se, and Zn, which are necessary for plant growth. Nonetheless, flora may get poisoned by a large concentration of these HMs (Kotyza et al., 2010; Lamine & Saunders, 2022). Five types of phytoremediation technologies are used to clean up contaminated water and soil.

2.1. Phytoextraction

Phytoextraction is the process of planting plants that are known to accumulate the highest quantities of pollutants (greater than 0.1% of DW) in their shoots. One potential selection criterion for these plants would be the extent to which the contamination has spread from the roots to the shoots (Adams et al., 2000). This strategy consists of two methods: continuous phytoextraction and triggered phytoextraction.

Permanent phytoextraction utilizes natural plants that naturally possess the ability to accumulate significant levels of heavy metals (hyper-accumulators). Induced phytoextraction boosted the accumulation of plant metals by introducing compounds such as chelates (Asante-Badu et al., 2020). Plants with fast development rates (high biomass) and deeper root systems are more adept at phytoextraction. Asante-Badu et al. (2020) also reported lead (Pb) in contaminated soils could be successfully eliminated by coffee weed or *Sesbania exaltata*. Ni was deposited at a higher level (1050 μg g−1 DW) in the above ground section of *Sesuvium portulacastrum*. Zn concentration in *Noccaea caerulescens* leaves exceeded 300 g Cd/ ha two months before implantation in the field conditions. *Melilotus ofcinalis* and *Amaranthus retrofexus* for their capacity to cure Cu- and Mo-contaminated soil. According to the results, *A. retrofexus* prefers Cu and Mo accumulation in shoots, while *M. Officinalis* prefers Zn storage in roots. A field study examined the ability of three cultivars of *Pennisetum purpureum Napier* grass to absorb zinc and copper. They found that in their shoots, *P. purpureum cv. Guiminyin,* accumulate the highest levels of Zn (5023.9 g ha−1) and Cd (197.5 g ha−1). *Alternanthera bettzickiana*'s phytoextraction potential for Ni and Cu using pot experiments. They found that this species accumulates twice as much Cu in shoots as the control after eight weeks of treatment (Ali et al., 2013).

2.2. Phytodegradation

The breakdown of pollutants absorbed by plants through metabolic processes, or the breakdown of pollutants external to the plant by root-produced enzymes such as peroxidases, nitroreductases, and dehalogenases, is known as phytodegradation or phytotransformation (Imron et al., 2019). The very toxic form of mercury, Hg^{2+}, can be changed to the far less dangerous form, Hg, in tissue cultures containing higher quantities of mercury thanks to genetic modification (Imron et al., 2019). Plants can convert pollutants into non-hazardous components using this process. *Vetiveria zizanioides* plants were able to remove 97% of TNT from the soil (Ali et al., 2013). *Nicotiana tabacum*'s roots produce the NfsA nitroreductase enzyme, which aids in the breakdown of TNT. It was revealed that Populus *deltoid* plants could convert RDX into metabolic components. Certain plants can offer the ideal conditions for mycorrhizae and bacteria to assemble, grow, and break down poisons. The breakdown's constituent parts volatilize or become part of the soil matrix (Ali et al., 2013). Fungi and bacterial populations are increased by the plant's production of exudates, which include sugars and organic acids (Asante-Badu et al., 2020). Favorable soil properties, such as soil aeration and moisture content, can promote rhizodegradation (Imron et al., 2019). In an attempt to restore the wetland contaminated by terbuthylazine (TER), Ali et al., 2013, conducted experiment culture and discovered that *Typha latifolia* rhizomes were a promising species for TER phytodegradation. Polycyclic aromatic hydrocarbons (PAHs) in contaminated sediment could be broken down by Rhizophora mangle mangrove linked to plant growth-promoting rhizobacteria (*Pseudomonas aeruginosa* and *Bacillus sp.*) (Imron et al., 2019).

2.3. Phytovolatilization

The process by which pollutants are absorbed and transformed into less dangerous volatile forms is known as phytovolatilization. According to this process, the pollutants enter the plant through the stomatal leaves, travel to the shoot, and then volatilize in the atmosphere. Indian mustard (*Brassica napus*) may be effective in phytovolatilizing selenium from soil. *Arabidopsis thaliana* was shown to convert organic Hg salts into their volatile elementals (Adams et al., 2000). It has been shown that hydroponically grown Brassica juncea can eliminate up to 95% of mercury by phytovolatilization and plant accumulation (Ali et al., 2013). Transgenic *Nicotiana tabacum* with the merA gene is capable of removing mercury from contaminated soils. In a recent study it was found that when plants were cultivated in sterile garden soil amended with 2, 10, and 20 mgL−1 of $NaAsO_2$, *Arundo donax* tended to volatilize arsenic with the help of plant growth-promoting rhizobacteria (*Stenotrophomonas maltophilia* and *Rhyzobacterium* sp.) (Asante-Badu et al., 2020).

2.4. Rhizofiltration

Through the precipitation of toxins onto surface roots or the absorption of soluble contaminants into the roots, rhizofiltration is a technique for removing pollutants from water and liquid waste (Ali et al., 2013). Large root surface areas and a fibrous root system are important components of this strategy (Adams et al., 2000). For this use, plants can absorb a lot of water from the soil. *Typha latifolia* may be an effective way to remove methyl parathion (MeP) from hydromorphic soils. The bean species *Phaseolus vulgaris* effectively eliminates cesium (Cs) and uranium (Ur) from groundwater. *Arundo donax* is an effective Poaceae to rhizofiltrate Cu from artificial wetlands (Asante-Badu et al., 2020). *Eichhornia crassipes, Salvinia molesta,* and *Pistia stratiotes* are potential terrestrial candidates for extracting heavy metals (Cu Cr, Cd, Ni, and Zn) from industrial sewage sludge (Ruley et al., 2020).

2.5. Phytostabilization

Pollutants are either precipitated by root exudates or stored by plant roots in this technique (Ali et al., 2013). As a result, this approach stops the pollutants from migrating to groundwater and reduces their movement. *Atriplex halimus* (Aramanthaceae family), a native halophyte of North Africa, was a promising candidate for phytostabilization of Cd (Asante-Badu et al., 2020). The same authors discovered that the roots of this species had higher Cd contents than the shoots (606.51 and 217.52 ppm DW, respectively). The two main qualities of plants that were good candidates for phytostabilization were their morphology and root depth. Many *fen* root system extents throughout the soil allowed for maximum interaction with the soil matrix (high surface area of the fibrous roots). The plant species, wetness, soil texture, calcareous encrust, dry conditions, and soil additives all have a substantial impact on root depth. Certain plant species can alter the solubility and/or mobility of metals through the secretion of root exudates. Other species, however, store a significant amount of HMs in their roots. By using pot tests, examined the remedial effects of *Sorghum bicolor* and *Carthamus tinctorius* on Ni, Pb, and Zn and found that these species could absorb these heavy metals. Moreover, roots gathered more metals than leaves did. *Erica australis* is thought to be a phytostabilization species because of its capacity to absorb Cu, Cd, and Pb through its roots without causing any harm (Ruley et al., 2020).

Helichrysum microphyllum root tissue has a substantially restricted uptake of zinc, copper, and lead, making it a good choice for phytostabilization. *Stigmatocarpum criniforum* and *Pelargonium hortorum* were recently chosen by scientists (Kotyza et al., 2010) as an evaluated species under Pb stress conditions, and their extraction ability was assessed. The findings showed that *P. hortorum*'s aboveground portion collects more lead in its roots than does *S. criniforum*. The best Cd accumulator in roots was discovered to be *Quercus robur*, making it a viable option for phytostabilizing Cd in

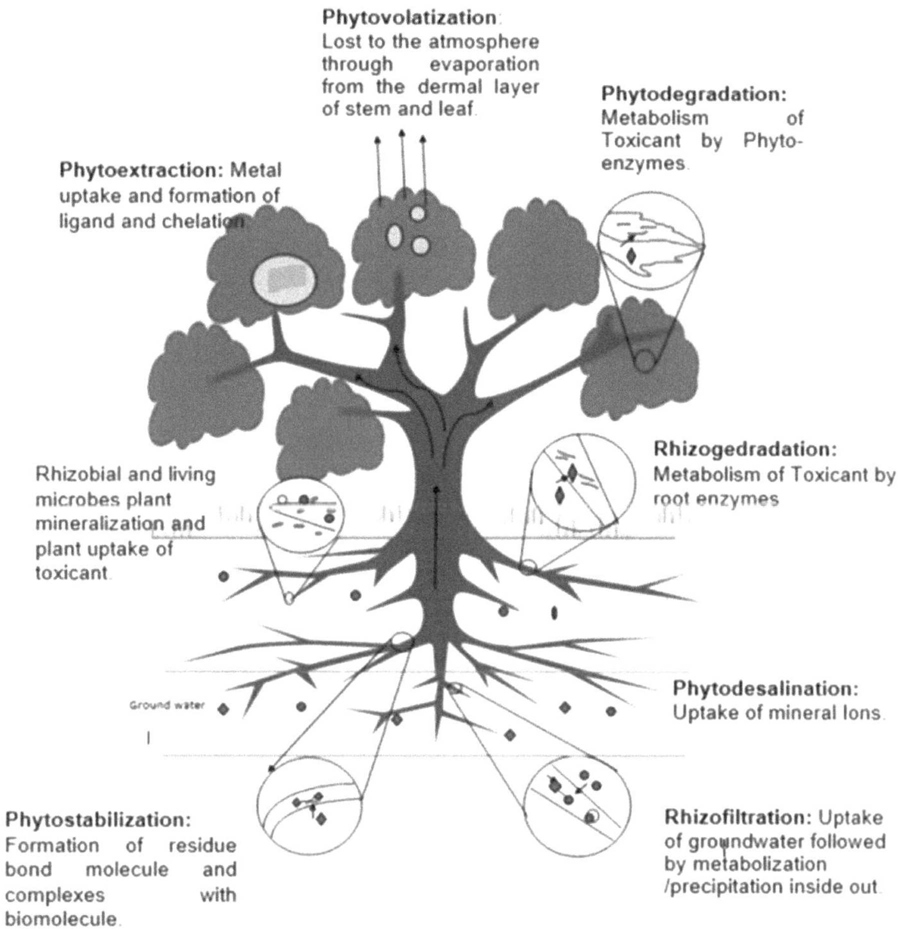

FIGURE 10.2 Different mechanisms for the removal of pollutants from wastewater by plants. Figure created by the authors.

contaminated soils. *Salix alba* was the helpful tree for Cd and Cu phytostabilization in the riparian soils of the Sava River under field circumstances (Article et al., 2022) (Figure 10.2).

3. REMOVAL OF POLLUTANTS FROM INDUSTRIAL WASTEWATER BY PLANT SPECIES

3.1. Phytoremediation of Dyes

The world population is growing, and this is causing significant growth in several sectors. Companies utilize a lot of dangerous chemicals to produce goods that the public wants, but the byproducts, such as pollutants, have an impact on the air, soil, and water bodies. In the textile industry, dyes are essential to the dyeing process and are in high demand. In the dyeing process, azo dyes are widely utilized; and 10% to 15% of the azo dyes are lost in the effluent, while 50% of other reactive dyes are used in the textile sector and end up as waste in the water. By destroying the ecology and releasing dye-containing wastewater into the surrounding area, the atmosphere is severely impacted and eventually affects water. Therefore, treating textile dye wastewater before releasing it into aquatic bodies presents a significant difficulty. Adsorption, coagulation, sedimentation, flocculation, filtration, photodegradation, and chemical oxidation are just a few of the numerous physical and chemical techniques available for controlling pollution brought on by textile dyes. These tactics have to do with excessive costs, poor output, large space requirements, and inconsistent performance. The development of an efficient, astute method for treating textile dyes is necessary in light of these problems (Kaur, 2020).

Biological methods are more effective than physical and chemical methods to treat textile dye wastewater. Bioremediation is a very efficient treatment for removing contaminants from textile dye effluent (Article et al., 2022). Different enzymes and microorganisms were used to remove the dyes from wastewater. *Phanerochaete chrysosporium*, the most spacious studied white-rot fungus, is used to metabolize and decolorize dyes colors, and compounds (Chandanshive et al., 2018). The biological decolorization of lignin-containing pulp and paper wastewater use white rot fungi *Tictoporia* species due to the high oxidative potential of many of the enzymes linked with white rot fungi as ligninase, laccase, Mn-peroxidase (Babu et al., 2021). *Aspergillus flavipes* fungus was reported to decolorize Xiron orange RHD 72 (Chandanshive et al., 2018). From the different bioremediation processes, phytoremediation is an energy-efficient, solar-driven process to remove contaminants from soil, air, and water. Phytoremediation is also used to remove pollutants from textile dye wastewater (Epps, 2006).

3.2. Phytoremediation of Heavy Metals

Furthermore, to pose major risks to human health, heavy metals pose significant risks to the environment. While several methods for eliminating heavy metals have been developed, they are more costly than using plants to remove harmful metals from industrial settings. In developing nations, heavy metal pollution is becoming a bigger issue. Many investigations have been done on less-priced and eco-friendly (plant-derived) adsorbents for removing heavy metals. The buildup of heavy metals in the air, soil, and water as a result of human activities like mining, urbanization, and industrialization has become a global problem. It is strongly recommended to use decorative plant species because of their many roles in the environment. Without proper treatment, industrial wastewater discharges contaminate land and water. Heavy metals found in these effluents are currently being removed using plant-based methods called phytoremediation. The pollution of wastewater with heavy metals poses a serious risk to aquatic life (Ahmed et al., 2021.) A long-term, environmentally benign, and reasonably priced approach is phytoremediation. For the removal of heavy metal contaminants, aquatic plants have high efficiency. A few other plants, including duckweed, are very effective metal accumulators. Adsorption of heavy metals by inexpensive, environmentally benign adsorbents such as plants is the favored method for treating wastewater. The best natural lignin residues for use as bio-adsorbents are those that have been separated using black liquor (Adams et al., 2000). Heavy metals and other harmful compounds found in the environment can be effectively removed using biosorption, a technique that has been extensively studied through multiple studies in the past. The utilization of plant-based waste for the removal of metals from industrial wastewater has garnered substantial attention due to its cost-effectiveness and notably higher removal rate, which can be attributed to several functional groups. Black oak bark and coconut wastes are effective at removing metals like lead, cadmium, and mercury (Article et al., 2022). *Trunk fiber* waste of several date palm kinds for the removal of heavy metals shows great results, but projects on vast scales are yet understudies owing to its cost-effectiveness and availability. Plants are employed in several industries to rehabilitate metal-affected habitats, remove heavy metals from polluted soil, and stop the environment's continued negative impacts on living things. The removal of heavy metals from soil and wastewater is more successful when green plants are used instead of other standard procedures (Kotyza et al., 2010). Heavy metals can be effectively removed from wastewater by a variety of techniques, such as membrane filtration, coagulation, ion exchange, electrolysis, and adsorption. Heavy metal removal is accomplished by chemicals that are derived from numerous microbes and plants. Atomic absorption spectroscopy's analysis of the hyacinth (*E. crassipes*) plant's metal-removing capacity reveals significant potential for arsenic, cadmium, and mercury. The dry Miscanthus plant, also known as silver grass, exhibits the highest degree of potential for the removal of heavy metals from wastewater when examined through a variety of analytical techniques, including elemental analysis, differential scanning calorimetry, X-ray diffraction, thermogravimetric analysis (TGA), inductively coupled plasma (ICP), and surface Brunauer-Emmett-Teller method (SBET). The frequent issues brought on by heavy metals in wastewater compelled researchers to look into a variety of heavy metal removal techniques for wastewater. In this case, phytoremediation demonstrated its usefulness in assisting with the heavy metal removal from wastewater (Tangahu et al., 2011).

Among the pollutants found in the environment are heavy metals. In addition to natural processes, practically every human activity has the potential to have unintended consequences that include the production of heavy metals. A few instances of events leading to ecosystem pollution include the migration of these contaminants into uncontaminated areas as dust or leachates via the soil and the spread of heavy metals-containing sewage sludge. There are now several techniques being used to rid the environment of these pollutants, but most of them are expensive and fall well short of their ideal levels of effectiveness. Chemical technologies lead to the production of huge volumes of sludge and are associated with higher costs (Ahmed, 2021). Chemical and thermal procedures are costly and technically complex, and they can all lead to the degradation of precious soil components (Tangahu et al., 2011). Traditionally, the process of cleaning up heavy-metal-contaminated soils has involved either management on-site or excavation followed by disposal in a landfill. This type of disposal only moves the contamination issue and the risks related to moving contaminated soil and allowing toxins to migrate from the landfill into the surrounding area to somewhere else. An alternative to excavation and landfill disposal for the removal of contaminated soil is soil washing. This process is quite expensive and leaves behind a heavy metal-rich residue that will need to be further processed. Furthermore, by eliminating all biological activity, these physio-chemical techniques for soil remediation turn the land into a substrate for plant growth (Ahmed, 2021). The creation of suitable technologies to evaluate the presence and mobility of metals in soil, water, and wastewater has been spurred by recent worries about environmental contamination. Currently, the extraction or removal of inert metals and metal contaminants from polluted soil can be accomplished economically and effectively through the application of phytoremediation. Using plants to remove contaminants from soils, sediments, and water is known as phytoremediation. This technique may be economical and environmentally beneficial (Kotyza et al., 2010). Hyperaccumulator plants are those that have a remarkable propensity to accumulate metals. Combined with the translocation, bioaccumulation, and pollutant degradation capacities of the entire plant body, phytoremediation leverages the special and targeted absorption capacities of plant root systems (Ahmed, 2021). Numerous plant species have demonstrated the ability to effectively absorb various pollutants from soils, including lead, cadmium, chromium, arsenic, and different radionuclides. Because phytoextraction is one of the categories of phytoremediation, heavy metals can be removed from soil by taking advantage of its capacity to absorb metals that are necessary for plant growth (Fe, Mn, Zn, Cu, Mg, Mo, and Ni). Additionally, certain metals (Cd, Cr, Pb, Co, Ag, Se, and Hg) with unclear biological functions might accumulate (Article et al., 2022). This paper's goals are to describe the potential of phytoremediation in treating heavy metal contamination, to give a brief overview of the mechanisms by which plants absorb heavy metals, to describe the abilities of various plant species to do so, and to explain the fate of heavy metals in plant tissue, with a focus on arsenic (As), lead (Pb), and mercury (Hg) (Tangahu et al., 2011).

3.3. Phytoremediation of Organic and Inorganic Elements

Due to the vast amount of garbage produced worldwide, there is a great need for affordable and environmentally suitable adsorbents. Large-scale industrial and residential trash releases a lot of toxic substances into the environment. Toxic substances found in wastewater are being eliminated through a variety of primary and secondary-stage wastewater treatment procedures (Article et al., 2022). For wastewater treatment, coagulation is the traditional approach utilized. Traditional therapies are insufficient and costly. Green technology offers the best practices in a healthier setting at an affordable price. *Opuntia ficus-indica* can sorb pollutants to the greatest extent possible. Because they are inexpensive and environmentally friendly wastewater treatments that may be used to remove harmful compounds on a larger scale, constructed wetlands are the best option. Phosphorus is primarily to blame for the issues with food security that are emerging in the 21st century. Plants with well-established wastewater treatment systems are important sources of inexpensive phosphorus recovery (Shikha & Gauba, 2016). One essential component of fertilizers is phosphate. Excess phosphate is being produced as wastewater by the agricultural sectors. Tomato plant roots are far more effective in eliminating surplus inorganic waste. Phytoremediation has made great strides in wastewater removal in the modern era. Microorganisms and plants play a major role in the elimination of nitrogen (Shikha & Gauba, 2016). In its natural state, biochar can adsorb phosphorus. Water bodies now have higher levels of contamination as a result of industrial and residential wastewater. Other plants, such as the water hyacinth (*E. crassipes*), are very effective in eliminating both organic and inorganic pollutants (Asante-Badu et al., 2020). Researchers have looked into using certain bacterial strains in plants like *Phragmites australis* to extract inorganic materials from wastewater and oil fields. Because of their high biomass efficiency, *Pistia stratiotes* and *S. molesta* are being widely introduced for use in large-scale enterprises. Several aquatic plants have more absorbent powder to eliminate hazardous inorganic components. Plants, bacteria, and fungi all have a significant part in phytoremediation. An essential part of the transformation of both organic and inorganic components is played by fungi (Article et al., 2022). The plants *Lactuca sativa L.*, *Centaurea cyanus L.*, *H. annuus*, and *Silybum marianum* have remarkable abilities to eliminate extremely dangerous radionuclides lethal to living organisms. Significant advancements have reduced wastewater's inorganic wastes (Babu et al., 2021). Even if many wastewater treatment systems are operating at maximum efficiency, there is still a need for inexpensive, environmentally friendly methods to eliminate harmful substances like nitrogen and phosphorus from wastewater (Ahmed, 2021).

Enzymes aid in the degradation process' catalysis. Many studies are being conducted to find out if plants can photodegrade both aromatic and chlorinated aliphatic chemical molecules (Adams et al., 2000). Certain enzymes within plant cells break down, mineralize, or metabolize organic contaminants. These enzymes are called nitroreductases, laccases (which break down anilines), and dehalogenases (which break down pesticides and chlorinated solvents) (degradation of nitro-aromatic compounds) (Imron et al., 2019). Examples of plants with these enzymatic systems are *Populus* species and *Myriophyllum spicatum*. Certain plants, like canna, can break down certain organic toxins, such as pesticides, explosives, solvents, industrial chemicals, and other xenobiotic compounds, making them benign through their metabolism. In other situations, though, these pollutants in soil or water may be broken down by microorganisms that reside in and attach themselves to plant roots. It is unable to decompose these resistant substances and pollutants into simpler molecules through their metabolism in plant tissue (carbon dioxide, water, etc.). Phytotransformation, then, refers to the alteration in the chemical composition of contaminants that occurred without a total disintegration of their complex structure (Shikha & Gauba, 2016).

3.4. Phytoremediation of Pesticides

The environment is being threatened and is seriously concerned about pesticide residues in soil and water. Thus, the most urgent concern among scientists and environmentalists is pesticide pollution. Pesticides can accumulate in plants, soil, and other creatures. These pollutants may find their way into the water, soil, and food chain. Moreover, pesticides may result in a decrease in the contaminated soil's productivity (Shikha & Gauba, 2016). Point and non-point pollution are the main causes of contamination of the environment and soil. Pesticide point source pollution originates from specific agricultural operations, spills, formulation facilities, agrochemical dealerships, and manufacturers. However, most people agree that the nonpoint source of pollution is the primary cause of pesticide contamination. Numerous research showed that by accumulating into the plant tissue, several plants can speed up the pesticides' rate of dissipation from contaminated soil. Furthermore, adding vegetation to the polluted soil may lessen leaching by reducing the pace at which organic pollutants diffuse and migrate through the soil's column. Significant research has been done in the past few years on the relationship between rhizosphere plants and microbes and its possible application in the rehabilitation of pesticide-contaminated media (Kumar et al., 2018). Further research on the phytoremediation of pesticides from contaminated soils and groundwater was carried out with *Kochia sp.* (Sites & Frazar, 2000). The findings demonstrated that plants have a great innate capacity to detoxify specific xenobiotic substances. However, to carry out mineralization or degradation, these plants typically need catabolic pathways. Thus, gene transfer from eukaryotes or bacteria to plants entails the degradation of xenobiotics. This will increase their ability to remove the pollutants even further (Article et al., 2022). The transgenic tobacco plant can eliminate pentachlorophenol by generating laccase and secreting it into the rhizosphere. Tobacco plants have been bred with a bacterial organophosphorus hydrolase (OPH) gene to eliminate and decontaminate organic phosphorus herbicide. The results of the investigation demonstrated that transgenic plants could break down over 99% of methyl parathion after growing for 14 days. The phytotransformation of xenobiotics in plant cells involves a wide range of plant enzymes, such as peroxidase, peroxygenases, glutathione S-transferases, cytochrome P450, carboxylesterases, O- and N-malonyl transferases, and O- and N-glucosyltransferases (Karavangeli et al., 2005). Additionally, it was observed that the transgenic potato plants' P450-dependent monooxygenase activity ranged from 3.5 to 4.2 times higher than that of the control plants (Kumar et al., 2018).

The usage of pesticide chemicals in agricultural contexts is the primary source of contaminated water. The environment and aquatic life are negatively impacted by its widespread use in agricultural areas. Developing state-of-the-art technology is essential to lowering pollution and purifying contaminated water. Today's water cleaning techniques are based on physical and electrochemical processes. Although these techniques are quite effective, they can also be expensive and produce hazardous consequences. Over the past ten years, phytoremediation has grown in popularity due to its low cost, capacity to remove numerous forms of contaminants in situ, and environmental friendliness. Phytoremediation is a four-step procedure that removes different types of pollution. Pollutants are directly absorbed into plant tissue and metabolized there. The transpiration mechanism, which also releases exudates from a different plant to eradicate various pollutants, allows separate volatile organic compounds to be released through the leaves. These exudates encourage the mycorrhizal fungi and microbial consortium, which are attached to plant roots, to remove different types of contaminants. Many aquatic plants, such as *Elodea canadensis, L. minor*, and *E. crassipes,* are used to treat water pollution due to their quick growth rates, ease of harvesting, high pollutant absorption rates, and photosynthetic activity. *E. canadensis* are aquatic macrophytes that are indigenous to North America, grow swiftly, and float freely in water. People consider them to be weeds (Srivastava & Giri, 2021) This plant's remarkable ability to eliminate water contaminants makes it a common tool in phytoremediation technologies used to treat pesticides. *E. canadensis* can eliminate three different pesticide classes: copper sulfate, dimethomorph, and flazasulfuron herbicides. The toxicity sequence for aquatic plants is the same for all of them, with dimethomorph being less toxic and flazasulfuron being more toxic than copper sulfate (Article et al., 2022). In four and two days, respectively, 16.5% of copper sulphates and 12% and 5.5% of dimethomorph were remediated. Water hyacinths, or *E. crassipes*, are aquatic macrophytes that grow swiftly in open ponds and at great depths, choking water bodies. Malathion and ethion are extracted from polluted water using an aqueous solution containing *E. crassipes*. It was discovered that 56% of 10 ppm, or around 250 mL, of malathion is broken down by *E. crassipes*. Recent research indicates that *E. crassipes* is better at absorbing and breaking down pesticides, indicating that this is a less involved but possibly efficient method of getting rid of pesticides from water. *L. minor* is another aquatic macrophyte known by the name duckweed. It thrives at pH 6 and can withstand frigid temperatures (1.7–35°C). It takes about one week to reach full growth. These plants have the potential to eliminate organic pollutants from the environment, including pesticides and heavy metals, by employing the cost-effective rhizofiltration technique (Kumar et al., 2018).

3.5. Phytoremediation of Pharmaceuticals

The most often discovered pharmaceutical wastes result from the incorrect disposal of medications used to treat people and animals. These include antibiotics, hormones, steroidal medications, analgesics, and therapeutic agents. The most common ways that these pollutants enter homes are through irrigation, sewage waste, and the disposal of unused and expired medications down the sink. They eventually find their way into groundwater and join the food chain (Shikha & Gauba, 2016).

3.5.1. Antibiotic waste

Ibuprofen (Article et al., 2022), ciprofloxacin, norfloxacin (Kotyza et al., 2010), oxytetracycline and tetracycline are among the antibiotics that persist in wastewater even after treatment. It is necessary to investigate technologies for their environmental cleanup, and some plants are capable of remediation. For example, Lupinus albus can remediate acetaminophen in hydroponic settings (Singh & Singh, 2017).

3.5.2. Hormones as wastes

It has been observed that human urine and wastewater from municipal wastes are the sources of hormones, particularly steroids (Nedjimi, 2021). In essence, these hormones are employed as treatments and contraceptives. Excreta and animal waste from homes and sewers combine to become hormonal waste in tainted water sources (Singh & Singh, 2017). It is known that they demolish fish and other aquatic creatures' endocrine systems. GC/MS tandem mass spectrometry (MS) and ELISA are the most innovative methods of separating these pollutants from water resources (Nedjimi, 2021). Very few studies have been done that mention the function of plants in removing hormone-contaminated wastewater from the environment. It has been observed that plants known as *Salix exigua* and *Arabidopsis thaliana* can remove synthetic estrogen and other derivatives from the solutions; this suggests that the plants have phytoremediation properties (Singh & Singh, 2017).

3.5.3. Inappropriate disposal of analgesics

In essence, analgesics, sometimes known as painkillers, are nonsteroidal anti-inflammatory medicines (NSAIDs), such as paracetamol. Even after treatment procedures, these bioactive substances and the products of their metabolism find their way into wastewater streams and river waters (Shikha & Gauba, 2016). Everyday medications that aren't needed, such as aspirin, paracetamol, etc., are thrown out with household waste and can have negative impacts on the water table and food chains, including hepatotoxicity and renal toxicity. Certain plants, such as *Phragmites australis*, are known to remove ibuprofen pollution from the environment through remediation (Nedjimi, 2021).

3.6. Phytoremediation of Hydrocarbons

In general, the uptake of hydrocarbons into plants, although possible, is not expected in great quantities given the compounds' chemical properties, including high molecular weights, relatively low solubilities in water, and hydrophobic nature (Dias et al., 2018). microbial degradation of hydrocarbons is possible, even without engineered vegetation. Many times more microorganisms are generally found in the plant rhizosphere than in unplanted soil, which suggests that hydrocarbon degradation could be enhanced by the presence of vegetation (Kafle et al., 2022). Numerous researchers have established that the primary mechanism for the disappearance of both petroleum hydrocarbons and PAHs is rhizodegradation. There is some indication that the presence of hydrocarbons may even encourage the proliferation of hydrocarbon-degrading microorganisms (Asante-Badu et al., 2020). Although the biodegradation of both petroleum hydrocarbons and PAHs can proceed under either aerobic or anaerobic conditions, the degradation rates will be faster in the presence of oxygen. The presence of plants can help with oxygen availability either by transporting oxygen or by creating void spaces in the subsurface that allow for greater oxygen diffusion from the atmosphere. All hydrocarbon constituents will not react in the same manner or degrade at the same rate. In general, smaller, less complex, or lower-molecular-weight hydrocarbons will be more readily degraded by microorganisms. This means that concentrations of straight-chain alkanes might disappear before branched alkanes, which will disappear more quickly in phytoremediation than aromatic hydrocarbons, including PAHs (Singh & Singh, 2017). Similarly, while low-molecular-weight PAHs are known to biodegrade under aerobic conditions, larger PAHs tend to be less amenable to biodegradation, making them more persistent in soil and groundwater (Kafle et al., 2022). Some researchers also believe that high-molecular-weight PAHs can degrade only metabolically, that is when lower-molecular-weight PAHs are present and can induce the production of enzymes required for PAH degradation (Singh & Singh, 2017), which would become less likely as smaller PAHs degrade and are no longer present to participate in the metabolic process. The persistence of larger (three rings or more) PAHs is also of concern because they may be more toxic or carcinogenic (Singh & Singh, 2017) than lighter PAHs. Some authors have asserted that the persistence of larger PAHs or other constituents remaining in weathered hydrocarbons in the presence of hydrocarbon-degrading microbes and the necessary enzymes may be due to low bio-availability. Furthermore, hydrocarbons tend to sorb to soil or root surfaces or may be incorporated into the organic material, as in humification. Humification is believed to further reduce the bioavailability of the incorporated compound. Because the presence of vegetation can cause increased humification, thereby further reducing both the bioavailability and the mobility of these compounds, this process can be thought of as 12 phytoremediation of petroleum hydrocarbons via phytostabilization, even though concentrations of the compounds might not decline (Imron et al., 2019). It should be noted that hydraulic control is also a feasible phytoremediation mechanism for the control of groundwater contamination in particular because the characteristics of the contaminants are not as relevant to the success of the technique.

4. ADVANTAGES OF PHYTOREMEDIATION

Compared to the existing physical and chemical process techniques, phytoremediation methods may also be less disruptive, more aesthetically beautiful, and more socially acceptable. This technology has several benefits, including low cost, broad applicability, efficaciousness in reducing toxins, and overall environmental friendliness. A few of the benefits of phytoremediation technology are shown in Figure 10.2. The ability of biomass-based heavy metal adsorption technology to effectively lower heavy metal ion concentrations to deficient levels and the utilization of low-cost biosorbent materials are its two main benefits (Adki et al., 2012).

Phytoremediation as possibly the cleanest and cheapest technology can be employed in the remediation of selected hazardous sites (Singh & Singh, 2017). Phytoremediation encompasses several different methods that can lead to contaminant degradation. Phytoremediation is a low-cost option and inexpensive approach for remediating environmental media. Several crops which carry out phytoremediation are *Alisma plantago-aquatica, Alternanthera philoxeroides, Amaranthus hybridus L., Amaranthus paniculata L., Brassica campestris L., Brassica carinata A. Br., Brassica juncea (L.) Czern., Brassica napus L., Brassica nigra (L.), Koch Brassica oleracea L.,*

Filipendula ulmaria, Helianthus annuus L., Nicotiana tabacum L., Oryza sativa L., Paulownia fortunei, Populus alba, Populus nigra, Populus tremula, Portulaca grandiflora, Potamogeton natans, Salix alba, Sanvitalia procumbens, Sorghum bicolor L., Triticum aestivum L., and *Zea mays L.* (Ruley et al., 2020).

Recently, this approach has drawn attention as a creative, affordable substitute for the more traditional treatment techniques employed in hazardous waste sites (Kotyza et al., 2010). Phytoremediation presents a promising low-cost solution to many of the present soil contamination issues. Compared to traditional physicochemical approaches, it is significantly less expensive (60–80%) because it doesn't require specialist personnel or expensive equipment. It is economical for big areas with low to moderately contaminated surface soils and for large volumes of water with low pollutant concentrations. It can be used to remediate a variety of environmental pollutants, including both organic and inorganic contaminants, and is effective against a wide spectrum of radioactive and hazardous metals (Lamine & Saunders, 2022).

Cleanup of contaminated soils, water, and ambient air is thought to be possible with the use of phytoremediation (Adams et al., 2000). Research on phytoremediation has the potential to remediate deficient soils, such as those with elevated levels of salt or aluminum (Asante-Badu et al., 2020). It can be used to remove secondary air or water-borne wastes, minimize environmental disruption, eliminate a variety of hazardous metals and radionuclides, and gain public approval (Article et al., 2022). Metals from contaminated soils can be removed in situ via phytoextraction, which is thought to be an environmentally favorable technique. This approach applies to considerably larger-scale cleanups and has been utilized for other heavy metals (Singh et. al 2017). This solar-powered cleansing system is aesthetically attractive, causes no environmental disturbance, and retains topsoil in place. When using in situ applications instead of traditional methods, less soil disturbance occurs. It can be completed with the least amount of disruption to the environment, leaving the topsoil in a workable state that can be recovered for agricultural purposes. The toxicity to the environment may be eliminated when the organic contaminants break down into CO_2 and H_2O (Tangahu et al., 2011). The much harsher remediation technologies of incineration, thermal vaporization, solvent washing, and other soil washing techniques can be substituted with phytoremediation. These methods essentially destroy the biological component of the soil and can significantly alter its chemical and physical characteristics, as well as produce solid waste that is relatively nonviable. Phytoremediation improves the soil and leaves behind a more functional ecosystem at a cost estimated to be a tenth of those of currently used technologies (Ahmed, 2021). Also referred to as "green technology," it is the most environmentally friendly method of cleaning contaminated soils. The creation of a recyclable plant residue rich in metals is another benefit of phytoremediation (Shikha & Gauba, 2016).

If the biomass created during the phytoremediation process can be profitably valorized as bioenergy, then phytoremediation may be a feasible solution for cleaning heavy metal-polluted soils. For this, using bioenergy crops that accumulate metals could be appropriate. Producing biodiesel from the resultant plant oil could be a practical way to produce bioenergy if heavy metal-contaminated soils are phytoremediation with oil crops (Adams et al., 2000). The potential energy that is stored can be used to produce thermal energy in large-scale applications (Article et al., 2022). The discovery of appropriate plant species that can hyperaccumulating heavy metals and generate significant amounts of biomass using accepted crop production and management techniques is essential to the success of the phytoextraction technique (Singh & Singh, 2017).

5. LIMITATIONS OF PHYTOREMEDIATION TECHNOLOGY

There are several restrictions with the phytoremediation system. These include the labor-intensive process, the volume of biomass generated, the depth of the roots, the chemistry and degree of contamination in the soil, the age of the plant, the concentration of contaminants, the effects of polluted vegetation, and the climate. A site may need to be cleaned up throughout multiple growing seasons if phytoremediation is to be successful. Plants may be cytotoxically affected by the intermediates resulting from those organic and inorganic pollutants (Babu et al., 2021). Additionally constrained by plant growth rate is phytoremediation. Comparatively speaking to other more conventional cleanup techniques, phytoremediation may take longer to complete a site. While phytoextraction or degradation may take years to complete, excavation, disposal, and incineration can be completed in a matter of weeks to months. Therefore, phytoremediation might not be the best remediation method for locations that present immediate threats to humans and other ecological receptors (Adki et al., 2012). In distant locations with little human interaction or where treatment of soil contamination is not urgently needed, phytoremediation may be the most appropriate approach (Babu et al., 2021). With irrigation and fertilization, overall biomass productivity can exceed 100 t/ha/y in the optimum climates. The trade-off between the accumulation of harmful elements and productivity is one of the outstanding challenges. In actuality, the maximum biomass production that might be harvested would probably be between 10 and 20 t/ha/y, especially for plants that accumulate heavy metals (Ali et al., 2013). Depending on the pollutant, plant species, climate, and other conditions, these values for biomass productivity and heavy metal content would restrict the yearly toxic element removal capability between roughly 10 to 400 kg/ha/y. This corresponds to an annual decrease in soil toxic element levels from 2.5 to 100 ppm for a target soil depth of 30 cm (4,000 t/ha) (Asante-Badu et al., 2020). In cases when the pollutant concentration may be reduced to a level that satisfies regulatory requirements, this is frequently an acceptable rate of contaminant removal that permits site cleanup over a few years to a few decades. Depending on the pollutant, plant species, climate, and other conditions, these biomass productivity and heavy metal concentration values would restrict the yearly toxic element removal capability between 10 and 400 kg/ha/y. Growing season, temperature, root depth, soil chemistry, and degree of pollution are a few examples of variables that may affect the effectiveness of phytoremediation (Imron et al., 2019). One of the main restrictions on the application of phytoremediation is root contact. Contaminants must come into contact with the plant roots for plant-based remediation to take place. The contaminated media must either be transferred to within the plants' reach or the plants must be able to extend their roots to the contaminants (Article et al., 2022). Only applicable to locations with minor contamination inside the rooting zone of remedial plants, the site's ground surface may need to be altered to stop erosion or floods (Article et al., 2022). A plant's physiological activity, particularly that of its roots, is significantly impacted by age. When two plants are of comparable size, the roots of a young plant often show a higher capacity to absorb ions than the roots of an old plant. Using young, healthy plants is essential for more effective plant removal. That being said, it is still possible to employ larger older plants, whose size may make up for their decreased physiological activity in comparison to younger, smaller plants (Ali et al., 2013). Elevated levels of pollutants might hinder the growth of plants, which could restrict their use in specific locations or segments of those locations. Because of this phytotoxicity, a remedial strategy that uses costly ex situ procedures to immediately reduce acute danger in high-concentration waste and long-term in situ phytoremediation to remove large volumes of lower contaminant concentrations may be adopted (Chandanshive et al., 2018). The highest concentration at which plants are capable of accumulating harmful substances is a significant barrier to phytoremediation. Known as "hyperaccumulators," plants with the highest concentrations of toxic metals typically show dry weight-based variations in toxic element concentrations ranging from above 2% for less toxic elements (Zn, Ni, and Cu) to roughly 2000 ppm (0.2%) for more toxic elements (Cd, Pb) (Kotyza et al., 2010). The treatment is often restricted to soils one meter below the surface and groundwater within a few meters of the surface, and soil amendments may be needed. It is only permitted at locations with low pollutant concentrations. Every time plants are utilized to interact with soil pollutants, there may be some ecological exposure. It is concerning what will happen to the metals in the biomass. While certain types of phytoremediation necessitate handling plant material embedded with metals and involve the accumulation of metals, most plants do not acquire considerable quantities of organic pollutants. Most phytoremediation plants don't need to be treated further or disposed of, but metal-accumulating plants must be harvested and recycled or disposed of by applicable legislation (Imron et al., 2019). Plant biomass that has been harvested using phytoextraction may be categorized as hazardous waste, hence disposal procedures must be followed. It is concerning when contaminated plant biomass is consumed because pollutants can still make their way into the food chain by being consumed by animals or insects that consume contaminated plant material. Hydrologic or climatic factors may limit how quickly plants grow (Tangahu et al., 2011).

ACKNOWLEDGMENTS

Authors Ms. Sadhana Y. More and Dr. Y. R Baste acknowledge M.V.P's K.S.K.W.A.S.C College, Uttam Nagar, CIDCO, Nashik and M.V.P's K.T.H.M College Nashik, Maharashtra, India, for providing the opportunity for research and development. Author Rohit Rai acknowledges the Science and Engineering Research Board (SERB), Department of Science & Technology, Government of India for providing financial assistance under the Start-up Research Grant (SRG/2021/001243) scheme, and also acknowledges the financial support received from the Research & Development Cell,

Lovely Professional University, India under the scheme of Matching Grant Seed Money (LPU/DRD/ SEED/SAC/090922/25274).

REFERENCES

Adams, N., Carroll, D., Madalinski, K., Rock, S., Wilson, T., & Pivetz, B. (2000). *Introduction to Phytoremediation. EPA/600/R*-(February), 104. https://clu-in.org/download/remed/introphyto.pdf

Adki, V. S., Jadhav, J. P., & Bapat, V. A. (2012). Exploring the phytoremediation potential of cactus (nopalea cochenillifera salm. dyck.) cell cultures for textile dye degradation. *International Journal of Phytoremediation*, 14(6), 554–569. https://doi.org/10.1080/15226514.2011.619226

Ahmed, A. (2021). Phytoremediation method for removal of selected heavy metals from pharmaceutical effluent. *Int J Waste Resour*, 11(9), 429. https://doi.org/10.35248/2252-5211.21.11.429

Ali, H., Khan, E., & Sajad, M. A. (2013). Phytoremediation of heavy metals-concepts and applications. *Chemosphere*, 91(7), 869–881. https://doi.org/10.1016/j.chemosphere.2013.01.075

Article, R., Khan, A. U., Khan, A. N., Waris, A., Ilyas, M., & Zamel, D. (2022). Phytoremediation of pollutants from wastewater: A concise review. *Open Life Sciences*, 17,488–17,496.

Asante-Badu, B., Kgorutla, L. E., Li, S. S., Danso, P. O., Xue, Z., & Qiang, G. (2020). Phytoremediation of organic and inorganic compounds in a natural and an agricultural environment: A review. In *Applied Ecology and Environmental Research*, 18(5). https://doi.org/10.15666/aeer/1805_68756904

Babu, S. M. O. F., Hossain, M. B., Rahman, M. S., Rahman, M., Ahmed, A. S. S., Hasan, M. M., … & Simal-Gandara, J. (2021). Phytoremediation of toxic metals: A sustainable green solution for a clean environment. *Applied Sciences (Switzerland)*, 11(21), 1–34. https://doi.org/10.3390/app112110348

Chandanshive, V. V, Kadam, S. K., Khandare, R. V, Kurade, M. B., Jeon, B., Jadhav, J. P., & Govindwar, S. P. (2018). In situ phytoremediation of dyes from textile wastewater using garden ornamental plants, effect on soil quality and plant growth. *Chemosphere*, *210*, 968-976. https://doi.org/10.1016/j.chemosphere.2018.07.064

Dias, S. M., Marisa, C., Almeida, R., & Gomes, C. R. (2018). *Phytoremediation of pharmaceuticals by estuarine salt marsh plants* [Master On Environmental Engineering, Universidade do Porto].

Imron, M. F., Kurniawan, S. B., Soegianto, A., & Wahyudianto, F. E. (2019). Phytoremediation of methylene blue using duckweed (Lemna minor). *Heliyon*, 5(8), e02206. https://doi.org/10.1016/j.heliyon.2019.e02206

Kafle, A., Timilsina, A., Gautam, A., Adhikari, K., Bhattarai, A., & Aryal, N. (2022). Phytoremediation: Mechanisms, plant selection and enhancement by natural and synthetic agents. *Environmental Advances*, *8*(2021), 100203. https://doi.org/10.1016/j.envadv.2022.100203

Kaur, N., & Kaushal, J. (2022). Screening the six plant species for phytoremediation of synthetic textile dye wastewater. *Materials Today: Proceedings*, *72*(2), 232–238.

Kotyza, J., Soudek, P., Kafka, Z., & Vaněk, T. (2010). Phytoremediation of pharmaceuticals-preliminary study. *International Journal of Phytoremediation*, *12*(3), 306–316. https://doi.org/10.1080/15226510903563900

Kumar, V., Kumar, M., & Prasad, R. (2018). Phytobiont and ecosystem restitution. In *Phytobiont and Ecosystem Restitution*. https://doi.org/10.1007/978-981-13-1187-1

Lamine, S., & Saunders, I. (2022). Phytoremediation of heavy-metals-contaminated soils: A short-term trial involving two willow species from gloucester willowbank in the UK. *Minerals*, *12*(5). https://doi.org/10.3390/min12050519

Nedjimi, B. (2021). Phytoremediation: a sustainable environmental technology for heavy metals decontamination. *SN Applied Sciences*, *3*(3), 1–19. https://doi.org/10.1007/s42452-021-04301-4

Ruley, J. A., Amoding, A., Tumuhairwe, J. B., Basamba, T. A., & Opolot, E. (2020). *Enhancing the phytoremediation of hydrocarbon-contaminated soils in the sudd wetlands, south sudan, using organic manure*.

Shikha, S., & Gauba, P. (2016). Phytoremediation of industrial and pharmaceutical pollutants. *Recent Advances in Biology and Medicine*, *02*, 113. https://doi.org/10.18639/rabm.2016.02.341789

Singh, T., & Singh, D. K. (2017). Phytoremediation of organochlorine pesticides: Concept, method, and recent developments. *International Journal of Phytoremediation*, *19*(9), 834–843. https://doi.org/10.1080/15226514.2017.1290579

Sites, P., & Frazar, C. (2000). *The bioremediation and phytoremediation of pesticide-contaminated sites* [National Network of Environmental Studies]. https://clu-in.org/download/studentpapers/frazar.pdf

Srivastava, P., & Giri, N. (2021). Role of plants in phytoremediation of industrial waste. *Bioprospecting of Plant Biodiversity for Industrial Molecules*, 73–89. https://doi.org/10.1002/9781119718017.ch4

Tangahu, B. V., Sheikh Abdullah, S. R., Basri, H., Idris, M., Anuar, N., & Mukhlisin, M. (2011). A review on heavy metals (As, Pb, and Hg) uptake by plants through phytoremediation. *International Journal of Chemical Engineering*, *2011*. https://doi.org/10.1155/2011/939161

Van Epps, A. (2006). *Phytoremediation of petroleum hydrocarbons*. U.S. Environmental Protection Agency. https://clu-in.org/download/studentpapers/a_van_epps-final.pdf

Role of Artificial Intelligence in Quality Assessment, Monitoring, and Implementation of Stringent Measures of Industrial Wastewater

11

R.V. Suganya and A.V. Senthil Kumar

1. INTRODUCTION

Managing industrial wastewaters presents a substantial challenge to the continuity of our environment. It is of utmost importance to ensure thorough quality evaluations, consistent monitoring, and the enforcement of stringent protocols since corporations play a significant role in polluting water sources with their effluents. Artificial intelligence (AI) is emerging as a revolutionary technology that completely transforms the approach to

DOI:10.1201/9781003496014-11

wastewater management. The management of industrial wastewater undergoes a remarkable change thanks to AI's capacity for data analysis and generation of valuable insights. Utilizing AI for quality assessments and monitoring, alongside the implementation of strict guidelines, can greatly enhance industries' environmental accountability. This overview delves into the profound influence that AI has exerted on industrial wastewater management, revolutionizing the field. It delves into the diverse applications of AI, ranging from real-time monitoring systems to predictive analytics for pollution control, and how these advancements are fostering sustainable methodologies in manufacturing industries worldwide. Despite covering 34% of the Earth's surface, merely 1% of the freshwater available worldwide is deemed appropriate for human consumption. Contaminants in aquatic environments are dispersing as a result of various influences such as industrial growth and the expanding global populace. As a result, continuous surveillance is essential for tracking the volume of water derived from conventional and unique origins. By examining the framework, one can acquire water that has been tested and assessed for its quality. Both physical and chemical parameters, along with organic information, are employed in the evaluation of water quality. The realm of AI has made significant strides since its inception in the 1950s within the field of computer science. Since then, it has progressed through numerous enhancements and upgrades, encompassing the integration of Internet of Things-enabled technologies, the utilization of artificial insights, and the traditional methodologies for assessing and appraising water quality. The scientific objective of the research is to demonstrate the potential use of various neural networks in determining water surface quality, thereby refining the strategies utilized at the test site where the input parameter is applied. Strict purity standards govern the vast array of water sources, from rivers to ponds and tributaries, ensuring their untainted state. Diverse water standards cater to specific needs, safeguarding various applications. Preserving ecosystems demands that irrigation water remains free of excessive salinity or harmful substances that might infiltrate the earth and vegetation. The criteria for industrial water quality varies according to the nature of the industrial process. Natural water reservoirs like surface water and groundwater are the preferred choices for human consumption. Human activities, engineering endeavours, and ecological interventions all play a role in potentially contaminating these vital resources. As a result, the quality of water has been deteriorating swiftly because of the expanding industrial activities. The influence of infrastructure, coupled with a lack of public awareness and limited sterilization measures, significantly impacts the quality of drinking water. Contaminated water sources pose a grave danger to human well-being, ecosystems, and community structures, making it an indisputable and significant issue.

2. STUDIES ON WATER QUALITY MONITORING INVOLVING AI

In consideration of the present situation of water scarcity and pollution, a model has been developed by us for monitoring of water quality via AI (Figure 11.1).

In the study conducted by Rana et al. (2023), a database with seven constraints to create models for predicting water quality was generated. It employs metrics like

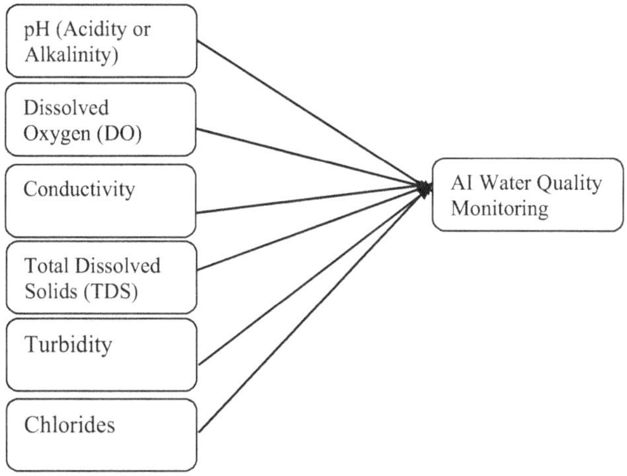

FIGURE 11.1 Conceptual model of the study for monitoring of water quality via AI.

MAE, MSE, and R2 to develop Artificial Neural Networks (ANN) and Long Short-Term Memory (LSTM) models. Results have shown that the LSTM model outperforms ANN with 95% accuracy. The research highlights the importance of using machine learning for water quality monitoring, especially in regions like Himachal Pradesh facing increased water contamination due to tourism.

Biraghi et al. (2021) explored AI for improvement of water monitoring by addressing the reliability of citizen science data. It focuses on the SIMILE project to create an open system for filtering Volunteer Geographic Information (VGI) related to global lake water monitoring. The aim is to automatically detect harmful phenomena (algae and foams) in images uploaded by citizen scientists to speed up the verification process. The data's diversity, consisting of geotagged pictures taken without specific instructions, makes the task challenging. The study test's various tools and deep learning techniques (Clarifai platform, CNN, and R-CNN) combine the SIMILE dataset with results from web searches and Flickr data. The study presents algorithm performance in detecting and labelling phenomena and discusses potential future improvements. Yang et al. (2022) discussed the importance of water-related factors in climate-change resilience; highlighted the use of remote sensing technology and AI for automating information related to water extraction and monitoring; and focused on water body extraction and water quality detection through remote sensing, addressing challenges and identifying research priorities. An interactive web application for reviewing relevant literature is also introduced.

Rasul and Chowdhury (2010) have explained that the appropriateness of water for human ingestion depends on its excellence, and is influenced by type and quantity of scums present in the sample. Al-Adhaileh and Alsaade (2021) estimated that one billion people globally will be grappling with a lack of clean water due to an ongoing reduction in the volume and purity of this essential element. Monitoring the quality of water is essential for effective management and distribution, especially as these statistics continue to climb. The Water Quality Index stands as the pinnacle in categorizing water quality (Elkiran & Ergil, 2006), and by employing water quality indices, one can gauge the quality of water accurately. Shalby et al. (2020) have described many strategies for managing water quality and utilized this tool to assess their effectiveness. Sharma (2011) has ascertained the adequacy of water samples for daily use, the figures of water quality indices were estimated. These values can be utilized and found to be beneficial in executing initiatives to enhance, thereby influencing the quality of water.

The concepts AI put forward by Solanki et al. (2015) encompass the exploration and enhancement of computer systems capable of emulating human intelligence across various areas such as visual interpretation, dialogue understanding, decision-making, and comprehension of everyday language. It was speculated that sectors like healthcare, finance, manufacturing, and transportation can be revolutionized by the crucial technology, and must also be easily accessible in the coming years (Marir et al., 2018). AI has evolved to trace back through a rich and prestigious journey. The emergence of programmable machines and various technological advancements during the 1900s has inspired the release of scholarly works detailing the concepts that could empower robots to imitate human cognitive functions (Zeilhofer et al., 2016). However, this advancement also sparks societal, financial, and moral dilemmas, such as its impact on employment and personal data security. Although AI would have rapid growth with the potential to transform multiple sectors, it also sparks crucial ethical, societal, and financial inquiries in need of resolution. It is also predicted that AI could influence individual cognition and decision-making. Hence the strategies for leveraging AI in constructive decision-making should be defined.

The utilization of a combination of physical-chemical and biological factors to evaluate the quality of surface water proved to be inadequately efficient by AI (Marir et al., 2018). Radar, wireless communication, and the exchange of data through the Internet of Things (IoT) are swiftly evolving realms, increasingly viewed as a promising choice for future control mechanisms. Pinto et al. (2018) designed and anticipated that a procedure could successfully distinguish between algae and foam within water. The procedure eliminates impurities and transforms them into waste, suitable for introduction into the water system or immediate retrieval. Diverse synthetic model variations, like adaptive network-based fuzzy inference system (ANFIS) and artificial neural network (ANN), have been assessed in research.

3. SMART TECHNOLOGY IN THE PRESENT SCENARIO

The basic objective of machine learning (ML) is to develop predictive models that can make accurate predictions or judgements using data-driven learning. ML is a strong data analysis technology that is commonly used to find patterns or produce predictions from massive amounts of data generated by different scenarios. Data collecting, method selection, model training, and validation must all be finalized before ML can be used in a real-world context. Whether or not the datasets contain labels is the primary determinant of which type is being compared. In this context, IoT is a system of interconnected physical items that are able to exchange data via various forms of electronic communication. In order to aid analytical procedures, including environmental sensing, these items are frequently outfitted with software, hardware, or other technology. The very essence of machine learning is the development of predictive abilities—the capacity to make exact predictions or decisions based on data-driven learning. Commonly used in water applications are internet-enabled systems that include sensors for water quality/characteristics, flow, pressure, and more. As a potent data analysis tool, ML is commonly employed to spot trends or make predictions from large datasets produced by a variety of sources. Supervised learning is employed to construct prediction functions with the help of labelled training datasets. Each training instance contains input values and expected output values; using the pertinent input data, supervised learning algorithms seek out correlations between the input and output values to build a prediction model that could forecast the outcome.

3.1. AI Enabled Reclamation of Wastewater

Water reclamation processes, which makes use of a wide variety of highly flexible and modular treatment approaches, are responsible for reclaiming volatile organic compounds (VOCs) and treating and reusing contaminated wastewater (Gadipelly et al., 2019). Prior to purification, a mix of sorption and extraction procedures is used to recover VOCs. Biodegradation and enhanced treatment (photocatalytical/photolytical degradation) are two supplementary methods used to treat wastewater. The budget, the specific wastewater needs, such as irrigation, reusing in the production cycle, additional treatment, etc., and the modular design of each process, determine their relative applicability and compatibility with one another. Data Science System (DSS) inputs include both static knowledge derived from domain expertise and experience as well as external data like weather conditions and the current price of irrigation water, as well as historical and real-time data. The latter comes from sensors that measure important wastewater parameters like temperature, total suspended solids (TSS), specific UV absorbance at 254 nm (SUVA254), total organic carbon (TOC), and pH. Respectively, the ±1%, 0.1%, 2%, 3%, 5%, and 0.05% accuracy of the sensors that were earlier indicated are valid. While the system is running, these are the most important metrics tracked. Simultaneously, the system's operation is regulated by a number of factors. These include, but are not limited to, the following: the quantity of coagulant, the flow rate of the treatment unit, the number of oxidizing agents in the advanced oxidation unit (AOP), the pH-changing flow rate of HCl or NaOH, the intensity of the irradiation source, and the concentration of dissolved oxygen in the mixture.

3.2. AI-Enabled Dual-Stage System for Chlorination

A Dual-Stage System (DSS) for chlorination prototype has been put into place to help with semi-automatic decision making. The DigiCirc accelerator programme was instrumental in developing the initial prototype. In this presentation, Zhang et al. (2015) build upon the prototype by incorporating various extended models that make use of external projections, like financial data and weather predictions.

Eliminating germs and viruses through the use of chlorine-based disinfectants is an essential step in water and wastewater treatment plants' disinfection processes.

While chlorination does a good job of disinfecting, it can be harmful to human health. Disinfectant by-products (DBP) are the result of chlorine's reactions with organic compounds and bromide in water systems.

More and more studies throughout the world are looking at these DBPs because of concerns that they could cause cancer in humans and interfere with reproduction. To manage the amounts of DBPs in drinking water and to forecast when they will be generated, ML technology shows potential. A number of studies have investigated pre-chlorination peroxide/ozonation, and many more have tested models of surface waters treated with chlorine, the principal disinfectant in drinking water plants. Taking into account parameters like water temperature, pH, contact duration, chlorine concentration, and total organic carbon concentrations as inputs to the model has allowed for the successful estimate of DBP concentrations in treated water distribution networks and consumer dwellings. Numerous applications involving support vector machines, fuzzy inference systems, evolutionary algorithms, and chlorination and DBP prediction have led to the widespread testing of ANNs as ML models. While SVMs do occasionally offer a little advantage when R2 is used as a benchmark, ANNs generally outperform GAs and SVMs in comparison studies. When it comes to machine learning models, ANNs have been the most successful in predicting chlorination and DBP. There are more uses for evolutionary algorithms, support vector machines, and fuzzy inference systems. Commonly modelled and anticipated DBPs include total trihalomethanes (TTHM) and total haloacetic acids (THAA). Studies have focused on certain DBP molecules, such as DCAA, TCM, BCAA, BDCM, TCAA, and DBCM. (Chinnappan et al., 2023; Hu et al., 2023).

3.3. Predictive Analytics for Sensors

There are two main goals of sensor forecasting:

- To make sure possible dangers can be prevented in due course. To illustrate the point, consider the potential consequences for the entire pipeline in the event that a variable reaches a critical level while processing.
- Making it easier to make decisions that take into consideration both present values and future trends by providing a water quality prediction module that calculates costs and optimal actions for the future. By considering the system's potential future evolution, the DSS can help uncover cost minima; these minima may be more efficient than local minima calculated using just the current values, which are collected by sensors (Thiyagarajan et al., 2017).

3.4. Machine Learning and AI Methods in Water and Sewage Treatment Settings

The use of AI and ML has been explored in a number of water and wastewater treatment applications. Combining AI models with more conventional methods and IoT architecture can improve WWT's intelligent design systems and its reusability. AI models are an effective tool for optimizing, simulating, and predicting the wastewater treatment process. They have found widespread application in several areas of WWT, such as the removal of dyes, metals, organic compounds, sediments, microbiological pollutants, minerals, and pesticides from water (Sarkar et al., 2023). AI models are most commonly employed in process design and research conducted on a laboratory scale. Optimization of process parameters and prediction of process performance are usually components of process design in practical applications (Kumari et al., 2024).

3.5. Application of AI in Methods Involving Adsorption

There are a number of contaminants that can be removed from water and wastewater using adsorption techniques, which are well-known as physical and chemical treatment alternatives. Adsorption is a technique that uses an exothermic mass transfer to move a target molecule from a fluid to a solid surface. But because of all the many interactions, it can be hard to estimate how well the adsorption process would work and to precisely calculate critical parameters. The adsorption process has been improved with the use of ML models, which have provided crucial predictions. ML has been utilized to model and forecast parameters in streams of water polluted with metals, industrial colours, and organic chemicals, and it can also aid operators in making judgements during adsorption procedures. Water temperature, initial adsorbate concentration, adsorbent dose, contact time, and pH are common variables in ML models of adsorption processes. Some models have also made use of additional parameters such bed height, agitation speed, system flow rate, and BET surface area, as well as adsorbent particle size. Research on the adsorption processes of diverse organic pollutants with distinct properties, like the target contaminant molar mass, has been conducted. The majority of research looked at adsorption efficiency, which is defined as the proportion of adsorbate that was successfully removed. Input water quality parameters, adsorption capacity, and non-dimensional effluent concentrations have all been tried to be predicted by some models. Research involving metal, organic, and industrial dye pollutants has commonly utilized ANN among the ML models. Additionally, RF, ANFIS, and SVM have all demonstrated usefulness in various contexts. The ANN, SVM, and RF ML models consistently achieve R2 values higher than 0.9 and, on rare occasions, even approach 0.99, which is indicative of impressive performance. The SVM models consistently produced more significant R2 and RMSE values than the ANN models, demonstrating their superior performance in most scenarios. In one case, the optimized ANFIS model outperformed competing success models for adsorption processes (Alnuwaiser et al., 2022).

4. UTILIZATION OF AI FOR POTABILITY OF WATER

The management of drinking water sources, distribution, and treatment processes, and decisions have all benefited greatly by the use of ML. To aid in early pollution detection and management, ML-based methods greatly improve forecasting and source water quality assessment (Rajapaksha et al. 2023). Bouamar and Ladjal (2012) conducted a study on real-time water quality monitoring using ANN (Artificial Neural Networks) and SVM (Support Vector Machines) algorithms, which were based on various sensors. Their findings showed that while both models performed well in differentiating between different types of water, the SVM algorithm demonstrated greater stability than the ANN. The importance of early risk assessment, decision-making, and warning systems to improve the quality of drinking water has been highlighted in Norwegian cities (Wu et al., 2012). Although most drinking water quality research has focused on physical and chemical properties, the SVM method has gained popularity for developing flocculation and disinfection plans due to its simplicity and robustness. While significant efforts have been made in fault monitoring, disaster prediction, and ensuring the proper operation of urban water supply systems, microbiological factors, especially those related to *E. coli*, have received little attention in terms of safeguarding the drinking water supply (Hall-Clifford et al., 2022). Even when water meets all standards at treatment facilities, it can become recontaminated while traveling through complex water distribution networks. One way to tackle this issue is by keeping an eye on indicators of biological stability and taking disinfection steps. Protecting groundwater is essential for human well-being because it is a major supply of drinkable water. ML has become an invaluable resource for a wide range of groundwater analytic activities, such as determining pollution sources, evaluating and forecasting groundwater quality, and more. Groundwater quality evaluations have seen a rise in the use of multivariate statistical methods including principal component analysis (PCA) and cluster analysis. Evaluation of groundwater quality has also made use of ML approaches such as DT, ANN, SVM, and RF. The bulk of groundwater quality-related research has examined the efficacy of several ML algorithms in assessing various challenges. Lee et al. (2019) utilized a self-organizing neural network in conjunction with fuzzy c-means clustering to assess the urban groundwater quality in Seoul, South Korea. In order to look into pollution patterns, groundwater samples were divided into three categories according to contamination levels using a self-organizing map algorithm. Next, the geographical distribution of these categories was examined. Five data-mining algorithms, including decision trees, random forests (RF), the automatic chi-square interaction detector, and iterative dichotomizer, have been applied to identify key factors influencing semi-arid groundwater and assess their impact on high-quality groundwater areas in Tabriz City, Iran (Jeihouni et al., 2020). Machine learning (ML) approaches have also been used to forecast future water quality and examine big regional datasets. Ahn et al. (2023) used support vector machines (SVM) and particle swarm optimization (PSO) to estimate and forecast groundwater quality index (WQI), which is especially beneficial for predicting nitrate and arsenic contamination levels.

The research investigates the use of evolutionary computational intelligence algorithms and self-tuning predictive models to calculate the water quality index (WQI). This novel approach intends to overcome the difficulties involved with reliably predicting WQI, which is critical for environmental monitoring and management. The literature survey is expected to cover a variety of computational intelligence techniques, including genetic algorithms, neural networks, and fuzzy logic systems, emphasizing their capacity to manage complicated, non-linear data interactions. It also examines the drawbacks of current WQI determination methods, such as large data pre-processing needs and adaptability concerns, which the proposed method aims to address. Furthermore, the survey investigates the use of self-tuning methods in predictive models, which are critical for enhancing model accuracy and dependability in dynamic situations such as water quality evaluation. The survey supports the argument for using evolutionary algorithms in environmental data analysis by citing successful applications in other domains, highlighting the research's potential significance (Abba et al., 2020).

A study was carried out to emphasize the need of assessing water quality, particularly in large bodies of water such as the Shatt Al-Arab River, utilizing the Water Quality Index (WQI) approach. It examines important physicochemical factors including pH, total dissolved solids, and total hardness, which are critical for determining water appropriateness for diverse applications. To assure water quality standards, the study uses the Weighted Arithmetic Water Quality Index Technique, which is based on World Health Organization (WHO) principles. It also underlines the importance of water quality to human health, emphasizing the need for effective water treatment if the water is unsafe to drink (Hamza, 2024).

Another study focused on the prediction of hydraulic performance in drain envelopes, which is critical for optimal water management in farm and urban settings. The work employs a K-means based multivariate adaptive regression spline (MARS) approach to improve prediction accuracy, addressing the constraints of classic empirical models that may fail to capture complex connections. The addition of K-means clustering to the MARS model enables more sophisticated examination of parameters influencing hydraulic performance, resulting in better forecasts. This unique strategy demonstrates the efficacy of integrating clustering techniques and regression models, and it has potential applications in other fields of engineering and environmental science. Overall, the study helps to improve drainage system design and administration (Adnan et al., 2021).

Feedforward artificial neural networks (FANN) were applied to estimate the WQI in Malaysia's Perak River basin, due to its effectiveness in dealing with nonlinear environmental systems. This study removes biological oxygen demand (BOD) and chemical oxygen demand (COD) as model inputs, demonstrating that their absence has no significant effect on forecast accuracy, allowing for real-time WQI prediction. It also discusses the idea of integrating multiple neural networks using bootstrap re-sampling to improve prediction resilience and generalization. Selective combination methods, such as forward selection and backward removal, enhance prediction accuracy over single or combined networks. The study emphasizes the necessity of real-time WQI prediction tools for effective environmental management, particularly in places that rely on river water quality (Ahmad et al., 2017).

The significance of understanding network traffic in predicting future demand and increasing network resource efficiency was determined. It examines several time series models for network traffic prediction, including statistical models, machine learning models, and hybrid systems, each addressing a distinct aspect of network performance. Deep learning methods such as convolutional and recurrent neural networks, particularly Long Short-Term Memory (LSTM), are noted for their ability to forecast traffic patterns. The study also discusses various models, including ON-OFF, ARIMA, FARIMA, and mobility models, and how they are used to analyze network properties and trends. Modern signal processing techniques are also discussed in terms of their ability to predict network activity. The report identifies research gaps and presents an intelligent hybrid methodology to improve network traffic prediction by combining clustering and machine intelligence (Aldhyani et al., 2020).

A study underlines the value of time series forecasting in a variety of disciplines, including internet traffic management, agricultural planning, catastrophe preparedness, and financial market investing. It examines the various features of time series data, such as seasonality or nonseasonality, volatility or stability, and linearity or nonlinearity. The paper discusses existing hybrid models that use ARIMA for linear components and ANN for nonlinear features, emphasizing their effectiveness in enhancing prediction accuracy. However, it also points out shortcomings in these models and proposes that a better knowledge of data volatility could improve predictions. The research offers a new hybrid ARIMA-ANN model that characterizes data volatility using a moving-average filter, with the goal of improving the effectiveness of ARIMA and ANN. This model is demonstrated to outperform earlier models, with higher accuracy in predicting various data sets (Babu & Reddy, 2014).

Due to the efforts of Ebenstein and Avraham (2012), a report delved deeply into the historical backdrop of water pollution in China, pointing out that present contamination levels are among the worst ever recorded, with major percentages of surface water unsuitable for human consumption. It concentrates on gut cancers, which are particularly common in China, accounting for two-thirds of all cancer cases and being less influenced by changes in smoking behaviours. The assessment highlights existing data linking water pollution to various malignancies, emphasizing that, while some studies imply a link, a comprehensive causal association has yet to be established. It compares the paper's findings to those of other research, demonstrating consistency but varied magnitude. The survey also addresses the possible health benefits of tougher environmental restrictions in developing nations, as well as the economic viability of such policies. Finally, it underlines the paper's methodological advancements, which involve using a nationally representative sample to better analyze the impact of environmental policies on public health in China. The critical role of Water Treatment Plants (WTPs) in protecting human health and the environment by assuring water purity and quality was emphasized. It introduces the idea of Smart Water Treatment Plants (SWTPs), which use software-based tools, data analytics, and predictive modelling to boost operational efficiency. The integration of IoT devices and cloud services is stressed, enabling real-time monitoring and analysis, improving response times, and lowering downtime. The report also emphasizes the need of Big Data and sensor monitoring in sustaining high water quality requirements while reducing energy expenditures. It also emphasizes the significance of effective risk management through constant evaluation and the use of IoT and cloud-based analytics to ensure operational efficiency (El Sayed et al., 2019).

The United States' federal Safe Drinking Water Act (SDWA) and the Environmental Protection Agency's Lead and Copper Rule, are critical in combating lead contamination in drinking water systems across the country. It examines studies and historical data on the prevalence of lead contamination, highlighting broad difficulties and using regional case studies such as Newark, New Jersey, to demonstrate these issues. The study examines the shortcomings of present rules, outlining their limitations in protecting public health and delving into the concept of environmental injustice, underlining how these failures disproportionately harm underprivileged populations. Comparative assessments with international standards and policy ideas for enhancing US rules are also discussed, offering a thorough picture of the difficulties and potential solutions to lead pollution in drinking water (Faherty & Alisha, 2021).

The evolution of water quality and pollution restrictions in India over the last seven decades was studied, and the focus has been on significant historical changes and their implications for current concerns. It outlines systemic difficulties, such as poor institutional frameworks, procedural inefficiencies, and a lack of coordination, that have hampered the effectiveness of water pollution controls. The survey also investigates growing regulatory, judicial, and legal solutions to both existing and emergent water quality challenges. It also covers various regulatory and legal measures that have been enacted over time, evaluating their success and identifying the challenges they confront. The importance of judicial interventions in creating policy and regulatory reforms is examined, shedding light on the dynamic interaction between legal frameworks and environmental governance in India (Sharma, 2024).

There have been reports emphasizing aquaculture's importance in ensuring food safety and nutritional security, as well as its rapid rise and concomitant environmental issues. It focuses on the largemouth bass in China, emphasizing its commercial worth since its introduction in 1983 and the scarcity of past research on its environmental implications. The study employs Life Cycle Assessment (LCA) to assess the environmental challenges associated with largemouth bass aquaculture, identifying electricity consumption and emissions during seed-rearing and pond-farming as important impact variables. The report examines upcoming technologies, such as clean energy usage and intelligent feeding tactics, as viable options to offset these effects. It also advocates for additional LCA study in China's aquaculture sector to develop methodology and promote sustainable practices around the world (Hou et al., 2023).

The research by Jaloree et al. (2014) investigated the implementation of a decision tree classification model to examine water quality data from the MAA Narmada River in Harda district in India using the WEKA software, a popular tool for data mining and machine learning. Decision trees are recognized as a well-established technique in data classification and prediction, which is popular in environmental data analysis due to their simplicity and interpretability. The study focuses on essential water quality measures such as nitrogen compounds (NH3_N, NO3_N), pH, temperature, Biological Oxygen Demand (BOD), and Chemical Oxygen Demand (COD), which are critical for classifying water pollutants. The usefulness of decision trees in managing complicated datasets with various variables has been well demonstrated in the literature, which supports their use in this context. The article is consistent with previous research that uses machine learning to increase environmental monitoring accuracy and efficiency, strengthening the trend of employing machine learning for sustainable environmental management. Although no extensive literature evaluation is offered, the research indirectly draws on known techniques in environmental studies, helping to develop water quality assessment methodologies.

A study by Kangabam et al. (2017) emphasizes the value of WQI as a tool for evaluating water quality since they provide a comprehensive estimate of numerous physicochemical and biological characteristics. It refers to earlier research that used various ways to assess water quality, highlighting the importance of indices in determining the efficacy of management strategies. The survey examines the selection of eleven important parameters for WQI computation, including pH, dissolved oxygen, and turbidity, and compares them to WHO and BIS criteria. It also investigates pollution sources in Loktak Lake using Pearson's correlation coefficient, with the goal of guiding effective conservation methods through the construction of a WQI for long-term management.

The study of Karn et al. (2021) focuses on worldwide water scarcity, emphasizing the importance of adequate wastewater treatment in meeting rising drinking water demand. It investigates the role of IoT technology in smart cities, highlighting successful applications across multiple disciplines and promoting its use in wastewater management. The research emphasizes the relevance of real-time monitoring and data analysis via IoT in improving the efficiency and reliability of wastewater treatment facilities. It also explores the use of complex event processing (CEP) to analyze data from IoT sensors, allowing for more prompt decision-making. A case study from South India highlights the practical benefits of the proposed IoT-based SMARTreat architecture, which are consistent with existing literature on the efficacy of such technologies in sustainable water management. The work of Lai et al. (2011) focuses on the creative integration of the Integrated Watershed Management Model (IWMM) and the Water Quality Analysis Simulation Program (WASP) to model river water quality, which is a novel way of integrating upstream land use changes with downstream water quality. It focuses on non-point source (NPS) pollution, which is a key contributor to water quality degradation, particularly through land use patterns such as orchards and farmlands. The study uses remote sensing and GIS methods to accurately identify land use and simulate pollution sources, while model calibration and validation with field data provide accurate depiction of the Kaoping River Basin. A sensitivity analysis evaluates model reliability by investigating the effect of parameter changes on outputs. Furthermore, the WASP model takes into account nutrient interactions as well as the impact of seasonal fluctuations on pollution loads, demonstrating the modelling approach's comprehensiveness.

The important need of precise groundwater level forecast in ensuring sustainable resource management, which is essential for drinking, agriculture, and industrial usage was emphasized by Maiti and Tiwari (2014). They investigated three sophisticated soft computing techniques: Artificial Neural Networks (ANN), Bayesian Neural Networks (BNN), and Adaptive Neuro-Fuzzy Inference System (ANFIS), which were chosen for their ability to tackle complicated, nonlinear issues. ANNs are known for their iterative learning skills, BNNs for their probabilistic approach and resilience to data noise, and ANFIS for their transparency and interpretability via fuzzy logic. The study compares these models using hydrometeorological data from a drought-prone area in Southern India, evaluating performance with statistical measures such as root mean square error and Pearson's correlation coefficient. The findings indicate that ANFIS performs best with noise-free data, whereas BNN.SCG performs better with noisy data, emphasizing the necessity of picking the appropriate model according on the data conditions.

The novel application of aquaponics as a climate-smart food production system that combines aquaculture with hydroponics, emphasizing its potential to recover nutrients from aquaculture wastewater, making it environmentally friendly was put forward by Kyle (2023). It emphasizes a significant difficulty in aquaponics: maintaining optimum dissolved oxygen (DO) levels, which are necessary for optimal system functioning.

To solve this, the research investigates the use of micronanobubbles (MNBs) for aeration, which are known for their small size and stability, resulting in improved oxygen mass transfer in water. The study compares MNB aeration to traditional methods, noting improvements in DO levels, plant production, and nitrogen cycling, as well as solids and nitrate reduction. Furthermore, it investigates the impact of MNBs on microbial community dynamics, concluding that, while they change microbial composition, they do not significantly disturb metabolic activities. This survey provides an in-depth analysis of the advantages and prospects of employing MNBs to improve aquaponic systems.

The study of Marie et al. (2012) describes the water quality legislation in Palestine and the evolution of legal frameworks over the last century; affected by numerous political regimes from the Ottoman era to the current Palestinian Authority. During the Ottoman time, religious rules affected water legislation, and there was no community involvement or enforcement mechanisms. The British Mandate period incorporated scientific findings into water regulation, whereas the Jordanian administration defined pollutants and developed a decentralized, penalty-based water management system that was relatively stable and effective. Military ordinances were enacted during the Israeli occupation to address water quality issues, but the Palestinian Authority has struggled to put these regulations into practice due to limited enforcement measures. The research focuses on how these administrations modified earlier laws to meet acute difficulties without long-term planning. It emphasizes the importance of strong legislative frameworks and enforcement to support Palestinian economic development, public health, and environmental quality, with a focus on equitable water resource management and community involvement.

The study on IoT-integrated systems for water quality monitoring underlines the growing need for innovative technology to manage urban water systems sustainably in the face of diminishing water conditions. It emphasizes the potential of wireless sensor networks (WSNs) for water quality monitoring, which offer substantial advantages over traditional centralized and manual techniques by enabling decentralized monitoring customized to dynamic urban water infrastructures. However, the paper identifies a market gap in the availability of low-cost wireless sensor nodes, which are critical for the cost effective deployment of these systems. It examines the integration of low-cost water quality monitoring devices with IoT platforms, which is required for real-time, decentralized monitoring and decision-making in wastewater treatment facilities. The paper also presents a novel ion chromatography detection method for analyzing nitrates and nitrites, which has been validated in a real-world setting, highlighting the advantages of decentralized systems that allow for real-time communication, preventive measures, and data analytics for better water quality management (Martínez et al., 2020).

A study by Castillo et al. (2024) described the river systems and their vital role in providing global ecosystem services, as well as the growing concern of water quality issues caused by contaminants. It emphasizes the use of machine learning models to improve the accuracy of water quality predictions, which is critical for effective environmental management. A significant difficulty mentioned is the lower performance of models trained on redundant data, emphasizing the importance of selecting representative and varying water quality scenarios to increase forecast accuracy. The survey investigates combining analytical methodologies, such as clustering and time series analysis, to improve data selection for machine learning applications. It examines specific models such as ANFIS, ANN, and SVM, observing that models trained on carefully selected subsets of data frequently outperform those trained on entire datasets. This emphasizes the significance of strategic data selection in improving machine learning-based water quality prediction.

Further, the vital role of river systems' vital role in providing global ecosystem services, as well as the growing concern of water quality issues caused by contaminants. It emphasizes the use of machine learning models to improve the accuracy of water quality predictions, which is critical for effective environmental management. A significant difficulty mentioned is the lower performance of models trained on redundant data, emphasizing the importance of selecting representative and varying water quality scenarios to increase forecast accuracy. The survey investigates combining analytical methodologies, such as clustering and time series analysis, to improve data selection for machine learning applications. It examines specific models such as ANFIS, ANN, and SVM, observing that models trained on carefully selected subsets of data frequently outperform those trained on entire datasets. This emphasizes the significance of strategic data selection in improving machine learning-based water quality prediction (Ray et al., 2023).

The study of Shen et al. (2024) discusses the problems and improvements in Electrochemical Recirculating Aquaculture Systems (ERAS) and addresses the high energy costs and negative consequences of residual chlorine created during electrochemical operations, which have hampered the use of ERAS in aquaculture. The study investigates how characteristics such as current density and geometric aspects of electrochemical systems can be tuned for effective control of inorganic nitrogen, microorganisms, and residual chlorine. It discusses pilot tests with shrimp culture, which demonstrate the systems' practical usefulness. The research describes the successful use of a 50-L electrochemical batch reactor to eradicate inorganic nitrogen and Vibrio bacteria while also controlling residual chlorine levels by air stripping techniques. The study reveals that by adjusting the timing of intermittent processes such as electrooxidation and air stripping, effective disinfection and energy cost reduction can be achieved, making ERAS more viable for wider aquaculture application. According to Krisht et al. (2024), the treated wastewater can be reused. It could underline the growing issue of water shortage caused by population growth, increased freshwater demand for agriculture, and reduced water availability due to climate change. In response, there is a global shift toward non-conventional water resources, which is consistent with Sustainable Development Goal (SDG) 6.3, which encourages water quality improvement as well as increased recycling and safe reuse. The study investigates the use of treated wastewater (TWW) for agricultural and domestic purposes as a technique of closing the water supply gap. It does, however, address a variety of TWW reuse concerns, including as cultural, socioeconomic, environmental, and public health difficulties. The survey includes case studies on successful wastewater reuse initiatives, as well as emerging research and technology breakthroughs aimed at enhancing TWW quality, lowering hazards, and increasing its utilization. This detailed analysis emphasizes the potential of treated wastewater as a solution to water constraint while also recognizing the hurdles that must be solved to assure its safe and sustainable use.

The research of Turcios and Papenbrock (2014) investigated the considerable issues faced by wastewater from aquaculture systems, which contain high amounts of suspended solids, nitrogen, and phosphorus as a result of rising worldwide fish demand and intensive fish production procedures. It examines traditional physical, chemical, and biological wastewater treatment technologies, focusing on artificial wetlands as a cost-effective alternative, particularly in recirculating aquaculture systems (RAS). The survey stresses the use of halophytic plants as biofilters in saline systems to degrade hazardous chemicals and successfully manage nutrient levels. Innovative techniques to sustainable aquaculture are presented, including enhancing RAS with biofiltering organisms and employing commercially valuable salt-tolerant plants. The assessment also emphasizes the importance of sustainable approaches for minimizing environmental consequences, such as increasing feed quality and management to reduce waste and increase nutrient bioavailability. Overall, this report describes both classic and novel approaches for controlling aquaculture wastewater and proposes future paths for the industry's sustainable practices.

Steel Plate Shear Walls (SPSWs) were utilized as useful structural solutions for resisting lateral pressures in buildings, focusing on their composition of thin web plates infilled within steel frames. It addresses the behaviour of unstiffened steel plates, noting that they are designed to buckle in shear at low stress levels, resulting in a diagonal tension field that improves strength, stiffness, and ductility while also contributing to significant energy dissipation, which is critical for seismic performance. The research investigates the inelastic structural response of SPSWs using both a modified strip model and a new simpler strip model for beam-connected SPSWs, with a focus on plasticity and nonlinear strip behaviour under compressive stresses. Furthermore, the study emphasizes essential features of SPSW seismic performance, such as energy dissipation capacity, ductility requirements, inter-story drift, and load demands on boundary elements. Finally, the results of these models are evaluated by comparisons to experimental and numerical data, resulting in a thorough investigation of SPSW behaviour in earthquake-resistant design (Totter et al., 2018). In the study carried out by Yadav et al. (2022), the application of Smart Aquaculture Management Systems (SAMS) was anticipated with the advent of IoT technology to increase aquaculture efficiency and productivity. Sensor nodes, AI, cloud computing, and data storage are critical components of these systems, which aid in the monitoring of water quality, animal health, and feeding behaviour. Recent improvements in IoT have dramatically improved farmers' benefit-cost ratios by enhancing aquaculture methods. The study also identifies future research directions for scaling digital technologies in aquaculture, with the goal of guiding stakeholders in embracing these innovations for long-term growth.

The vital necessity for accurate water quality forecast models as freshwater resources are under increasing demand was underlined by Jianzhuo (2019). Traditional Backpropagation (BP) neural networks, which are typically employed for such predictions, frequently encounter issues such as delayed convergence and are becoming locked in local minima, reducing their accuracy. To address these limitations, the paper develops a hybrid model that optimizes BP networks using both Particle Swarm Optimization (PSO) and Genetic Algorithms (GA). This method improves prediction ability, particularly for time series data. According to the study, the hybrid model outperforms standard and individually tuned BP networks in terms of accuracy and robustness when estimating water quality. The use of spectrometry as a novel method for predicting water quality, with a specific emphasis on the Total Organic Carbon (TOC) criterion was investigated by Yan-jun and Qian (2012). Their technique stands out for its speed, environmental friendliness, and potential in soft measurement technology. A major difficulty encountered was the accuracy and resilience of prediction models, which are heavily reliant on the quality of training samples. To improve model performance, the research suggests utilizing affinity propagation (AP) clustering based on spectral data to choose the best training samples. Following clustering, TOC is evaluated using a nonlinear modelling method called a least squares support vector machine (LSSVM). The experimental results show that the AP-LSSVM model outperforms existing models in terms of accuracy and robustness for predicting water quality.

The research of Zeilhofer et al. (2007) delves into the use of Geographic Information Systems (GIS) for mapping and spatial modelling of urban water quality in Cuiabá, Brazil. It employs spatial analysis tools to investigate water quality issues and the danger of waterborne infections in a lower-middle-income metropolitan area. Logistic regression models were used to examine factors such as turbidity, *Escherichia coli,* total coliforms, and intestinal parasite infection, demonstrating that water contamination can occur from both public infrastructure and in-house sanitary equipment. GIS was successfully utilized to generate spatial datasets, build risk maps, and examine the spatial distribution of water quality issues and associated health consequences. The study also included a socio-economic breakdown of homes, emphasizing the importance of hygiene behaviours and infrastructure in water quality. The findings show that GIS can improve logistic regression analysis by adding spatial context and increasing the depiction of health hazards, which is critical for environmental planning and public health interventions.

5. FUTURISTIC IMPACT OF AI IN WASTEWATER MANAGEMENT

The different aspects in which AI is expected to impact the future of water and infrastructure globally are as follows.

5.1. Condition of the Water Supply

In order to detect changes in water quality that could indicate pollution or other issues, AI can analyze past results and data from sensors to predict future patterns. Now that they have plans in place, the local agency can respond quickly to issues like harmful algal blooms and other forms of pollution (Park et al., 2024).

5.2. Predictive Upkeep of Water Supply Systems

Adefemi et al. (2023) explored AI's ability for detection of equipment problems and maintenance needs. A number of US-based municipal governments and agencies have implemented AI and ML systems to monitor equipment health in real-time and alert maintenance teams to potential issues before they escalate.

5.3. Flood Risk Prediction for the Future

In the context of floods, the natural world could be exceedingly complex. Historical data, sometimes known as "past performance," was traditionally used to estimate future flood risk (Waqas et al., 2023).

5.4. Sustainable Energy and Minimizing Energy Waste

Given the high energy consumption of water distribution and purification, these processes are ideal candidates for optimization using AI and ML. AI can analyze historical water consumption patterns to predict purification and distribution network needs. Furthermore, AI can optimize the distribution system by analyzing pressure, flow rates, and other data to identify under- or over-utilized areas within the network. Implementing a decision-support system that leverages these insights can help reduce energy consumption and carbon emissions by adjusting the network and ensuring efficient water distribution (Muhammad et al., 2024).

5.5. Asset and Water Resource Management

Prioritizing and managing infrastructure assets with the help of AI helps ensure that they are properly maintained and replaced when needed. Additionally, it can enhance irrigation and reduce waste, both of which help conserve water (Nova, 2023).

5.6. Decarbonization and Decentralization of Wastewater

Based on the interventions implemented over the last 30 years to progressively evolve water systems, the five primary characteristics of the future are decentralized, environmentally friendly, circular, carbon neutral, and autonomous. Instead of being mutually exclusive, the development routes that lead to these traits are interdependent on one another, since they all contribute to making water systems that are both robust and sustainable through increasing the system's capacity, performance, and efficiency. Consider swales as an example of a greener road management solution. Not only can they reduce energy consumption and greenhouse gas emissions, but they can also decouple storm water services from centralized sewer systems, which might pave the way for decarbonization and decentralization.

When it comes to decentralized systems, AI can be a huge help when it comes to planning and controlling things like water supply, stormwater management, water recycling, and wastewater treatment. This way, resources like water, energy, and materials aren't wasted. Because decentralized systems must be interconnected to create a whole that can deliver essential water and wastewater services while simultaneously maximizing environmental and ecological benefits, the optimal planning of such a system is inherently more difficult than that of a centralized system. According to Garrido-Baserba et al. (2020), it is possible under certain conditions to develop decentralized systems that eliminate the need for centralized water supply or waste collection using existing technologies. To maximize the benefits of any decentralized system, AI technologies are essential for identifying reliable and robust operations and developing predictive maintenance strategies. In recent years, citizen science programs have increasingly leveraged AI and machine learning (ML) to enhance engagement and efficiently distribute tasks. By recruiting participants, targeting specific demographics, and analyzing results, AI and ML can support studies that explore public perceptions of decentralized systems.

5.7. Transition to a Circular Economy

WWTPs, being energy intensive, vital for nutrient recovery, water reuse, and renewable energy generation employing a range of technologies, are the backbone of the circular economy plan (Chmielewski et al., 2024). They also account for 25% of the energy utilized in the water sector. Using process-based models has improved water treatment systems' energy and nutrient recovery rates. More and more, vital process parameters and nutritional qualities are being monitored in real-time due to a particular real-world challenge in extremely changeable water streams. The incorporation of machine learning allows for precise simulation of system dynamics and online learning and control in real-time for optimal resource recovery.

5.8. Automated Process

Sewage and water treatment plants make extensive use of automation as a result of developments in sensing, communication, and computing technology. However, its main use is in managing a single process unit or a small number of units in a system. While research into integrated control of UWI systems has made great strides in recent decades, few real-world applications have resulted. There are a lot of obstacles to overcome when trying to automate water systems or build fully autonomous ones. These include the complex interdependencies between different parts of the system, the future's unpredictability, the explainability of control strategies and the credibility of AI, and the software and monitoring costs. Integrating machine learning into system control approaches can enhance control performance. Reports indicate that AI integration leads to notable improvements in water quality, system reliability, and energy efficiency. Using predicted demands to regulate water treatment plant discharges is one example of these advantages. Pressure, water demand, water depth, flow, and pollution load forecasts have all been much improved by recent advances in deep learning technology (Zhu & Qiu, 2017; Zhu et al., 2020; Sarkar et al., 2023).

6. CONCLUSION AND FUTURE PERSPECTIVE

This study has offered an analysis and summary of numerous ML models, in addition to cutting-edge technology and methodologies used in various water-related domains. Wastewater treatment, water-based agriculture, and natural system monitoring and management have all benefited greatly from the use of AI and ML techniques for process improvement, modelling, and automation. It is believed that by combining AI and ML, we may improve water-based applications, cut costs, and find computer-assisted solutions to difficult problems in physical and biological processes and water chemistry. Water and wastewater treatment facilities, natural systems, and water-based agriculture are just a few examples of the many water-related sectors and processes that have benefited greatly from the use of ML and AI technologies for prediction, modelling, automation, and optimization. Research into developing soft sensors for use in water treatment facilities and the use of artificial intelligence and machine learning to enhance the resilience of water infrastructure (as measured by water quality) should be prioritized in the future.

REFERENCES

Abba, S. I., Hadi, S. J., Sammen, S. S., Salih, S., Abdulkadir, R. A., Pham, Q. B., & Yaseen, Z. M. (2020). Evolutionary computational intelligence algorithm coupled with self-tuning predictive model for water quality index determination. *Journal of Hydrology, 587*. doi: 10.1016/J.JHYDROL.2020.124974

Adefemi, A., Ukpoju, E. A., Adekoya, O., Abatan, A., & Adegbite, A. O. (2023). Artificial intelligence in environmental health and public safety: A comprehensive review of USA strategies. *World Journal of Advanced Research and Reviews, 20*(3), 1420–1434.

Adnan, R.M., Khosravinia, P., Karimi, B., Ozgur, Kisi., & Kisi, O. (2021). Prediction of hydraulics performance in drain envelopes using Kmeans based multivariate adaptive regression spline. *Applied Soft Computing, 100*, 107008. doi: 10.1016/J.ASOC.2020.107008

Ahmad, Z., Rahim, N. A., Bahadori, A., & Zhang, J. (2017). Improving water quality index prediction in Perak River Basin Malaysia through the combination of multiple neural networks. *International Journal of River Basin Management, 15*(1), 79 –87. doi: 10.1080/15715124.2016.1256297

Al-Adhaileh, M. H., & Alsaade, F. W. (2021). Modelling and prediction of water quality by using artificial intelligence. *Sustainability, 13*(8), 4259.

Aldhyani, T. H. H., Alrasheedi, M., Alqarni, A. A., Alzahrani, M. Y., & Bamhdi, A. M. (2020). Intelligent hybrid model to enhance time series models for predicting network traffic. IEEE Access, 8, 130431–130451. doi: 10.1109/ACCESS.2020.3009169

Alnuwaiser, M. A., Javed, M. F., Khan, M. I., Ahmed, M. W., & Galal, A. M. (2022). Support vector regression and ANN approach for predicting the ground water quality. *Journal of the Indian Chemical Society, 99*(7), 100538, ISSN 0019-4522.https://doi.org/10.1016/j.jics.2022.100538

Babu, C. N., B., & Reddy, B. E. (2014). A moving-average filter based hybrid ARIMA-ANN model for forecasting time series data. *Applied Soft Computing, 23*, 27–38. doi: 10.1016/J.ASOC.2014.05.028

Biraghi, C. A., Lotfian, M., Carrion, D., & Brovelli, M. A. (2021). AI in support to water quality monitoring. *The International Archives of the Photogrammetry, Remote Sensing and Spatial Information Sciences*, 167–174.

Bouamar M., & Ladja, M. (2012). Performance and evaluation of three pattern classification techniques used for water quality monitoring. *International Journal of Computational Intelligence and Applications, 11*(02), 1250013. https://doi.org/10.1142/S1469026812500137

Castillo, A. F. D, Garibay, M. V., Díaz-Vázquez, D., Yebra-Montes, C., Brown, L. E., Johnson, A., … & Gradilla-Hernández, M. G. (2024). Improving river water quality prediction with hybrid machine learning and temporal analysis. *Ecological Informatics,* 102655–102655. doi: 10.1016/j.ecoinf.2024.102655

Chinnappan, C. V., John William, A. D., Nidamanuri, S. K. C., Jayalakshmi, S., Bogani, R., Thanapal, P., … & Syed Masood, J. A. I. (2023). IoT-enabled chlorine level assessment and prediction in water monitoring system using machine learning. *Electronics, 12*, 1458. https://doi.org/10.3390/electronics12061458

Chmielewski, A. G., Sudlitz, M., & Żubrowska-Sudoł, M. (2024). Advanced technology for energy, plant nutrients and water recovery at wastewater treatment plants. *Energies, 17*(11), 2749.

Ebenstein, A. (2012). The consequences of industrialization: Evidence from water pollution and digestive cancers in China. *The Review of Economics and Statistics, 94*(1), 186–201. doi: 10.1162/REST_A_00150

El Sayed, H. Y., Al-Kady, M., & Siddik, Y. (2019). *Management of smart water treatment plant using IoT cloud services.* 2024 International Conference on Smart Communications and Networking (SmartNets). doi: 10.1109/SMARTNETS48225.2019.9069763

Elkiran, G., & Ergil, M. (2006). The assessment of a water budget of North Cyprus. *Building and Environment, 41*(12), 1671–1677.

Faherty, A. (2021). Tapped out: How Newark, New Jersey's lead drinking water crisis illuminates the inadequacy of the federal drinking water regulatory scheme and fuels environmental injustice throughout the nation. *Environmental Claims Journal, 33*(4), 304-327. doi: 10.1080/10406026.2020.1848078

Gadipelly, C., Pérez-González, A., Yadav, G. D., Ortiz, I., Ibáñez, R., Rathod, V. K., & Marathe, K. V. (2014). Pharmaceutical industry wastewater: review of the technologies for water treatment and reuse. *Industrial & Engineering Chemistry Research, 53*(29), 11571–11592.

Garrido-Baserba, M., Corominas, L., Ulises C., Rosso D., & Poch, M. (2020). The fourth-revolution in the water sector encounters the digital revolution *Environmental Science & Technology, 54*(8). https://pubs.acs.org/doi/10.1021/acs.est.9b0425

Hall-Clifford, R., Arzu, A., Contreras, S., Croissert Muguercia, M. G., de Leon Figueroa, D. X., Ochoa Elias, M. V., … & Pennington, P. (2022). Toward co-design of an AI solution for detection of diarrheal pathogens in drinking water within resource-constrained contexts. *PLOS Global Public Health, 2*(8), e0000918. https://doi.org/10.1371/journal.pgph.0000918

Hamza, Z. A. H. (2024). Application of water quality index in assessing water quality: A case study of the Shatt Al Arab River. *Technium.* doi: 10.47577/technium.v21i.10737

Hou, H., Ren, A., Yu, L., Ma, Z., Zhang, Y., & Liu, Y.(2023). An environmental impact assessment of largemouth bass (*Micropterus salmoides*) aquaculture in Hangzhou, China. *Sustainability*. doi: 10.3390/su151612368

Hu, G., Mian, H. R., Mohammadiun, S., Rodriguez, M. J., Hewage, K., & Sadiq, R. (2023). Appraisal of machine learning techniques for predicting emerging disinfection byproducts in small water distribution networks. *Journal of Hazardous Materials*, *446*, 130633, ISSN 0304-3894, https://doi.org/10.1016/j.jhazmat.2022.130633.

Jaloree, S., Rajput, A., & Gour, S. (2014). Decision tree approach to build a model for water quality. doi:10.5138/BJDMN.V4I1.1563

Jeihouni, M., Toomanian, A., & Mansourian, A. (2020). Decision tree-based data mining and rule induction for identifying high quality groundwater zones to water supply management: A novel hybrid use of data mining and GIS. *Water Resources Management, 34*, 10.1007/s11269-019-02447-w.

Jianzhuo, Y., Zongbao, X., Yongchuan, Y., Hongxia, X., & Kaili, G. (2019). Application of a Hybrid Optimized BP network model to estimate water quality parameters of Beihai Lake in Beijing. *Applied Sciences, 9*(9), 1863. doi: 10.3390/APP9091863

Kangabam, R. D., Bhoominathan, S. D., Kanagaraj, S., & Govindaraju, M. (2017). Development of a water quality index (WQI) for the Loktak Lake in India. *Applied Water Science, 7*(6), 2907–2918. doi: 10.1007/S13201-017-0579-4

Karn, A. L., Pandya, S., Mehbodniya, A., Arslan, F., Sharma, D. K., Phasinam, K., … & Sengan S. (2021). An integrated approach for sustainable development of wastewater treatment and management system using IoT in smart cities. *Soft Computing*, *27*. doi: 10.1007/S00500-021-06244-9

Krisht, G., Said, R. B., Aboujaoude, L., Hajjar, T., Kamaleddine, F., Soufi, A. R., … & Dare, A. (2024). Irrigating with treated wastewater. 635–651. doi: 10.1016/b978-0-323-90386-8.00091-7

Kumari, S., Chowdhry, J., & Chandra Garg, M. (2024). AI-enhanced adsorption modeling: Challenges, applications, and bibliographic analysis. *Journal of Environmental Management, 351*, 119968. https://doi.org/10.1016/j.jenvman.2023.119968

Kyle, R., Marcelino, S., Wongkiew., T., Shitanaka., K. C., Surendra., B., Song., & Samir, K. (2023). Micronanobubble aeration enhances plant yield and nitrification in aquaponic systems. doi: 10.1021/acsestengg.3c00344

Lai, Y. C., Yang, C. P., Hsieh, C. Y., Wu, C. Y., & Kao, C. M. (2011). Evaluation of non-point source pollution and river water quality using a multimedia two-model system. *Journal of Hydrology, 409*(3), 583–595. doi: 10.1016/J.JHYDROL.2011.08.040

Lee, K. J., Yun, S. T., Yu, S., Kim, K. H., Lee, J. H., & Lee, S.H.(2019). The combined use of self-organizing map technique and fuzzy c-means clustering to evaluate urban groundwater quality in Seoul metropolitan city. *South Korea Journal of Hydrology*, *569*, 685–697. ISSN 0022-1694, https://doi.org/10.1016/j.jhydrol.2018.12.031

Maiti, S., & Tiwari, R. K. (2014). A comparative study of artificial neural networks, Bayesian neural networks and adaptive neuro-fuzzy inference system in groundwater level prediction. *Environmental Earth Sciences, 71*(7), 3147-3160. doi: 10.1007/S12665-013-2702-7

Marie, A., Khayat, S., & Dajani, M. (2012). Water quality legislation in Palestine over the past century. *Environmental Sciences Europe, 24*(1), 15. doi: 10.1186/2190-4715-24-15

Marir, N., Wang, H., Feng, G., Li, B., & Jia, M. (2018). Distributed abnormal behavior detection approach based on deep belief network and ensemble svm using spark. *IEEE Access*, *6*, 59657–59671. https://doi.org/10.1109/ACCESS.2018.2875045

Martínez, R., Vela, N., el Aatik, A., Murray, E., Roche, P., & Navarro, J.M.(2020). On the use of an IoT integrated system for water quality monitoring and management in wastewater treatment plants. *Water, 12*(4), 1096. doi: 10.3390/W12041096

Muhammad, V. F., Kadaria, U., & Purnaini, R. (2024). Planning for the development of a drinking water distribution system network in Sungai Pinyuh. *Jurnal Teknik Sipil, 24*(1), 1059–1067. https://doi.org/10.26418/jts.v24i2.78461

Nova, K. (2023). AI-enabled water management systems: an analysis of system components and interdependencies for water conservation. *Eigenpub Review of Science and Technology, 7*(1), 105–124.

Park, J., Patel, K., & Lee, W. H. (2024). Recent advances in algal bloom detection and prediction technology using machine learning. *Science of the Total Environment, 938*, 173546, ISSN 0048-9697.https://doi.org/10.1016/j.scitotenv.2024.173546.

Pinto, M. M. S. C., Marinho-Reis, A. P., Almeida, A., Ordens, C. M., Silva, M. M. V. G., Freitas, S., ... & Condesso De Melo, M. T. (2018). Human predisposition to cognitive impairment and its relation with environmental exposure to potentially toxic elements. *Environmental Geochemistry and Health, 40*(5), 1767–1784. https://doi.org/10.1007/s10653-017-9928-3

Rajapaksha, R. M. L. S., Karunasiri, R. P. W. N., Perera, T. A. N. T., & Sirisena, K. (2023, December 15th-17th). *AI-based machine learning algorithms for water quality analysis: A review.* The 14th International conference on sustainable Built environment, Kandy, Sri Lanka.

Rana, R., Kalia, A., Boora, A., Alfaisal, F. M., Alharbi, R. S., Berwal, P., ... & Qamar, O. (2023). Artificial intelligence for surface water quality evaluation, monitoring and assessment. *Water, 15*(22), 3919. https://doi.org/10.3390/w15223919

Rasul, G., & Chowdhury, A. K. M. J. U. (2010). *Equity and social justice in water resource management in Bangladesh.* Gatekeeper. https://iwrmactionhub.org/system/files/2021-11/bangladesh.-equity-and-social-justice-in-water-resource-management-413.pdf

Ray, S., Verma, R., Singh, A., Ganesapillai, M., & Kwon, Y.N. (2023). A holistic review on how artificial intelligence has redefined water treatment and seawater desalination processes. *Desalination, 546,* 116221–116221. doi: 10.1016/j.desal.2022.116221

Sarkar, O., Dey, A. K., & Malik, T. (2023). Modernized management of biomedical waste assisted with artificial intelligence. *International Journal of Biomedical and Clinical Analysis, 3*(2), 69–86. https://researchlakejournals.com/index.php/IJBCA/article/view/265/229

Shalby, A., Elshemy, M., & Zeidan, B. A. (2020). Assessment of climate change impacts on water quality parameters of Lake Burullus, Egypt. *Environmental Science and Pollution Research, 27*(26), 32157–32178.

Sharma, A. (2024). Water quality and water pollution, 611–630. doi: 10.1093/oxfordhb/9780198884682.013.35

Sharma, D., & Kansal, A. (2011). Water quality analysis of River Yamuna using water quality index in the national capital territory, India (2000–2009). *Applied Water Science, 1*(3–4), 147–157. https://doi.org/10.1007/s13201-011-0011-4

Shen, Y., Chen, C., Li, P., Huang, X., & Li, Y. (2024). Application of a smart pilot electrochemical system for recycling aquaculture seawater. *Aquaculture International, 32*(1), 197–213.

Solanki, A., Agrawal, H., & Khare, K. (2015). Predictive analysis of water quality parameters using deep learning. *Int. J. Comput. Appl.,* 125, 29–34

Thiyagarajan, K., Kodagoda, S., & Nguyen, L.V. (2017). *Predictive analytics for detecting sensor failure using autoregressive integrated moving average model.* 12th IEEE Conference on Industrial Electronics and Applications (ICIEA), Siem Reap, Cambodia. doi: 10.1109/ICIEA.2017.8283153.

Totter, E., Formisano, A., Mazzolani, F.M., & Crisafulli, F. (2018). Strip model analysis for steel plate shear walls in earthquake resistant structures. *Key Engineering Materials, 763,* 743–750. doi: 10.4028/WWW.SCIENTIFIC.NET/KEM.763.743

Turcios, A. E., & Papenbrock, J. (2014). Sustainable treatment of aquaculture effluents—What can we learn from the past for the future? *Sustainability, 6*(2), 836–856. doi: 10.3390/SU6020836

Waqas, M., Bonnet, S., Wannasing, U. H., Hlaing, P. T., Lin, H. A. & Hashim, S. (2023). *Assessment of Advanced Artificial Intelligence Techniques for Flood Forecasting.* 2023 International Multi-disciplinary Conference in Emerging Research Trends (IMCERT), Karachi, Pakistan.

Wu, Y., Zhang, X., Xiao, Y., & Feng, J. (2020). Attention neural network for water image classification under IoT environment. *Applied Sciences, 10*(3), 909.

Yadav, A., Noori, M. T., Biswas, A., & Min, B. (2022). A concise review on the recent developments in the Internet of Things (IoT)-based smart aquaculture practices. *Reviews in Fisheries Science & Aquaculture, 31*(1), 103-118. doi: 10.1080/23308249.2022.2090228

Yang, H., Kong, J., Hu, H., Du, Y., Gao, M., & Chen, F. (2022). A review of remote sensing for water quality retrieval: Progress and challenges. *Remote Sens., 14*, 1770. https://doi.org/10.3390/rs14081770

Yan-jun, L., & Qian, M. (2012, July). AP-LSSVM modeling for water quality prediction. In *Proceedings of the 31st Chinese Control Conference* (pp. 6928–6932). IEEE.

Zeilhofer, P., Calheiros, D. F., de Oliveira, M. D., de Carvalho Dores, E. F. G., Lima, G. A. R., & Fantin-Cruz, I. (2016). Temporal patterns of water quality in the Pantanal floodplain and its contributing Cerrado upland rivers: Implications for the interpretation of freshwater integrity. *Wetlands Ecology and Management, 24*, 697–716.

Zeilhofer, P., Zeilhofer, L.V.A.C., Hardoim, E.L., de Lima, Z.M., & Oliveira, C.S. (2007). GIS applications for mapping and spatial modeling of urban-use water quality: A case study in District of Cuiabá, Mato Grosso, Brazil. Cadernos De Saude Publica, 23(4), 875–884. doi: 10.1590/S0102-311X2007000400015

Zhang, D., Chen, X., & Yao, H. (2015). Development of a prototype web-based decision support system for watershed management. *Water, 7*(2), 780–793.

Zhu, H., & Qiu, X. (2017). The application of PLC in sewage treatment. *Journal of Water Resource and Protection, 9*, 841–850. doi:10.4236/jwarp.2017.97056.

Zhu, W., Wang, Z., & Zhang, Z. (2020). Renovation of automation system based on industrial internet of things: A case study of a sewage treatment plant. *Sensors, 20*, 2175. doi:10.3390/s20082175

Index

Note: Page numbers in **bold** indicate a table; *italics* indicate a figure.

Kolibri 360 GmbH
Ferdinand-Jühlke-Straße 7
99095 Erfurt, Deutschland
produktsicherheit@kolibri360.de